Mobile Computing

Mobile Computing

Edited by
Ian Phillips

Larsen & Keller
www.larsen-keller.com

Mobile Computing
Edited by Ian Phillips
ISBN: 978-1-63549-187-6 (Hardback)

Published by Larsen and Keller Education,
5 Penn Plaza,
19th Floor,
New York, NY 10001, USA

Cataloging-in-Publication Data

Mobile computing / edited by Ian Phillips.
     p. cm.
Includes bibliographical references and index.
ISBN 978-1-63549-187-6
1. Mobile computing. 2. Cell phone systems. 3. Wireless
communication systems. I. Phillips, Ian.
QA76.59 .M63 2017
004--dc23

The publisher's policy is to use permanent paper from mills that operate a sustainable forestry policy. Furthermore, the publisher ensures that the text paper and cover boards used have met acceptable environmental accreditation standards.

Printed and bound in the United States of America.

For more information regarding Larsen and Keller Education and its products, please visit the publisher's website www.larsen-keller.com

Table of Contents

Preface

This book provides comprehensive insights into the field of mobile computing and technology. It describes in detail the various techniques and applications of this science in the present day scenario. Mobile computing and technology refers to the technology of human and computer interaction. It allows the transfer of data, video and voice. Mobile software includes applications and operating systems while, hardware includes mobile devices and device components. This text attempts to understand the multiple branches that fall under the discipline of mobile computing and technology and how such concepts have practical applications. Fundamental theories and concepts related to mobile computing and related technologies are related in this book. The topics included in it on this subject are of utmost significance and bound to provide incredible insights to readers. It aims to serve as a resource guide for students and experts alike and contribute to the growth of the discipline.

To facilitate a deeper understanding of the contents of this book a short introduction of every chapter is written below:

Chapter 1- The main focus of mobile computing is on the design and technology used in mobiles. Mobile computing includes mobile communication, hardware and also the software of any mobile. This chapter is an overview of the subject matter incorporating all the major aspects of mobile computing and technology.

Chapter 2- The various mobile computing devices discussed within this section are wearable computer, laptop, tablet computer, smart card and personal digital assistant. Wearable computers are devices that are small in size and are usually worn on the body, for example an Apple Watch. The section strategically encompasses and incorporates the various mobile computing devices, proving a complete understanding.

Chapter 3- Cellular networks are networks where the last link of the network uses wireless technology. Code division multiple access is the network used by radio technologies. Alternatively, the networks also used in mobile computing are code division multiple access, public switched telephone network, 4G, 5G and general packet radio service. The major networks of mobile computing are discussed in this chapter.

Chapter 4- Mobile computing management helps in the administration of mobile devices. It is a way of ensuring the productivity of employees and making sure they do not violate corporate policies. The section serves as a source to understand the major aspects of mobile computing management.

Chapter 5- Mobile phones are portable phones that a person can carry wherever they want to, and can receive and make calls while moving around. The applications of mobile phones mentioned in this text are text messaging, camera phone, Internet access, Bluetooth, mobile cloud computing etc. This section is an overview of the subject matter incorporating all the major aspects of mobile phones.

Chapter 6- The main concerns and challenges of mobile computing are mobile malware, spyware and ghost push. Mobile malware causes the system of a mobile to collapse whereas spyware is the software that collects information of people or companies without their knowledge. Spyware is usually used for malicious purposes, and is mostly used without the knowledge of the user. The aspects elucidated in this chapter are of vital importance, and provides a better understanding of mobile computing.

Chapter 7- Mobile security has become an alarming concern in the past few decades. The aspects explained in this text are browser security, wireless security camera and mobile secure gateway. This chapter helps the reader in developing an in depth understanding on the issue of mobile security.

I would like to share the credit of this book with my editorial team who worked tirelessly on this book. I owe the completion of this book to the never-ending support of my family, who supported me throughout the project.

Editor

Introduction to Mobile Computing

The main focus of mobile computing is on the design and technology used in mobiles. Mobile computing includes mobile communication, hardware and also the software of any mobile. This chapter is an overview of the subject matter incorporating all the major aspects of mobile computing and technology.

Mobile Computing

Mobile computing is human–computer interaction by which a computer is expected to be transported during normal usage, which allows for transmission of data, voice and video. Mobile computing involves mobile communication, mobile hardware, and mobile software. Communication issues include ad hoc networks and infrastructure networks as well as communication properties, protocols, data formats and concrete technologies. Hardware includes mobile devices or device components. Mobile software deals with the characteristics and requirements of mobile applications.

The Galaxy Nexus, capable of web browsing, e-mail access, video playback, document editing, image editing, among many other tasks common on smartphones. A smartphone is a tool of mobile computing.

Principles of Mobile Computing

Portability

Facilitates movement of device(s) within the mobile computing environments.

Connectivity

Ability to continuously stay connected with minimal amount of lag/downtime, without being affected by movements of the device.

Social Interactivity

Maintaining the connectivity to collaborate with other users, at least within the same environment.

Individuality

Adapting the technology to suit individual needs.

Devices

Some of the most common forms of mobile computing devices are as follows.

- portable computers, compacted lightweight units including a full character set keyboard and primarily intended as hosts for software that may be parameterized, as laptops, notebooks, notepads, etc.
- *mobile phones* including a restricted key set primarily intended but not restricted to for vocal communications, as smartphones, cell phones, feature phones, etc.
- Smart cards that can run multiple applications but typically payment, travel and secure area access
- *wearable computers*, mostly limited to functional keys and primarily intended as incorporation of software agents, as watches, wristbands, necklaces, keyless implants, etc.

The existence of these classes is expected to be long lasting, and complementary in personal usage, none replacing one the other in all features of convenience.

Other types of mobile computers have been introduced since the 1990s including the:

- Portable computer (discontinued)
- Personal digital assistant/Enterprise digital assistant (discontinued)
- Ultra-Mobile PC (discontinued)
- Laptop
- Smartphone
- Robots
- Tablet computer
- Wearable computer
- Carputer
- Application-specific computer

Limitations

- Range & Bandwidth: Mobile Internet access is generally slower than direct cable connections, using technologies such as GPRS and EDGE, and more recently HSDPA, HSUPA, 3G and 4G networks and also the upcoming 5G network. These networks are usually available within range of commercial cell phone towers. High speed network wireless LANs are inexpensive but have very limited range.

- Security standards: When working mobile, one is dependent on public networks, requiring careful use of VPN. Security is a major concern while concerning the mobile computing standards on the fleet. One can easily attack the VPN through a huge number of networks interconnected through the line.

- Power consumption: When a power outlet or portable generator is not available, mobile computers must rely entirely on battery power. Combined with the compact size of many mobile devices, this often means unusually expensive batteries must be used to obtain the necessary battery life.

- Transmission interferences: Weather, terrain, and the range from the nearest signal point can all interfere with signal reception. Reception in tunnels, some buildings, and rural areas is often poor.

- Potential health hazards: People who use mobile devices while driving are often distracted from driving and are thus assumed more likely to be involved in traffic accidents. (While this may seem obvious, there is considerable discussion about whether banning mobile device use while driving reduces accidents or not.) Cell phones may interfere with sensitive medical devices. Questions concerning mobile phone radiatiom and health have been raised.

- Human interface with device: Screens and keyboards tend to be small, which may make them hard to use. Alternate input methods such as speech or handwriting recognition require training.

In-vehicle Computing and Fleet Computing

The Compaq Portable - Circa 1982 pre-laptop

Many commercial and government field forces deploy a rugged portable computer with their fleet of vehicles. This requires the units to be anchored to the vehicle for driver safety, device security, and ergonomics. Rugged computers are rated for severe vibration associated with large service vehicles and off-road driving and the harsh environmental conditions of constant professional use such as in emergency medical services, fire, and public safety.

Other elements affecting function in vehicle:

- Operating temperature: A vehicle cabin can often experience temperature swings from -20F to +140F. Computers typically must be able to withstand these temperatures while operating. Typical fan-based cooling has stated limits of 95F-100F of ambient temperature, and temperatures below freezing require localized heaters to bring components up to operating temperature (based on independent studies by the SRI Group and by Panasonic R&D).

- Vibration can decrease the life expectancy of computer components, notably rotational storage such as HDDs.

- Visibility of standard screens becomes an issue in bright sunlight.

- Touchscreen users easily interact with the units in the field without removing gloves.

- High-temperature battery settings: Lithium ion batteries are sensitive to high temperature conditions for charging. A computer designed for the mobile environment should be designed with a high-temperature charging function that limits the charge to 85% or less of capacity.

- External antenna connections go through the typical metal cabins of vehicles which would block wireless reception, and take advantage of much more capable external communication and navigation equipment.

Security Issues Involved in Mobile

Mobile security or mobile phone security has become increasingly important in mobile computing. It is of particular concern as it relates to the security of personal information now stored on the smartphone.

More and more users and businesses use smartphones as communication tools but also as a means of planning and organizing their work and private life. Within companies, these technologies are causing profound changes in the organization of information systems and therefore they have become the source of new risks. Indeed, smartphones collect and compile an increasing amount of sensitive information to which access must be controlled to protect the privacy of the user and the intellectual property of the company.

All smartphones, as computers, are preferred targets of attacks. These attacks exploit weaknesses related to smartphones that can come from means of communication like SMS, MMS, wifi networks, and GSM. There are also attacks that exploit software vulnerabilities from both the web browser and operating system. Finally, there are forms of malicious software that rely on the weak knowledge of average users.

Different security counter-measures are being developed and applied to smartphones, from security in different layers of software to the dissemination of information to end users. There are good practices to be observed at all levels, from design to use, through the development of operating systems, software layers, and downloadable apps.

Portable Computing Devices

Several categories of portable computing devices can run on batteries but are not usually classified as laptops: portable computers, PDAs, ultra mobile PCs (UMPCs), tablets and smartphones.

- A portable computer (discontinued) is a general-purpose computer that can be easily moved from place to place, but cannot be used while in transit, usually because it requires some "setting-up" and an AC power source. The most famous example is the Osborne 1. Portable computers are also called a "transportable" or a "luggable" PC.

- A personal digital assistant (PDA) (discontinued) is a small, usually pocket-sized, computer with limited functionality. It is intended to supplement and to synchronize with a desktop computer, giving access to contacts, address book, notes, e-mail and other features.

A Palm TX PDA

- An ultra mobile PC (discontinued) is a full-featured, PDA-sized computer running a general-purpose operating system.

- A tablet computer that lacks a keyboard (also known as a non-convertible tablet) is shaped like a slate or a paper notebook. Instead a physical keyboard it has a touchscreen with some combination of virtual keyboard, stylus and/or handwriting recognition software. Tablets may not be best suited for applications requiring a physical keyboard for typing, but are otherwise capable of carrying out most of the tasks of an ordinary laptop.

- A smartphone has a wide range of features and install-able applications.

- A carputer is installed in an automobile. It operates as a wireless computer, sound system, GPS, and DVD player. It also contains word processing software and is bluetooth compatible.

- A |Pentop (discontinued) is a computing device the size and shape of a pen. It functions as a writing utensil, MP3 player, language translator, digital storage device, and calculator.

- An application-specific computer is one that is tailored to a particular application. For example, Ferranti introduced a handheld application-specific mobile computer (the MRT-100) in the form of a clipboard for conducting opinion polls.

Boundaries that separate these categories are blurry at times. For example, the OQO UMPC is also a PDA-sized tablet PC; the Apple eMate had the clamshell form factor of a laptop, but ran PDA software. The HP Omnibook line of laptops included some devices small more enough to be called ultra mobile PCs. The hardware of the Nokia 770 internet tablet is essentially the same as that of a PDA such as the Zaurus 6000; the only reason it's not called a PDA is that it does not have PIM software. On the other hand, both the 770 and the Zaurus can run some desktop Linux software, usually with modifications.

Mobile Data Communication

Wireless data connections used in mobile computing take three general forms so. Cellular data service uses technologies such as GSM, CDMA or GPRS, 3G networks such as W-CDMA, EDGE or CDMA2000. and more recently 4G networks such as LTE, LTE-Advanced. These networks are usually available within range of commercial cell towers. Wi-Fi connections offer higher performance, may be either on a private business network or accessed through public hotspots, and have a typical range of 100 feet indoors and up to 1000 feet outdoors. Satellite Internet access covers areas where cellular and Wi-Fi are not available and may be set up anywhere the user has a line of sight to the satellite's location, which for satellites in geostationary orbit means having an unobstructed view of the southern sky. Some enterprise deployments combine networks from multiple cellular networks or use a mix of cellular, Wi-Fi and satellite. When using a mix of networks, a mobile virtual private network (mobile VPN) not only handles the security concerns, but also performs the multiple network logins automatically and keeps the application connections alive to prevent crashes or data loss during network transitions or coverage loss.

Human–computer Interaction

Human–computer interaction (commonly referred to as HCI) researches the design and use of computer technology, focused on the interfaces between people (users) and computers. Researchers in the field of HCI both *observe* the ways in which humans interact with computers and *design* technologies that let humans interact with computers in novel ways.

As a field of research, human-computer interaction is situated at the intersection of computer science, behavioral sciences, design, media studies, and several other fields of study. The term was popularized by Stuart K. Card, Allen Newell, and Thomas P. Moran in their seminal 1983 book, *The Psychology of Human-Computer Interaction*, although the authors first used the term in 1980 and the first known use was in 1975. The term connotes that, unlike other tools with only limited uses (such as a hammer, useful for driving nails but not much else), a computer has many uses and this takes place as an open-ended dialog between the user and the computer. The notion of dialog likens human-computer interaction to human-to-human interaction, an analogy which is crucial to theoretical considerations in the field.

Introduction

Humans interact with computers in many ways; and the interface between humans and the computers they use is crucial to facilitating this interaction. Desktop applications, internet browsers, handheld computers, and computer kiosks make use of the prevalent graphical user interfaces (GUI) of today. Voice user interfaces (VUI) are used for speech recognition and synthesising systems, and the emerging multi-modal and gestalt User Interfaces (GUI) allow humans to engage with embodied character agents in a way that cannot be achieved with other interface paradigms. The growth in human-computer interaction field has been in quality of interaction, and in different branching in its history. Instead of designing regular interfaces, the different research branches have had different focus on the concepts of multimodality rather than unimodality, intelligent adaptive interfaces rather than command/action based ones, and finally active rather than passive interfaces

The Association for Computing Machinery (ACM) defines human-computer interaction as "a discipline concerned with the design, evaluation and implementation of interactive computing systems for human use and with the study of major phenomena surrounding them". An important facet of HCI is the securing of user satisfaction (or simply End User Computing Satisfaction). "Because human–computer interaction studies a human and a machine in communication, it draws from supporting knowledge on both the machine and the human side. On the machine side, techniques in computer graphics, operating systems, programming languages, and development environments are relevant. On the human side, communication theory, graphic and industrial design disciplines, linguistics, social sciences, cognitive psychology, social psychology, and human factors such as computer user satisfaction are relevant. And, of course, engineering and design methods are relevant." Due to the multidisciplinary nature of HCI, people with different backgrounds contribute to its success. HCI is also sometimes termed *human–machine interaction* (HMI), *man–machine interaction* (MMI) or *computer–human interaction* (CHI).

Poorly designed human-machine interfaces can lead to many unexpected problems. A classic example of this is the Three Mile Island accident, a nuclear meltdown accident, where investigations concluded that the design of the human–machine interface was at least partly responsible for the disaster. Similarly, accidents in aviation have resulted from manufacturers' decisions to use non-standard flight instrument or throttle quadrant layouts: even though the new designs were proposed to be superior in basic human–machine interaction, pilots had already ingrained the "standard" layout and thus the conceptually good idea actually had undesirable results.

Leading academic research centers include CMU's Human-Computer Interaction Institute, GVU Center at Georgia Tech, and the University of Maryland Human–Computer Interaction Lab.

Goals

Human–computer interaction studies the ways in which humans make, or don't make, use of computational artifacts, systems and infrastructures. In doing so, much of the research in the field seeks to *improve* human-computer interaction by improving the *usability* of computer interfaces. How usability is to be precisely understood, how it relates to other social and cultural values and when it is, and when it may not be a desirable property of computer interfaces is increasingly debated.

Much of the research in the field of human-computer interaction takes an interest in:

- Methods for designing novel computer interfaces, thereby optimizing a design for a desired property such as, e.g., learnability or efficiency of use.

- Methods for implementing interfaces, e.g., by means of software libraries.

- Methods for evaluating and comparing interfaces with respect to their usability and other desirable properties.

- Methods for studying human computer use and its sociocultural implications more broadly.

- Models and theories of human computer use as well as conceptual frameworks for the design of computer interfaces, such as, e.g., cognitivist user models, Activity Theory or ethnomethodological accounts of human computer use.

- Perspectives that critically reflect upon the values that underlie computational design, computer use and HCI research practice.

Visions of what researchers in the field seek to achieve vary. When pursuing a cognitivist perspective, researchers of HCI may seek to align computer interfaces with the mental model that humans have of their activities. When pursuing a post-cognitivist perspective, researchers of HCI may seek to align computer interfaces with existing social practices or existing sociocultural values.

Researchers in HCI are interested in developing new design methodologies, experimenting with new devices, prototyping new software and hardware systems, exploring new interaction paradigms, and developing models and theories of interaction.

Differences with Related Fields

HCI differs from human factors and ergonomics as HCI focuses more on users working specifically with computers, rather than other kinds of machines or designed artifacts. There is also a focus in HCI on how to implement the computer software and hardware mechanisms to support human–computer interaction. Thus, *human factors* is a broader term; HCI could be described as the human factors of computers – although some experts try to differentiate these areas.

HCI also differs from human factors in that there is less of a focus on repetitive work-oriented tasks and procedures, and much less emphasis on physical stress and the physical form or industrial design of the user interface, such as keyboards and mouse devices.

Three areas of study have substantial overlap with HCI even as the focus of inquiry shifts. In the study of personal information management (PIM), human interactions with the computer are placed in a larger informational context – people may work with many forms of information, some computer-based, many not (e.g., whiteboards, notebooks, sticky notes, refrigerator magnets) in order to understand and effect desired changes in their world. In computer-supported cooperative work (CSCW), emphasis is placed on the use of computing systems in support of the collaborative work of a group of people. The principles of human interaction management (HIM) extend the scope of CSCW to an organizational level and can be implemented without use of computers.

Design

Principles

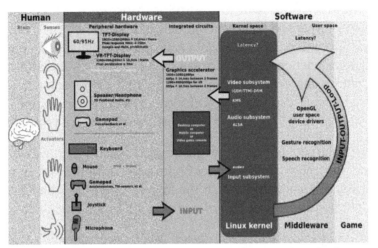

The user interacts directly with hardware for the human *input* and *output* such as displays, e.g. through a graphical user interface. The user interacts with the computer over this software interface using the given input and output (I/O) hardware. Software and hardware must be matched, so that the processing of the user input is fast enough, the latency of the computer output is not disruptive to the workflow.

When evaluating a current user interface, or designing a new user interface, it is important to keep in mind the following experimental design principles:

- Early focus on user(s) and task(s): Establish how many users are needed to perform the task(s) and determine who the appropriate users should be; someone who has never used the interface, and will not use the interface in the future, is most likely not a valid user. In addition, define the task(s) the users will be performing and how often the task(s) need to be performed.

- Empirical measurement: Test the interface early on with real users who come in contact with the interface on a daily basis. Keep in mind that results may vary with the performance level of the user and may not be an accurate depiction of the typical human-computer interaction. Establish quantitative usability specifics such as: the number of users performing the task(s), the time to complete the task(s), and the number of errors made during the task(s).

- Iterative design: After determining the users, tasks, and empirical measurements to include, perform the following iterative design steps:

  1. Design the user interface

  2. Test

  3. Analyze results

  4. Repeat

Repeat the iterative design process until a sensible, user-friendly interface is created.

Methodologies

A number of diverse methodologies outlining techniques for human–computer interaction design have emerged since the rise of the field in the 1980s. Most design methodologies stem from a model for how users, designers, and technical systems interact. Early methodologies, for example, treated users' cognitive processes as predictable and quantifiable and encouraged design practitioners to look to cognitive science results in areas such as memory and attention when designing user interfaces. Modern models tend to focus on a constant feedback and conversation between users, designers, and engineers and push for technical systems to be wrapped around the types of experiences users want to have, rather than wrapping user experience around a completed system.

- Activity theory: used in HCI to define and study the context in which human interactions with computers take place. Activity theory provides a framework to reason about actions in these contexts, analytical tools with the format of checklists of items that researchers should consider, and informs design of interactions from an activity-centric perspective.

- User-centered design: user-centered design (UCD) is a modern, widely practiced design philosophy rooted in the idea that users must take center-stage in the design of any computer system. Users, designers and technical practitioners work together to articulate the wants, needs and limitations of the user and create a system that addresses these elements. Often, user-centered design projects are informed by ethnographic studies of the environments in which users will be interacting with the system. This practice is similar but not identical to participatory design, which emphasizes the possibility for end-users to contribute actively through shared design sessions and workshops.

- Principles of user interface design: these are seven principles of user interface design that may be considered at any time during the design of a user interface in any order: tolerance, simplicity, visibility, affordance, consistency, structure and feedback.

- Value sensitive design: Value Sensitive Design (VSD) is a method for building technology that account for the values of the people who use the technology directly, as well as those who the technology affects, either directly or indirectly. VSD uses an iterative design process that involves three types of investigations: conceptual, empirical and technical. Conceptual investigations aim at understanding and articulating the various stakeholders of the technology, as well as their values and any values conflicts that might arise for these stakeholders through the use of the technology. Empirical investigations are qualitative or quantitative design research studies used to inform the designers' understanding of the users' values, needs, and practices. Technical investigations can involve either analysis of how people use related technologies, or the design of systems to support values identified in the conceptual and empirical investigations.

Display Designs

Displays are human-made artifacts designed to support the perception of relevant system variables and to facilitate further processing of that information. Before a display is designed, the task that the display is intended to support must be defined (e.g. navigating, controlling, decision making, learning, entertaining, etc.). A user or operator must be able to process whatever information that a system generates and displays; therefore, the information must be displayed according to principles in a manner that will support perception, situation awareness, and understanding.

Thirteen Principles of Display Design

Christopher Wickens et al. defined 13 principles of display design in their book *An Introduction to Human Factors Engineering*.

These principles of human perception and information processing can be utilized to create an effective display design. A reduction in errors, a reduction in required training time, an increase in efficiency, and an increase in user satisfaction are a few of the many potential benefits that can be achieved through utilization of these principles.

Certain principles may not be applicable to different displays or situations. Some principles may seem to be conflicting, and there is no simple solution to say that one principle is more important than another. The principles may be tailored to a specific design or situation. Striking a functional balance among the principles is critical for an effective design.

Perceptual Principles

1. Make displays legible (or audible). A display's legibility is critical and necessary for designing a usable display. If the characters or objects being displayed cannot be discernible, then the operator cannot effectively make use of them.

2. Avoid absolute judgment limits. Do not ask the user to determine the level of a variable on the basis of a single sensory variable (e.g. colour, size, loudness). These sensory variables can contain many possible levels.

3. Top-down processing. Signals are likely perceived and interpreted in accordance with what is expected based on a user's experience. If a signal is presented contrary to the user's expectation, more physical evidence of that signal may need to be presented to assure that it is understood correctly.

4. Redundancy gain. If a signal is presented more than once, it is more likely that it will be understood correctly. This can be done by presenting the signal in alternative physical forms (e.g. colour and shape, voice and print, etc.), as redundancy does not imply repetition. A traffic light is a good example of redundancy, as colour and position are redundant.

5. Similarity causes confusion: Use discriminable elements. Signals that appear to be similar will likely be confused. The ratio of similar features to different features causes signals to be similar. For example, A423B9 is more similar to A423B8 than 92 is to 93. Unnecessary similar features should be removed and dissimilar features should be highlighted.

Mental Model Principles

6. Principle of pictorial realism. A display should look like the variable that it represents (e.g. high temperature on a thermometer shown as a higher vertical level). If there are multiple elements, they can be configured in a manner that looks like it would in the represented environment.

7. Principle of the moving part. Moving elements should move in a pattern and direction compatible with the user's mental model of how it actually moves in the system. For example, the moving element on an altimeter should move upward with increasing altitude.

Principles Based on Attention

8. Minimizing information access cost. When the user's attention is diverted from one location to another to access necessary information, there is an associated cost in time or effort. A display design should minimize this cost by allowing for frequently accessed sources to be located at the nearest possible position. However, adequate legibility should not be sacrificed to reduce this cost.

9. Proximity compatibility principle. Divided attention between two information sources may be necessary for the completion of one task. These sources must be mentally integrated and are defined to have close mental proximity. Information access costs should be low, which can be achieved in many ways (e.g. proximity, linkage by common colours, patterns, shapes, etc.). However, close display proximity can be harmful by causing too much clutter.

10. Principle of multiple resources. A user can more easily process information across different resources. For example, visual and auditory information can be presented simultaneously rather than presenting all visual or all auditory information.

Memory Principles

11. Replace memory with visual information: knowledge in the world. A user should not need to retain important information solely in working memory or retrieve it from long-term memory. A menu, checklist, or another display can aid the user by easing the use of their memory. However, the use of memory may sometimes benefit the user by eliminating the need to reference some type of knowledge in the world (e.g., an expert computer operator would rather use direct commands from memory than refer to a manual). The use of knowledge in a user's head and knowledge in the world must be balanced for an effective design.

12. Principle of predictive aiding. Proactive actions are usually more effective than reactive actions. A display should attempt to eliminate resource-demanding cognitive tasks and replace them with simpler perceptual tasks to reduce the use of the user's mental resources. This will allow the user to focus on current conditions, and to consider possible future conditions. An example of a predictive aid is a road sign displaying the distance to a certain destination.

13. Principle of consistency. Old habits from other displays will easily transfer to support processing of new displays if they are designed consistently. A user's long-term memory will trigger actions that are expected to be appropriate. A design must accept this fact and utilize consistency among different displays.

Human–computer Interface

The human–computer interface can be described as the point of communication between the human user and the computer. The flow of information between the human and computer is defined as the *loop of interaction*. The loop of interaction has several aspects to it, including:

- Visual Based :The visual based human computer inter-action is probably the most wide-spread area in HCI research.

- Audio Based : The audio based interaction between a computer and a human is another important area of in HCI systems. This area deals with information acquired by different audio signals.

- *Task environment*: The conditions and goals set upon the user.

- *Machine environment*: The environment that the computer is connected to, e.g. a laptop in a college student's dorm room.

- *Areas of the interface*: Non-overlapping areas involve processes of the human and computer not pertaining to their interaction. Meanwhile, the overlapping areas only concern themselves with the processes pertaining to their interaction.

- *Input flow*: The flow of information that begins in the task environment, when the user has some task that requires using their computer.

- *Output*: The flow of information that originates in the machine environment.

- *Feedback*: Loops through the interface that evaluate, moderate, and confirm processes as they pass from the human through the interface to the computer and back.

- *Fit*: This is the match between the computer design, the user and the task to optimize the human resources needed to accomplish the task.

Current Research

Topics in HCI include:

User Customization

End-user development studies how ordinary users could routinely tailor applications to their own needs and use this power to invent new applications based on their understanding of their own domains. With their deeper knowledge of their own knowledge domains, users could increasingly be important sources of new applications at the expense of generic systems programmers (with systems expertise but low domain expertise).

Embedded Computation

Computation is passing beyond computers into every object for which uses can be found. Embedded systems make the environment alive with little computations and automated processes, from computerized cooking appliances to lighting and plumbing fixtures to window blinds to automobile braking systems to greeting cards. To some extent, this development is already taking place. The expected difference in the future is the addition of networked communications that will allow many of these embedded computations to coordinate with each other and with the user. Human interfaces to these embedded devices will in many cases be very different from those appropriate to workstations.

Augmented Reality

A common staple of science fiction, augmented reality refers to the notion of layering relevant information into our vision of the world. Existing projects show real-time statistics to users performing difficult tasks, such as manufacturing. Future work might include augmenting our social interactions by providing additional information about those we converse with.

Social Computing

In recent years, there has been an explosion of social science research focusing on interactions as the unit of analysis. Much of this research draws from psychology, social psychology, and sociology. For example, one study found out that people expected a computer with a man's name to cost more than a machine with a woman's name. Other research finds that individuals perceive their interactions with computers more positively than humans, despite behaving the same way towards these machines.

Knowledge-driven Human-computer Interaction

In human and computer interactions, there usually exists a semantic gap between human and computer's understandings towards mutual behaviors. Ontology (information science), as a formal representation of domain-specific knowledge, can be used to address this problem, through solving the semantic ambiguities between the two parties.

Factors of Change

Traditionally, as explained in a journal article discussing user modeling and user-adapted interaction, computer use was modeled as a human-computer dyad in which the two were connected by a narrow explicit communication channel, such as text-based terminals. Much work has been done to make the interaction between a computing system and a human. However, as stated in the introduction, there is much room for mishaps and failure. Because of this, human-computer interaction shifted focus beyond the interface (to respond to observations as articulated by D. Engelbart: "If ease of use was the only valid criterion, people would stick to tricycles and never try bicycles."

The means by which humans interact with computers continues to evolve rapidly. Human–computer interaction is affected by the forces shaping the nature of future computing. These forces include:

- Decreasing hardware costs leading to larger memory and faster systems

- Miniaturization of hardware leading to portability

- Reduction in power requirements leading to portability

- New display technologies leading to the packaging of computational devices in new forms

- Specialized hardware leading to new functions

- Increased development of network communication and distributed computing

- Increasingly widespread use of computers, especially by people who are outside of the computing profession

- Increasing innovation in input techniques (e.g., voice, gesture, pen), combined with lowering cost, leading to rapid computerization by people formerly left out of the *computer revolution.*

- Wider social concerns leading to improved access to computers by currently disadvantaged groups

The future for HCI, based on current promising research, is expected to include the following characteristics:

- *Ubiquitous computing and communication.* Computers are expected to communicate through high speed local networks, nationally over wide-area networks, and portably via infrared, ultrasonic, cellular, and other technologies. Data and computational services will be portably accessible from many if not most locations to which a user travels.

- *High-functionality systems.* Systems can have large numbers of functions associated with them. There are so many systems that most users, technical or non-technical, do not have time to learn them in the traditional way (e.g., through thick manuals).

- *Mass availability of computer graphics.* Computer graphics capabilities such as image processing, graphics transformations, rendering, and interactive animation are becoming widespread as inexpensive chips become available for inclusion in general workstations and mobile devices.

- *Mixed media.* Commercial systems can handle images, voice, sounds, video, text, formatted data. These are exchangeable over communication links among users. The separate fields of consumer electronics (e.g., stereo sets, VCRs, televisions) and computers are merging partly. Computer and print fields are expected to cross-assimilate.

- *High-bandwidth interaction.* The rate at which humans and machines interact is expected to increase substantially due to the changes in speed, computer graphics, new media, and new input/output devices. This can lead to some qualitatively different interfaces, such as virtual reality or computational video.

- *Large and thin displays.* New display technologies are finally maturing, enabling very large displays and displays that are thin, lightweight, and low in power use. This is having large effects on portability and will likely enable developing paper-like, pen-based computer interaction systems very different in feel from desktop workstations of the present.

- *Information utilities.* Public information utilities (such as home banking and shopping) and specialized industry services (e.g., weather for pilots) are expected to proliferate. The rate of proliferation can accelerate with the introduction of high-bandwidth interaction and the improvement in quality of interfaces.

Scientific Conferences

One of the main conferences for new research in human-computer interaction is the annually held Association for Computing Machinery's (ACM) *Conference on Human Factors in Computing Systems*, usually referred to by its short name CHI (pronounced *kai*, or *khai*). CHI is organized by ACM Special Interest Group on Computer–Human Interaction (SIGCHI). CHI is a large conference, with thousands of attendants, and is quite broad in scope. It is attended by academics, practitioners and industry people, with company sponsors such as Google, Microsoft, and PayPal.

There are also dozens of other smaller, regional or specialized HCI-related conferences held around the world each year, including:

- ASSETS: ACM International Conference on Computers and Accessibility
- CSCW: ACM conference on Computer Supported Cooperative Work
- CC: Aarhus decennial conference on Critical Computing
- DIS: ACM conference on Designing Interactive Systems
- ECSCW: European Conference on Computer-Supported Cooperative Work
- GROUP: ACM conference on supporting group work
- HRI: ACM/IEEE International Conference on Human–robot interaction
- ICMI: International Conference on Multimodal Interfaces
- ITS: ACM conference on Interactive Tabletops and Surfaces
- MobileHCI: International Conference on Human–Computer Interaction with Mobile Devices and Services
- NIME: International Conference on New Interfaces for Musical Expression
- OzCHI: Australian Conference on Human-Computer Interaction
- TEI: International Conference on Tangible, Embedded and Embodied Interaction
- Ubicomp: International Conference on Ubiquitous computing
- UIST: ACM Symposium on User Interface Software and Technology
- i-USEr: International Conference on User Science and Engineering
- INTERACT: IFIP TC13 Conference on Human-Computer Interaction

Mobile Interaction

Mobile interaction is the study of interaction between mobile users and computers. Mobile interaction is an aspect of human–computer interaction that emerged when computers became small enough to enable mobile usage around 1990's.

Mobile devices are a pervasive part of our everyday lives. People use mobile phones, PDAs, and portable media players almost everywhere. These devices are the first truly pervasive interaction devices that are currently used for a huge variety of services and applications. Mobile devices affect the way we interact, share, and communicate with others. They are growing in diversity and complexity, featuring new interaction paradigms, modalities, shapes, and purposes (e.g., e-readers, portable media players, handheld game consoles). The strong differentiating factors that characterize mobile devices from traditional personal computing (e.g., desktop computers), are their ubiquitous use, usual small size, and mixed interaction modalities.

Mobile Phone Device

The history of mobile interaction includes different design trends. The main six design trends are Portability, Miniaturization, Connectivity, Convergence, Divergence, and Apps. The main reason behind those trends is to understand the requirements and needs of mobile users which is the main goal for mobile interaction. Mobile interaction is a multidisciplinary area with various academic subjects making contributions to it. The main disciplines involved in mobile interaction are Psychology, Computer Science, Sociology, Design, and Information Systems. The processes in mobile interaction design includes three main activities: understanding users, developing prototype designs, and evaluation.

History

The history of mobile interaction can be divided into a number of eras, or waves, each characterized by a particular technological focus, interaction design trends, and by leading to fundamental changes in the design and use of mobile devices. Although not strictly sequential, they provide a good overview of the legacy on which current mobile computing research and design is built.

1. Portability

   One of the first work in the mobile interaction discipline was the concept of the Dynabook by Alan Kay in 1968. However, at that time the necessary hardware to build such system was not available. When the first laptops were built in the early 1980s they were seen as transportable desktop computers.

2. Miniaturization

   By the early 1990s, many types of handheld devices were introduced such as labelled palmtop computers, digital organizers, or personal digital assistants (PDAs).

3. Connectivity

    By 1973, Martin Cooper worked at Motorola developed a handheld mobile phone concept, which later on by 1983, led to the introduction of the first commercial mobile phone called the DynaTAC 8000X.

Apple iPhone

4. Convergence

    During this era, different types of specialized mobile devices started to converge into new types of hybrid devices with primarily different form factors and interaction designs. On 1992, the first device of such technique, the "smartphones" was introduced. The first smart phone was the IBM Simon and it was used for making phone calls, calendars, addresses, notes, e-mail, fax and games.

5. Divergence

    During the 2000s, a trend toward a single function many devices started to spread. the basic idea behind divergence is that specialized tools facilitate optimization of functionality over time and enhancement of use. The most famous device of this era was the Apple iPod on 2001.

6. Apps

    During 2007, Apple Inc. introduced the first truly "smart" cellular phone; the iPhone. It was a converged mobile device with different features functionality. The most important thing is that it represents a significant rethinking of the design of mobile interactions and a series of notable interaction design choices. In less than a decade Apple Inc. would sell over one-billion iPhones

Goals

With the evolution of both software and hardware on the mobile devices, the users are becoming more demanding of the user interface that provide both functionality and pleasant user experience. The goal of mobile interaction researches is to understand the requirements and needs of mobile users. Compared with stationary devices mobile devices have specific, often restricted, input and output requirements. A goal that is often named is to overcome the limitations of mobile devices. However, exploiting the special opportunities of mobile usage can also be seen as a central goal.

Disciplines Involved in Mobile Interaction

Mobile interaction is a multidisciplinary area with various academic subjects making contributions. This is a reflection of the complicated nature of an individual's interaction with a computer system. This includes factors such as an understanding of the user and the task the user wants to perform with the system, understanding of the design tools, software packages that are needed to achieve this and an understanding of software engineering tools. The following are the main disciplines involved in mobile interaction:

1. Psychology

Many of the research methods and system evaluation techniques currently used in mobile Human Computer Interaction research are borrowed from Psychology. As well as attitude measures, performance measures that are used in mobile Human Computer Interaction research studies come from the area of experimental psychology. Understanding users and their needs is a key aspect in the design of mobile systems, devices, and applications so that they will be easy and enjoyable to use. Individual user characteristics such as age, or personality physical disabilities such as blindness, all have an affect on users' performance when they are using mobile applications and systems, and these individual differences can also affect people's attitude towards the mobile service or device that they interact with.

2. Computer Science

Computer Science (along with Software Engineering) is responsible for providing software tools to develop the interfaces that users need to interact with system. These include the software development tools.

3. Sociology

Sociologists working in this area are responsible for looking at socio-technical aspects of Human Computer Interaction. They bring methods and techniques from the social sciences (e.g., observational studies, ethnography) that can be used in the design and evaluation of mobile devices and applications.

4. Design

People working in this area are concerned with looking at the design layout of the interface (e.g., colors, positioning of text or graphics on a screen of a PDA). This is a crucial area of mobile Human Computer Interaction research due to the limited screen space available for most mobile devices. Therefore, it is crucial that services and applications reflect this limitation by reducing information complexity to fit the parameters of the mobile device, without losing any substantial content.

5. Information Systems

People who work in this area are interested in investigating how people interact with information and technologies in an organisational, managerial, and business context. In an organisational context, information system professionals and researchers are interested in looking at ways in which mobile technologies and mobile applications can be used to make an organisation more effective in conducting its business on a day-to-day business.

Mobile Interaction Design

Mobile interaction design is part of the interaction design which heavily focused on satisfying the needs and desires of the majority of people who will use the product. The processes in mobile interaction design are in the following main types of activity:

1. Understanding users - having a sense of people's capabilities and limitations; gaining a rich picture of what makes up the detail of their lives, the things they do and use.

2. Developing prototype designs - representing a proposed interaction design in such a way that it can be demonstrated, altered, and discussed.

3. Evaluation - each prototype is a stepping stone to the next, better, refined design. Evaluation techniques identify the strengths and weaknesses of a design but can also lead the team to propose a completely different approach, discarding the current line of design thinking for a radical approach.

Mobile Technology

Mobile technology is the technology used for cellular communication. Mobile code division multiple access (CDMA) technology has evolved rapidly over the past few years. Since the start of this millennium, a standard mobile device has gone from being no more than a simple two-way pager to being a mobile phone, GPS navigation device, an embedded web browser and instant messaging client, and a handheld game console. Many experts argue that the future of computer technology rests in mobile computing with wireless networking. Mobile computing by way of tablet computers are becoming more popular. Tablets are available on the 3G and 4G networks.

4G Networking

One of the most important features in the 4G mobile networks is the domination of high-speed packet transmissions or burst traffic in the channels. The same codes used in the 2G-3G networks is applied to 4G mobile or wireless networks, the detection of very short bursts will be a serious problem due to their very poor partial correlation properties. Recent study has indicated that traditional multilayer network architecture based on the Open Systems Interconnection (OSI) model may not be well suited for 4G mobile network, where transactions of short packets will be the major part of the traffic in the channels. As the packets from different mobiles carry completely different channel characteristics, the receiver should execute all necessary algorithms, such as channel estimation, interactions with all upper layers and so on, within a very short period of time.

Operating Systems

Many types of mobile operating systems (OS) are available for smartphones, including: Android, BlackBerry OS, webOS, iOS, Symbian, Windows Mobile Professional (touch screen), Windows Mobile Standard (non-touch screen), and Bada. Among the most popular are the Apple iPhone, and the newest – Android. Android is a mobile operating system (OS) developed by Google. Android is the first completely open source mobile OS, meaning that it is free to any cell phone mobile network.

Since 2008 customizable OSs allow the user to download applications ("apps") like games, GPS, Utilities, and other tools. Any user can also create their own Apps and publish them e.g. to Apple's App Store. The Palm Pre using webOS has functionality over the Internet and can support Internet-based programming languages such as Cascading Style Sheets (CSS), HTML, and JavaScript. The Research In Motion (RIM) BlackBerry is a smartphone with a multimedia player and third-party software installation. The Windows Mobile Professional Smartphones (Pocket PC or Windows Mobile PDA) are like that of a personal digital assistant (PDA) and have touchscreen abilities. The Windows Mobile Standard does not have a touch screen but uses a trackball, touchpad, rockers, etc.

Channel Hogging and File Sharing

There will be a hit to file sharing, the normal web surfer would want to look at a new web page every minute or so at 100 kbs a page loads quickly. Because of the changes to the security of wireless networks users will be unable to do huge file transfers because service providers want to reduce channel use. AT&T claimed that they would ban any of their users that they caught using *peer-to-peer* (P2P) file sharing applications on their 3G network. It then became apparent that it would keep any of their users from using their iTunes programs. The users would then be forced to find a Wi-Fi hotspot to be able to download files. The limits of wireless networking will not be cured by 4G, as there are too many fundamental differences between wireless networking and other means of Internet access. If wireless vendors do not realize these differences and bandwidth limits, future wireless customers will find themselves disappointed and the market may suffer setback

Future of Smartphone

The next generation of smartphones are going to be context-aware, taking advantage of the growing availability of embedded physical sensors and data exchange abilities. One of the main features applying to this is that the phones will start keeping track of your personal data, but adapt to anticipate the information you will need based on your intentions. There will be all-new applications coming out with the new phones, one of which is an X-Ray device that reveals information about any location at which you point your phone. One thing companies are developing software to take advantage of more accurate location-sensing data. How they described it was as wanting to make the phone a virtual mouse able to click the real world. An example of this is where you can point the phone's camera while having the live feed open and it will show text with the building and saving the location of the building for use in the future.

Along with the future of a smart phone comes the future of another device. Omnitouch is a device in which applications can be viewed and used on your hand, arm, wall, desk, or any other everyday surface. The device uses a sensor touch interface, which enables the user to access all the functions through the use of finger touch. It was developed at Carnegie Mellon University. This device uses a projector and camera that is worn on the person's shoulder, with no controls other than the user's fingers.

References

- Kjeldskov, Jesper (2014). Mobile Interactions in Context: A Designerly Way Toward Digital Ecology. Morgan & Claypool. ISBN 9781627052269.

- Cooper, Alan; Reimann, Robert; Cronin, Dave (2007). About Face 3: The Essentials of Interaction Design. Indianapolis, Indiana: Wiley. p. 610. ISBN 978-0-470-08411-3.

- Zimmermann, Andreas; Henze, Niels; Righetti, Xavier; Rukzio, Enrico (2009). "Mobile Interaction with the Real World". MobileHCI'09, Article No. 106. doi:10.1145/1613858.1613980. ISBN 978-1-60558-281-8.

- Grudin, Jonathan (1992). "Utility and usability: research issues and development contexts". Interacting with Computers. 4 (2): 209–217. doi:10.1016/0953-5438(92)90005-z. Retrieved 7 March 2015.

- Chalmers, Matthew; Galani, Areti. "Seamful interweaving: heterogeneity in the theory and design of interactive systems". Proceedings of the 5th conference on Designing interactive systems: processes, practices, methods, and techniques: 243–252. Retrieved 7 March 2015.

- Barkhuus, Louise; Polichar, Valerie E. (2011). "Empowerment through seamfulness: smart phones in everyday life". Personal and Ubiquitous Computing. 15 (6): 629–639. doi:10.1007/s00779-010-0342-4. Retrieved 7 March 2015.

- Rogers, Yvonne (2012). "HCI Theory: Classical, Modern, and Contemporary". Synthesis Lectures on Human-Centered Informatics. 5: 1–129. doi:10.2200/S00418ED1V01Y201205HCI014. Retrieved 7 March 2015.

- Suchman, Lucy (1987). Plans and Situated Action. The Problem of Human-Machine Communication. New York, Cambridge: Cambridge University Press. Retrieved 7 March 2015.

- Hewett; Baecker; Card; Carey; Gasen; Mantei; Perlman; Strong; Verplank. "ACM SIGCHI Curricula for Human-Computer Interaction". ACM SIGCHI. Retrieved 15 July 2014.

Various Mobile Computing Devices

The various mobile computing devices discussed within this section are wearable computer, laptop, tablet computer, smart card and personal digital assistant. Wearable computers are devices that are small in size and are usually worn on the body, for example an Apple Watch. The section strategically encompasses and incorporates the various mobile computing devices, proving a complete understanding.

Wearable Computer

Wearable computers, also known as body-borne computers or wearables are miniature electronic devices that are worn under, with or on top of clothing. This class of wearable technology has been developed for general or special purpose information technologies. It is also used in media development. Wearable computers are especially useful for applications that require more complex computational support, such as accelerometers or gyroscopes, than just hardware coded logic.

The Apple Watch, released in 2015.

Wearable computing devices are variously defined. For example, consumers often refer to wearable computers as computers that can be easily carried on the body, or systems with a heads-up display or speech activated. This contrasts with academics that define wearables as a system that can perform a set of functions without being constrained by the physical hardware of the system. Merchandiser often use the broadest definition, as any computing device worn on the body. This article page will use the broadest definition.

Smartwatches and the Fitbit system are the most common form, worn on the wrist. Google Glass is an optical head-mounted display supplying a augmented reality perspective, controlled by novel gestural movements.

One common feature of wearable computers is their persistence of activity. There is constant interaction between the wearable and user, so there is no need to turn the device on or off. Another feature is the ability to multi-task. When using a wearable computer, there is no need to stop what one is doing to use the device; its functionality blends seamlessly into all other user actions. These devices can be used by the wearer to act as a prosthetic. It may therefore be an extension of the user's mind or body.

Many issues are common to wearables as with mobile computing, ambient intelligence and ubiquitous computing research communities. These include power management and heat dissipation, software architectures, wireless and personal area networks, and data management, all of which are essential for overall data quality and trust in the device.

Areas of Applications

In many applications, the user's body is actively engaged as the device's interface. This usually includes: the skin, hands, voice, eyes, and arms. Wearables are also receptive to any motion or attention.

Wearable computer items have been developed and applied in the following:

- sensory integration, e.g. to help people see better or understand the world better (whether in task-specific applications like camera-based welding helmets or for everyday use like computerized "digital eyeglass"),

- behavioral modeling,

- health care monitoring systems,

- service management

- mobile phones

- smartphones

- electronic textiles

- fashion design

Today "wearable computing" is still a topic of active research, with areas of study including user interface design, augmented reality, and pattern recognition. The use of wearables for specific applications, for compensating disabilities or supporting elderly people steadily increases. The application of wearable computers in fashion design is evident through Microsoft's prototype of "The Printing Dress" at the International Symposium on Wearable Computers in June 2011.

History

Due to the varied definitions of "wearable" and "computer", the first wearable computer could be as early as the first abacus on a necklace, a 16th-century abacus ring, the first wristwatch made by Breguet for the Queen of Naples in 1810, or the covert timing devices hidden in shoes to cheat at roulette by Thorp and Shannon in the 1960s and 1970s.

However, a computer is not merely a time-keeping or calculating device, but rather a user-programmable item for complex algorithms, interfacing, and data management. By this definition, the wearable computer was invented by Steve Mann, in the late 1970s:

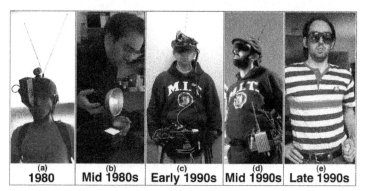

Evolution of Steve Mann's *WearComp* wearable computer from backpack based systems of the 1980s to his current covert systems.

Steve Mann, a professor at the University of Toronto, was hailed as the father of the wearable computer and the ISSCC's first virtual panelist, by moderator Woodward Yang of Harvard University (Cambridge Mass.).

— IEEE ISSCC 8 Feb. 2000

The development of wearable items has taken several steps of miniaturization from discrete electronics over hybrid designs to fully integrated designs, where just one processor chip, a battery and some interface conditioning items make the whole unit.

1500s

Queen Elizabeth I of England received the wristwatch from Robert Dudley in 1571 as a new year's present.

1600s

The Qing Dynasty saw the introduction of a fully functional abacus on a ring, which could be used while it was being worn.

1800s

The first wearable timepiece was made by watchmaker Breguet for the Queen of Naples in 1810. It was a small ladies' pocket watch on a bracelet chain. A wristwatch is a "wearable computer" in the sense that it can be worn and that it also computes time. But it is not a general-purpose computer in the sense of the modern word.

Girard-Perregaux made wristwatches for the German Imperial Navy after an artillery officer complained that it was not convenient to use both hands to operate a pocket watch while timing his bombardments. The officer had strapped a pocket watch onto his wrist and his superiors liked his solution, and thus asked La Chaux-de-Fonds to travel to Berlin to begin production of small pocket watches attached to wrist bracelets.

Early acceptance of wristlets by men serving in the military was not widespread, though:

Wristlets, as they were called, were reserved for women, and considered more of a passing fad than a serious timepiece. In fact, they were held in such disdain that many a gentlemen were actually quoted to say they "would sooner wear a skirt as wear a wristwatch".

—International Watch Magazine

1960s

In 1961, mathematicians Edward O. Thorp and Claude Shannon built some computerized timing devices to help them cheat at a game of roulette. One such timer was concealed in a shoe and another in a pack of cigarettes. Various versions of this apparatus were built in the 1960s and 1970s. Detailed pictures of a shoe-based timing device can be viewed at www.eyetap.org.

Thorp refers to himself as the inventor of the first "wearable computer" In other variations, the system was a concealed cigarette-pack sized analog computer designed to predict the motion of roulette wheels. A data-taker would use microswitches hidden in his shoes to indicate the speed of the roulette wheel, and the computer would indicate an octant of the roulette wheel to bet on by sending musical tones via radio to a miniature speaker hidden in a collaborator's ear canal. The system was successfully tested in Las Vegas in June 1961, but hardware issues with the speaker wires prevented it from being used beyond test runs. This was not a wearable computer, because it could not be re-purposed during use; rather it was an example of task-specific hardware. This work was kept secret until it was first mentioned in Thorp's book *Beat the Dealer* (revised ed.) in 1966 and later published in detail in 1969.

1970s

The 1970s saw the rise of special purpose hardware timing devices, similar to the ones from the 1960s, such as roulette prediction devices using next-generation technology. In particular, a group known as Eudaemonic Enterprises used a CMOS 6502 microprocessor with 5K RAM to create a shoe computer with inductive radio communications between a data-taker and bettor.

Another early wearable system was a camera-to-tactile vest for the blind, published by C.C. Collins in 1977, that converted images into a 1024-point, 10-inch square tactile grid on a vest. On the consumer end, 1977 also saw the introduction of the HP-01 algebraic calculator watch by Hewlett-Packard.

1980s

The 1980s saw the rise of more general-purpose wearable computers that fit the modern definition of "computer" by going beyond task-specific hardware to more general-purpose (e.g. re-programmable by the user) devices. In 1981, Steve Mann designed and built a backpack-mounted 6502-based wearable multimedia computer with text, graphics, and multimedia capability, as well as video capability (cameras and other photographic systems). Mann went on to be an early and active researcher in the wearables field, especially known for his 1994 creation of the Wearable Wireless Webcam, the first example of Lifelogging.

Seiko Epson released the RC-20 Wrist Computer in 1984. It was an early smartwatch, powered by a computer on a chip.

Though perhaps not technically "wearable," in 1986 Steve Roberts built Winnebiko-II, a recumbent bicycle with on-board computer and chorded keyboard. Winnebiko II was the first of Steve Roberts' forays into nomadic computing that allowed him to type while riding.

In 1989, Reflection Technology marketed the Private Eye head-mounted display, which scans a vertical array of LEDs across the visual field using a vibrating mirror. This display gave rise to several hobbyist and research wearables, including Gerald "Chip" Maguire's IBM / Columbia University Student Electronic Notebook, Doug Platt's Hip-PC, and Carnegie Mellon University's VuMan 1 in 1991.

The Student Electronic Notebook consisted of the Private Eye, Toshiba diskless AIX notebook computers (prototypes), a stylus based input system and a virtual keyboard. It used direct-sequence spread spectrum radio links to provide all the usual TCP/IP based services, including NFS mounted file systems and X11, which all ran in the Andrew Project environment.

The Hip-PC included an Agenda palmtop used as a chording keyboard attached to the belt and a 1.44 megabyte floppy drive. Later versions incorporated additional equipment from Park Engineering. The system debuted at "The Lap and Palmtop Expo" on 16 April 1991.

VuMan 1 was developed as part of a Summer-term course at Carnegie Mellon's Engineering Design Research Center, and was intended for viewing house blueprints. Input was through a three-button unit worn on the belt, and output was through Reflection Tech's Private Eye. The CPU was an 8 MHz 80188 processor with 0.5 MB ROM.

1990s

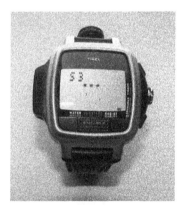

Timex Datalink USB Dress edition with *Invasion* video game. The watch crown (*icontrol*) can be used to move the defender left to right and the fire control is the Start/Split button on the lower side of the face of the watch at 6 o' clock.

In 1993, the Private Eye was used in Thad Starner's wearable, based on Doug Platt's system and built from a kit from Park Enterprises, a Private Eye display on loan from Devon Sean McCullough, and the Twiddler chording keyboard made by Handykey. Many iterations later this system became the MIT "Tin Lizzy" wearable computer design, and Starner went on to become one of the founders of MIT's wearable computing project. 1993 also saw Columbia University's augmented-reality system known as KARMA (Knowledge-based Augmented Reality for Maintenance Assistance).

Users would wear a Private Eye display over one eye, giving an overlay effect when the real world was viewed with both eyes open. KARMA would overlay wireframe schematics and maintenance instructions on top of whatever was being repaired. For example, graphical wireframes on top of a laser printer would explain how to change the paper tray. The system used sensors attached to objects in the physical world to determine their locations, and the entire system ran tethered from a desktop computer.

In 1994, Edgar Matias and Mike Ruicci of the University of Toronto, debuted a "wrist computer." Their system presented an alternative approach to the emerging head-up display plus chord keyboard wearable. The system was built from a modified HP 95LX palmtop computer and a Half-QWERTY one-handed keyboard. With the keyboard and display modules strapped to the operator's forearms, text could be entered by bringing the wrists together and typing. The same technology was used by IBM researchers to create the half-keyboard "belt computer. Also in 1994, Mik Lamming and Mike Flynn at Xerox EuroPARC demonstrated the Forget-Me-Not, a wearable device that would record interactions with people and devices and store this information in a database for later query. It interacted via wireless transmitters in rooms and with equipment in the area to remember who was there, who was being talked to on the telephone, and what objects were in the room, allowing queries like "Who came by my office while I was on the phone to Mark?". As with the Toronto system, Forget-Me-Not was not based on a head-mounted display.

Also in 1994, DARPA started the Smart Modules Program to develop a modular, *humionic* approach to wearable and carryable computers, with the goal of producing a variety of products including computers, radios, navigation systems and human-computer interfaces that have both military and commercial use. In July 1996, DARPA went on to host the "Wearables in 2005" workshop, bringing together industrial, university, and military visionaries to work on the common theme of delivering computing to the individual. A follow-up conference was hosted by Boeing in August 1996, where plans were finalized to create a new academic conference on wearable computing. In October 1997, Carnegie Mellon University, MIT, and Georgia Tech co-hosted the IEEE International Symposium on Wearables Computers (ISWC) in Cambridge, Massachusetts. The symposium was a full academic conference with published proceedings and papers ranging from sensors and new hardware to new applications for wearable computers, with 382 people registered for the event.

2000s

Dr. Bruce H Thomas and Dr. Wayne Piekarski developed the Tinmith wearable computer system to support augmented reality. This work was first published internationally in 2000 at the ISWC conference. The work was carried out at the Wearable Computer Lab in the University of South Australia.

In 2002, as part of Kevin Warwick's Project Cyborg, Warwick's wife, Irena, wore a necklace which was electronically linked to Warwick's nervous system via an implanted electrode array The color of the necklace changed between red and blue dependent on the signals on Warwick's nervous system.

GoPro released their first product, the GoPro HERO 35mm, which began a successful franchise of wearable cameras. The cameras can be worn atop the head or around the wrist and are shock and waterproof. GoPro cameras are used by many athletes and extreme sports enthusiasts, a trend that became very apparent during the early 2010s.

In the late 2000s, various Chinese companies began producing mobile phones in the form of wristwatches, the descendants of which as of 2013 include the i5 and i6, which are GSM phones with 1.8 inch displays, and the ZGPAX s5 Android wristwatch phone.

2010s

LunaTik, a machined wristband attachment for the 6th-generation iPod Nano.

Standardization with IEEE, IETF, and several industry groups (e.g. Bluetooth) lead to more various interfacing under the WPAN (wireless personal area network). It also lead the WBAN (Wireless body area network) to offer new classification of designs for interfacing and networking.

The 6th-generation iPod Nano, released in September 2010, has a wristband attachment available to convert it into a wearable wristwatch computer.

The development of wearable computing spread to encompass rehabilitation engineering, ambulatory intervention treatment, life guard systems, and defense wearable systems.

Sony produced a wristwatch called Sony SmartWatch that must be paired with an Android phone. Once paired, it becomes an additional remote display and notification tool.

Fitbit released several wearable fitness trackers and the Fitbit Surge, a full smartwatch that is compatible with Android and iOS.

On April 11, 2012, Pebble launched a Kickstarter campaign to raise $100,000 for their initial smartwatch model. The campaign ended on May 18 with $10,266,844, over 100 times the fundraising target. Pebble has released several smartwatches since, including the Pebble Time and the Pebble Round.

Google Glass launched their optical head-mounted display (OHMD) to a test group of users in 2013, before it became available to the public on May 15, 2014. Google's mission was to produce a mass-market ubiquitous computer that displays information in a smartphone-like hands-free format that can interact with the Internet via natural language voice commands. Google Glass received criticism over privacy and safety concerns. On January 15, 2015, Google announced that it would stop producing the Google Glass prototype but would continue to develop the product.

According to Google, Project Glass was ready to "graduate" from Google X, the experimental phase of the project.

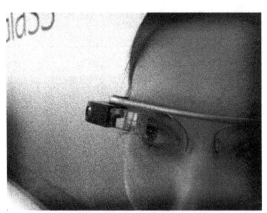

Google Glass, Google's head-mounted display, which was launched in 2013

Thync, a headset launched in 2014, is a wearable that stimulates the brain with mild electrical pulses, causing the wearer to feel energized or calm based on input into a phone app. The device is attached to the temple and to the back of the neck with an adhesive strip.

In January 2015, Intel announced the sub-miniature Intel Curie for wearable applications, based on its Intel Quark platform. As small as a button, it features a 6-axis accelerometer, a DSP sensor hub, a Bluetooth LE unit, and a battery charge controller. It was scheduled to ship in the second half of the year.

On April 24, 2015, Apple released their take on the smartwatch, known as the Apple Watch. The Apple Watch features a touchscreen, many applications, and a heart-rate sensor.

Commercialization

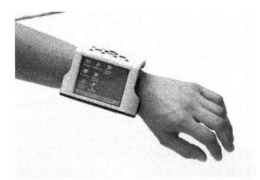

Image of the ZYPAD wrist wearable computer from Arcom Control Systems

The commercialization of general-purpose wearable computers, as led by companies such as Xybernaut, CDI and ViA, Inc. has thus far been met with limited success. Publicly traded Xybernaut tried forging alliances with companies such as IBM and Sony in order to make wearable computing widely available, and managed to get their equipment seen on such shows as The X-Files, but in 2005 their stock was delisted and the company filed for Chapter 11 bankruptcy protection amid financial scandal and federal investigation. Xybernaut emerged from bankruptcy protection in January, 2007. ViA, Inc. filed for bankruptcy in 2001 and subsequently ceased operations.

In 1998, Seiko marketed the Ruputer, a computer in a (fairly large) wristwatch, to mediocre returns. In 2001, IBM developed and publicly displayed two prototypes for a wristwatch computer running Linux. The last message about them dates to 2004, saying the device would cost about $250, but it is still under development. In 2002, Fossil, Inc. announced the Fossil Wrist PDA, which ran the Palm OS. Its release date was set for summer of 2003, but was delayed several times and was finally made available on January 5, 2005. Timex Datalink is another example of a practical wearable computer. Hitachi launched a wearable computer called Poma in 2002. Eurotech offers the ZYPAD, a wrist wearable touch screen computer with GPS, Wi-Fi and Bluetooth connectivity and which can run a number of custom applications. In 2013, a wearable computing device on the wrist to control body temperature was developed at MIT.

Evidence of weak market acceptance was demonstrated when Panasonic Computer Solutions Company's product failed. Panasonic has specialized in mobile computing with their Toughbook line for over 10 years and has extensive market research into the field of portable, wearable computing products. In 2002, Panasonic introduced a wearable brick computer coupled with a handheld or a touchscreen worn on the arm. The "Brick" Computer is the CF-07 Toughbook, dual batteries, screen used same batteries as the base, 800 x 600 resolution, optional GPS and WWAN. Has one M-PCI slot and one PCMCIA slot for expansion. CPU used is a 600 MHz Pentium 3 factory under clocked to 300 MHz so it can stay cool passively as it has no fan. Micro DIM RAM is upgradeable. The screen can be used wirelessly on other computers. The brick would communicate wirelessly to the screen, and concurrently the brick would communicate wirelessly out to the internet or other networks. The wearable brick was quietly pulled from the market in 2005, while the screen evolved to a thin client touchscreen used with a handstrap.

Google has announced that it has been working on a head-mounted display-based wearable "augmented reality" device called Google Glass. An early version of the device was available to the US public from April 2013 until January 2015. Despite ending sales of the device through their Explorer Program, Google has stated that they plan to continue developing the technology.

LG and iriver produce earbud wearables measuring heart rate and other biometrics, as well as various activity metrics.

Greater response to commercialization has been found in creating devices with designated purposes rather than all-purpose. One example is the WSS1000. The WSS1000 is a wearable computer designed to make the work of inventory employees easier and more efficient. The device allows workers to scan the barcode of items and immediately enter the information into the company system. This removed the need for carrying a clipboard, removed error and confusion from hand written notes, and allowed workers the freedom of both hands while working; the system improves accuracy as well as efficiency.

Popular Culture

Many technologies for wearable computers derive their ideas from science fiction. There are many examples of ideas from popular movies that have become technologies or are technologies currently being developed.

- 3D User Interface: Devices to display usable interfaces wherever you are. Ex. Minority Report, Gate Workers in Zion (Matrix), etc.

- Intelligent Textiles: Clothing that can relay and collect information. Ex. Tron and also many Sci-Fi Military movies.

- Threat Glasses: Scan others in vicinity and assess threat-to-self level. Ex. Terminator 2 or 'Threep' Technology in Lock-In

- Computerized Contact Lenses: Ex. Mission Impossible 4

- Combat Suit Armor: Ex. Iron Man Suit

- Brain Nano-Bots to Store Memories in the Cloud: Ex. Total Recall

- Infrared Headsets: Can help identify suspects and see through walls. Ex. Robocop.

- Wrist worn Computers: Various abilities such as wearer's stats, area maps, inventory, flashlight, geiger counter (Pip Boy 3000) or communication, laser, detect poison in food, tracking device (Leela's Wrist Device). Ex. Pip Boy 3000 from Fallout games and Leela's Wrist Device from Futurama

Military Use

The wearable computer was introduced to the US Army in 1989 as a small computer that was meant to assist soldiers in battle. Since then, the concept has grown to include the = Land Warrior program and proposal for future systems. The most extensive military program in the wearables arena is the US Army's Land Warrior system, which will eventually be merged into the Future Force Warrior system.

F-INSAS is an Indian Military Project, designed largely with wearable computing.

Laptop

A laptop, often called a notebook or "notebook computer", is a small, portable personal computer with a "clamshell" form factor, an alphanumeric keyboard on the lower part of the "clamshell" and a thin LCD or LED computer screen on the upper portion, which is opened up to use the computer. Laptops are folded shut for transportation, and thus are suitable for mobile use. Although originally there was a distinction between laptops and notebooks, the former being bigger and heavier than the latter, as of 2014, there is often no longer any difference. Laptops are commonly used in a variety of settings, such as at work, in education, and for personal multimedia and home computer use.

A laptop combines the components, inputs, outputs, and capabilities of a desktop computer, including the display screen, small speakers, a keyboard, pointing devices (such as a touchpad or trackpad), a processor, and memory into a single unit. Most 2016-era laptops also have integrated webcams and built-in microphones. Some 2016-era laptops have touchscreens. Laptops can be powered either from an internal battery or by an external power supply from an AC adapter. Hardware specifications, such as the processor speed and memory capacity, significantly vary between different types, makes, models and price points. Design elements, form factor, and construction can also vary significantly between models depending on intended use. Examples of specialized models of laptops include rugged notebooks for use in construction or military applications, as well as low production cost laptops such as those from the One Laptop per Child organization, which incorporate features like solar charging and semi-flexible components not found on most laptop computers.

An Apple MacBook Pro

Portable computers, which later developed into modern laptops, were originally considered to be a small niche market, mostly for specialized field applications, such as in the military, for accountants, or for traveling sales representatives. As portable computers evolved into the modern laptop, they became widely used for a variety of purposes.

Term Variants

The terms *laptop* and *notebook* are used interchangeably to describe a portable computer in English, although in some parts of the world one or the other may be preferred. There is some question as to the original etymology and specificity of either term—the term *laptop* appears to have been coined in the early 1980s to describe a mobile computer which could be used on one's lap, and to distinguish these devices from earlier, much heavier, portable computers (often called "luggables"). The term "notebook" appears to have gained currency somewhat later as manufacturers started producing even smaller portable devices, further reducing their weight and size and incorporating a display roughly the size of A4 paper; these were marketed as *notebooks* to distinguish them from bulkier laptops. Regardless of the etymology, by the late 1990s, the terms were interchangeable.

History

Alan Kay with his 1972 "Dynabook" prototype (photo: 2008 in Mountain View, California)

The Epson HX-20 was first sold to the public in 1981

As the personal computer (PC) became feasible in 1971, the idea of a portable personal computer soon followed. A "personal, portable information manipulator" was imagined by Alan Kay at Xerox PARC in 1968, and described in his 1972 paper as the "Dynabook". The IBM Special Computer APL Machine Portable (SCAMP) was demonstrated in 1973. This prototype was based on the IBM PALM processor. The IBM 5100, the first commercially available portable computer, appeared in September 1975, and was based on the SCAMP prototype.

As 8-bit CPU machines became widely accepted, the number of portables increased rapidly. The Osborne 1, released in 1981, used the Zilog Z80 and weighed 23.6 pounds (10.7 kg). It had no battery, a 5 in (13 cm) CRT screen, and dual 5.25 in (13.3 cm) single-density floppy drives. In the same year the first laptop-sized portable computer, the Epson HX-20, was announced. The Epson had an LCD screen, a rechargeable battery, and a calculator-size printer in a 1.6 kg (3.5 lb) chassis. Both Tandy/RadioShack and HP also produced portable computers of varying designs during this period. The first laptops using the flip form factor appeared in the early 1980s. The Dulmont Magnum was released in Australia in 1981–82, but was not marketed internationally until 1984–85. The US$8,150 (US$20,020 today) GRiD Compass 1101, released in 1982, was used at NASA and by the military, among others. The Sharp PC-5000, Ampere and Gavilan SC released in 1983. The Gavilan SC was the first computer described as a "laptop" by its manufacturer, while the Ampere had a modern clamshell design.

From 1983 onward, several new input techniques were developed and included in laptops, including the touchpad (Gavilan SC, 1983), the pointing stick (IBM ThinkPad 700, 1992), and handwriting recognition (Linus Write-Top, 1987). Some CPUs, such as the 1990 Intel i386SL, were designed to use minimum power to increase battery life of portable computers and were supported by dynamic power management features such as Intel SpeedStep and AMD PowerNow! in some designs.

Displays reached 640x480 (VGA) resolution by 1988 (Compaq SLT/286), and color screens started becoming a common upgrade in 1991, with increases in resolution and screen size occurring frequently until the introduction of 17" screen laptops in 2003. Hard drives started to be used in portables, encouraged by the introduction of 3.5" drives in the late 1980s, and became common in laptops starting with the introduction of 2.5" and smaller drives around 1990; capacities have typically lagged behind physically larger desktop drives. Optical storage, read-only CD-ROM followed by writeable CD and later read-only or writeable DVD and Blu-ray players, became common in laptops early in the 2000s.

Classification

Apple MacBook Pro, a laptop with a traditional design

Sony VAIO P series, variant of a subnotebook

A Samsung Chromebook, typical netbook

Asus Transformer Pad, a hybrid tablet, powered by Android OS

Microsoft Surface Pro 3, a prominent 2-in-1 detachable

Alienware desktop replacement gaming laptop

Panasonic Toughbook CF-M34, a rugged laptop/subnotebook

Since the introduction of portable computers during late 1970s, their form has changed significantly, spawning a variety of visually and technologically differing subclasses. Except where there is a distinct legal trademark around a term (notably Ultrabook), there are rarely hard distinctions between these classes and their usage has varied over time and between different sources.

Traditional Laptop

The form of the traditional laptop computer is a clamshell, with a screen on one of its inner sides and a keyboard on the opposite, facing the screen. It can be easily folded to conserve space while traveling. The screen and keyboard are inaccessible while closed. Devices of this form are commonly called a 'traditional laptop' or notebook, particularly if they have a screen size of 11 to 17 inches measured diagonally and run a full-featured operating system like Windows 10, OS X, or

Linux. Traditional laptops are the most common form of laptops, although Chromebooks, Ultrabooks, convertibles and 2-in-1s (described below) are becoming more common, with similar performance being achieved in their more portable or affordable forms.

Subnotebook

A *subnotebook* or an *ultraportable*, is a laptop designed and marketed with an emphasis on portability (small size, low weight, and often longer battery life). Subnotebooks are usually smaller and lighter than standard laptops, weighing between 0.8 and 2 kg (2-5 lb), with a battery life exceeding 10 hours. Since the introduction of *netbooks* and *ultrabooks*, the line between *subnotebooks* and either category has blurred. Netbooks are a more basic and cheap type of subnotebook, and while some ultrabooks have a screen size too large to qualify as subnotebooks, certain ultrabooks fit in the subnotebook category. One notable example of a subnotebook is the Apple MacBook Air.

Netbook

The netbook was an inexpensive, light-weight, energy-efficient form of laptop, especially suited for wireless communication and Internet access. Netbooks first became commercially available around 2008, weighing under 1 kg, with a display size of under 9". The name *netbook* (with *net* short for *Internet*) is used as "the device excels in web-based computing performance". Netbooks were initially sold with light-weight variants of the Linux operating system, although later versions often have the Windows XP or Windows 7 operating systems. The term "netbook" is largely obsolete, although machines that would have once been called netbooks—small, inexpensive, and low powered—never ceased being sold, in particular the smaller Chromebook models.

Convertible, Hybrid, 2-in-1

The latest trend of technological convergence in the portable computer industry spawned a broad range of devices, with a combined features of several previously separate device types. The *hybrids*, *convertibles* and *2-in-1s* emerged as crossover devices, which share traits of both tablets and laptops. All such devices have a touchscreen display designed to allow users to work in a *tablet* mode, using either multi-touch gestures or a stylus/digital pen.

Convertibles are devices with the ability to conceal a hardware keyboard. Keyboards on such devices can be flipped, rotated, or slid behind the back of the chassis, thus transforming from a laptop into a tablet. *Hybrids* have a keyboard detachment mechanism, and due to this feature, all critical components are situated in the part with the display. *2-in-1s* can have a hybrid or a convertible form, often dubbed *2-in-1 detachables* and *2-in-1 convertibles* respectively, but are distinguished by the ability to run a desktop OS, such as Windows 10. 2-in-1s are often marketed as *laptop replacement tablets*.

2-in-1s are often very thin, around 10 millimetres (0.39 in), and light devices with a long battery life. 2-in-1s are distinguished from mainstream tablets as they feature an x86-architecture CPU (typically a low- or ultra-low-voltage model), such as the Intel Core i5, run a full-featured desktop OS like Windows 10, and have a number of typical laptop I/O ports, such as USB 3 and Mini DisplayPort.

2-in-1s are designed to be used not only as a media consumption device, but also as valid desktop or laptop replacements, due to their ability to run *desktop* applications, such as Adobe Photoshop. It is possible to connect multiple peripheral devices, such as a mouse, keyboard, and a number of external displays to a modern 2-in-1.

Microsoft Surface Pro-series devices and Surface Book are examples of modern 2-in-1 detachables, whereas Lenovo Yoga-series computers are a variant of 2-in-1 convertibles. While the older Surface RT and Surface 2 have the same chassis design as the Surface Pro, their use of ARM processors and Windows RT do not classify them as 2-in-1s, but as hybrid tablets. Similarly, a number of hybrid laptops run a mobile operating system, such as Android. These include Asus's Transformer Pad devices, example of hybrids with a detachable keyboard design, which not fall in the category of 2-in-1s.

Desktop Replacement

A desktop-replacement laptop is a class of large device which is not intended primarily for mobile use. They are bulkier and not as portable as other laptops, and are intended for use as compact and transportable alternatives to a desktop computer. Desktop replacements are larger and typically heavier than other classes of laptops. They are capable of containing more powerful components and have a 15-inch or larger display. Desktop replacement laptops' operation time on batteries is typically shorter than other laptops; in rare cases they have no battery at all. In the past, some laptops in this class used a limited range of desktop components to provide better performance for the same price at the expense of battery life, although this practice has largely died out. The names *Media Center Laptops* and *Gaming Laptops* are used to describe specialized notebook computers, often overlapping with the desktop replacement form factor.

Rugged Notebook

A rugged laptop is designed to reliably operate in harsh usage conditions such as strong vibrations, extreme temperatures, and wet or dusty environments. Rugged laptops are usually designed from scratch, rather than adapted from regular consumer laptop models. Rugged laptops are bulkier, heavier, and much more expensive than regular laptops, and thus are seldom seen in regular consumer use.

The design features found in rugged laptops include a rubber sheeting under the keyboard keys, sealed port and connector covers, passive cooling, very bright displays easily readable in daylight, cases and frames made of magnesium alloys that are much stronger than plastics found in commercial laptops, and solid-state storage devices or hard disc drives that are shock mounted to withstand constant vibrations. Rugged laptops are commonly used by public safety services (police, fire, and medical emergency), military, utilities, field service technicians, construction, mining, and oil drilling personnel. Rugged laptops are usually sold to organizations rather than individuals, and are rarely marketed via retail channels.

Components

The basic components of laptops function identically to their desktop counterparts. Traditionally they were miniaturized and adapted to mobile use, although desktop systems increasingly use the same smaller, lower-power parts which were originally developed for mobile use. The design restrictions on power, size, and cooling of laptops limit the maximum performance of laptop parts compared to that of desktop components, although that difference has increasingly narrowed.

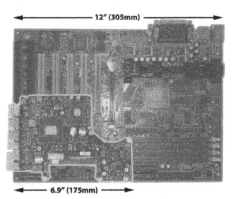

Miniaturization: a comparison of a desktop computer motherboard (ATX form factor) to a motherboard from a 13" laptop (2008 unibody Macbook)

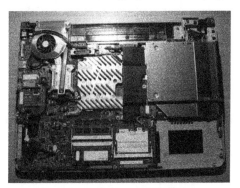

Inner view of a Sony VAIO laptop

A SODIMM memory module

In general, laptop components are not intended to be replaceable or upgradable, with the exception of components which can be detached, such as a battery or CD/CDR/DVD drive. This restriction is one of the major differences between laptops and desktop computers, because the large "tower" cases used in desktop computers are designed so that new motherboards, hard disks, sound cards, RAM, and other components can be added. In a very compact laptop, such as laplets, there may be no upgradeable components at all.

Intel, Asus, Compal, Quanta, and some other laptop manufacturers have created the Common Building Block standard for laptop parts to address some of the inefficiencies caused by the lack of standards and inability to upgrade components. The following sections summarizes the differences and distinguishing features of laptop components in comparison to desktop personal computer parts.

Display

Most modern laptops feature a 13 inches (33 cm) or larger color active matrix display based on LED lighting with resolutions of 1280×800 (16:10) or 1366×768 (16:9) pixels and above. Models with LED-based lighting offer lesser power consumption, and often increased brightness. Netbooks with a 10 inches (25 cm) or smaller screen typically use a resolution of 1024×600, while netbooks and subnotebooks with a 11.6 inches (29 cm) or 12 inches (30 cm) screen use standard notebook resolutions. Having a higher resolution display allows more items to fit onscreen at a time, improving the user's ability to multitask, although at the higher resolutions on smaller screens, the resolution may only serve to display sharper graphics and text rather than increasing the usable area. Since the introduction of the MacBook Pro with Retina display in 2012, there has been an increase in the availability of very-high resolution (1920×1080 and higher) displays, even in relatively small systems, and in typical 15-inch screens resolutions as high as 3200×1800 are available. External displays can be connected to most laptops, and models with a Mini DisplayPort can handle up to three.

Central Processing Unit

A laptop's central processing unit (CPU) has advanced power-saving features and produces less heat than one intended purely for desktop use. Typically, laptop CPUs have two processor cores, although 4-core models are also available. For low price and mainstream performance, there is no longer a significant performance difference between laptop and desktop CPUs, but at the high end, the fastest 4-to-8-core desktop CPUs still substantially outperform the fastest 4-core laptop processors, at the expense of massively higher power consumption and heat generation; the fastest laptop processors top out at 56 watts of heat, while the fastest desktop processors top out at 150 watts.

There have been a wide range of CPUs designed for laptops available from both Intel, AMD, and other manufacturers. On non-x86 architectures, Motorola and IBM produced the chips for the former PowerPC-based Apple laptops (iBook and PowerBook). Many laptops have removable CPUs, although this has become less common in the past few years as the trend has been towards thinner and lighter models. In other laptops the CPU is soldered on the motherboard and is non-replaceable; this is nearly universal in ultrabooks.

In the past, some laptops have used a desktop processor instead of the laptop version and have had high performance gains at the cost of greater weight, heat, and limited battery life, but the practice was largely extinct as of 2013. Unlike their desktop counterparts, laptop CPUs are nearly impossible to overclock. A thermal operating mode of laptops is very close to its limits and there is almost no headroom for an overclocking–related operating temperature increase. The possibility of improving a cooling system of a laptop to allow overclocking is extremely difficult to implement.

Graphical Processing Unit

On most laptops a graphical processing unit (GPU) is integrated into the CPU to conserve power and space. This was introduced by Intel with the Core i-series of mobile processors in 2010, and similar APU processors by AMD later that year. Prior to that, lower-end machines tended to use graphics processors integrated into the system chipset, while higher end machines had a separate graphics processor. In the past, laptops lacking a separate graphics processor were limited in their utility for gaming and professional applications involving 3D graphics, but the capabilities of CPU-integrated graphics have converged with the low-end of dedicated graphics processors in the past few years. Higher-end laptops intended for gaming or professional 3D work still come with dedicated, and in some cases even dual, graphics processors on the motherboard or as an internal expansion card. Since 2011, these almost always involve switchable graphics so that when there is no demand for the higher performance dedicated graphics processor, the more power-efficient integrated graphics processor will be used. Nvidia Optimus is an example of this sort of system of switchable graphics.

Memory

Most laptops use SO-DIMM (small outline dual in-line memory module) memory modules, as they are about half the size of desktop DIMMs. They are sometimes accessible from the bottom of the laptop for ease of upgrading, or placed in locations not intended for user replacement. Most laptops have two memory slots, although some of the lowest-end models will have only one, and some high end models (usually mobile engineering workstations and a few high-end models intended for gaming) have four slots. Most mid-range laptops are factory equipped with 4–6 GB of RAM. Netbooks are commonly equipped with only 1–2 GB of RAM and are generally only expandable to 2 GB, if at all. Due to the limitation of DDR3 SO-DIMM of a maximum of 8 GB per module, most laptops can only be expanded to a total of 16 GB of memory, until systems using DDR4 memory start becoming available. Laptops may have memory soldered to the motherboard to conserve space, which allows the laptop to have a thinner chassis design. Soldered memory cannot be upgraded.

Internal Storage

Traditionally, laptops had a hard disk drive (HDD) as a main non-volatile storage, but these proved inefficient for use in mobile devices due to high power consumption, heat production, and a presence of moving parts, which can cause damage to both the drive itself and the data stored when a laptop is unstable physically, e.g. during its use while transporting it or after its accidental drop. With the advent of flash memory technology, most mid- to high-end laptops opted for more compact, power efficient, and fast solid-state drives (SSD), which eliminated the hazard of drive and data corruption caused by a laptop's physical impacts. Most laptops use 2.5-inch drives, which are a smaller version of a 3.5-inch desktop drive form factor. 2.5-inch HDDs are more compact, power efficient, and produce less heat, while at the same time have a smaller capacity and a slower data transfer rate. Some very compact laptops support even smaller 1.8-inch HDDs. For SSDs, however, these miniaturization-related trade-offs are nonexistent, because SSDs were designed to have a very small footprint. SSDs feature a traditional 2.5- or 1.8-inch or a laptop-specific mSATA or M.2 card's form factor. SSDs have a higher data transfer rate, lower power consumption, lower failure rate, and a larger capacity compared to HDDs. However, HDDs have a significantly lower cost.

Most laptops can contain a single 2.5-inch drive, but a small number of laptops with a screen wider than 17 inches can house two drives. Some laptops support a hybrid mode, combining a 2.5-inch drive, typically a spacious HDD for data, with an mSATA or M.2 SDD drive, typically having less capacity, but a significantly faster read/write speed. The operating system partition would be located on the SSD to increase laptop I/O performance. Another way to increase performance is to use a smaller SSD of 16-32 GB as a cache drive with a compatible OS. Some laptops may have very limited drive upgradeability when the SSD used has a non-standard shape and/or requires a proprietary daughter card. Some laptops have very limited space on the installed SSD, instead relying on availability of cloud storage services for storing of user data; Chromebooks are a prominent example of this approach. A variety of external HDDs or NAS data storage servers with support of RAID technology can be attached to virtually any laptop over such interfaces as USB, FireWire, eSATA, or Thunderbolt, or over a wired or wireless network to further increase space for the storage of data. Many laptops also incorporate a card reader which allows for use of memory cards, such as those used for digital cameras, which are typically SD or microSD cards. This enables users to download digital pictures from an SD card onto a laptop, thus enabling them to delete the SD card's contents to free up space for taking new pictures.

Removable Media Drive

Optical disc drives capable of playing CD-ROMs, compact discs (CD), DVDs, and in some cases, Blu-ray Discs (BD), were nearly universal on full-sized models in the 2010s. A disc drive remains fairly common in laptops with a screen wider than 15 inches (38 cm), although the trend towards thinner and lighter machines is gradually eliminating these drives and players; these drives are uncommon in compact laptops, such as subnotebooks and netbooks. Laptop optical drives tend to follow a standard form factor, and usually have a standard mSATA connector. It is often possible to replace an optical drive with a newer model. In certain laptop models there is a possibility to replace an optical drive with a second hard drive, using a caddy that fills the extra space the optical drive would have occupied.

Inputs

Closeup of a touchpad on an Acer laptop

An alphanumeric keyboard is used to enter text and data and make other commands (e.g., function keys). A touchpad (also called a trackpad), a pointing stick, or both, are used to control the position of the cursor on the screen, and an integrated keyboard is used for typing. An external keyboard and/or mouse may be connected using a USB port or wirelessly, via Bluetooth or similar technology. With the advent of ultrabooks and support of touch input on screens by 2010-era operating systems, such as Windows 8.1, multitouch touchscreen displays are used in many models. Some models have webcams and microphones, which can be used to communicate with other people with both moving images and sound, via Skype, Google Chat and similar software. Laptops typically have USB ports and a microphone jack, for use with an external mic. Some laptops have a card reader for reading digital camera SD cards.

Input/Output (I/O) Ports

On a typical laptop there are several USB ports, an external monitor port (VGA, DVI, HDMI or Mini DisplayPort), an audio in/out port (often in form of a single socket) is common. It is possible to connect up to three external displays to a 2014-era laptop via a single Mini DisplayPort, utilizing multi-stream transport technology. Apple, in a 2015 version of its MacBook, transitioned from a number of different I/O ports to a single USB Type-C port. This port can be used both for charging and connecting a variety of devices through the use of aftermarket adapters. Google, with its updated version of Chromebook Pixel, shows a similar transition trend towards USB Type-C, although keeping older USB Type-A ports for a better compatibility with older devices. Although being common until the end of the 2000s decade, Ethernet network port are rarely found on modern laptops, due to widespread use of wireless networking, such as Wi-Fi. Legacy ports such as a PS/2 keyboard/mouse port, serial port, parallel port, or Firewire are provided on some models, but they are increasingly rare. On Apple's systems, and on a handful of other laptops, there are also Thunderbolt ports, but Thunderbolt 3 uses USB Type-C. Laptops typically have a headphone jack, so that the user can connect external headphones or amplified speaker systems for listening to music or other audio.

Expansion Cards

In the past, a PC Card (formerly PCMCIA) or ExpressCard slot for expansion was often present on laptops to allow adding and removing functionality, even when the laptop is powered on; these are becoming increasingly rare since the introduction of USB 3.0. Some internal subsystems such as: ethernet, Wi-Fi, or a Wireless cellular modem can be implemented as replaceable internal expansion cards, usually accessible under an access cover on the bottom of the laptop. The standard for such cards is PCI Express, which comes in both mini and even smaller M.2 sizes. In newer laptops, it is not uncommon to also see Micro SATA (mSATA) functionality on PCI Express Mini or M.2 card slots allowing the use of those slots for SATA-based solid state drives.

Battery and Power Supply

2016-era laptops use lithium ion batteries, with some thinner models using the flatter lithium polymer technology. These two technologies have largely replaced the older nickel metal-hydride batteries. Battery life is highly variable by model and workload, and can range from one hour to nearly a day. A battery's performance gradually decreases over time; substantial reduction in capacity is typically evident after one to three years of regular use, depending on the charging and discharging pattern and the design of the battery. Innovations in laptops and batteries have seen situations in which the battery can provide up to 24 hours of continued operation, assuming average power consumption levels. An example is the HP EliteBook 6930p when used with its ultra-capacity battery.

A laptop's battery is charged using an external power supply which is plugged into a wall outlet.

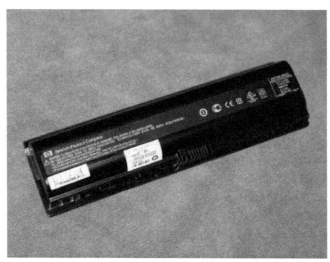

Almost all laptops use smart batteries

The power supply outputs a DC voltage typically in the range of 7.2—24 volts. The power supply is usually external, and connected to the laptop through a DC connector cable. In most cases, it can charge the battery and power the laptop simultaneously. When the battery is fully charged, the laptop continues to run on power supplied by the external power supply, avoiding battery use. The battery charges in a shorter period of time if laptop is turned off or sleeping. The charger typically adds about 400 grams (0.88 lb) to the overall transporting weight of a laptop, although some models are substantially heavier or lighter. Most 2016-era laptops use a smart battery, a rechargeable battery pack with a built-in battery management system (BMS). The smart battery can internally measure voltage and current, and deduce charge level and SoH (State of Health) parameters, indicating the state of the cells.

Cooling

Waste heat from operation is difficult to remove in the compact internal space of a laptop. Early laptops used heat sinks placed directly on the components to be cooled, but when these hot components are deep inside the device, a large space-wasting air duct is needed to exhaust the heat. Modern laptops instead rely on heat pipes to rapidly move waste heat towards the edges of the device, to allow for a much smaller and compact fan and heat sink cooling system. Waste heat is usually exhausted away from the device operator towards the rear or sides of the device. Multiple air intake paths are used since some intakes can be blocked, such as when the device is placed on a soft conforming surface like a chair cushion. It is believed that some designs with metal cases, like Apple's aluminum MacBook Pro and MacBook Air, also employ the case of the machine as a heat sink, allowing it to supplement cooling by dissipating heat out of the device core. Secondary device temperature monitoring may reduce performance or trigger an emergency shutdown if it is unable to dissipate heat, such as if the laptop were to be left running and placed inside a carrying case. Such a condition has the potential to melt plastics or ignite a fire. Aftermarket cooling pads with external fans can be used with most laptops to reduce operating temperatures.

Docking Station

A docking station (sometimes referred to simply as a *dock*) is a laptop accessory that contains multiple ports, and in some cases expansion slots and/or bays for fixed or removable drives. A laptop connects and disconnects to a docking station, typically through a single large proprietary connector. A docking station is an especially popular laptop accessory in a corporate computing environment, due to a possibility of a docking station to transform a laptop into a full-featured desktop replacement, yet allowing for its easy release. This ability can be advantageous to "road warrior" employees who have to travel frequently for work, and yet who also come into the office. If more ports are needed, or their position on a laptop is inconvenient, one can use a cheaper passive device known as a port replicator. These devices mate to the connectors on the laptop, such as through USB or FireWire.

Docking station and laptop

Charging Trolleys

Laptop charging trolleys, also known as laptop trolleys or laptop carts, are mobile storage containers to charge multiple laptops, netbooks, and tablet computers at the same time. The trolleys are used in schools that have replaced their traditional static computer labs suites of desktop equipped with "tower" computers, but do not have enough plug sockets in an individual classroom to charge all of the devices. The trolleys can be wheeled between rooms and classrooms so that all students and teachers in a particular building can access fully charged IT equipment.

Laptop charging trolleys are also used to deter and protect against opportunistic and organized theft. Schools, especially those with open plan designs, are often prime targets for thieves who steal high-value items. Laptops, netbooks, and tablets are among the highest–value portable items in a school. Moreover, laptops can easily be concealed under clothing and stolen from buildings. Many types of laptop–charging trolleys are designed and constructed to protect against theft. They are generally made out of steel, and the laptops remain locked up while not in use. Although the trolleys can be moved between areas from one classroom to another, they can often be mounted or locked to the floor or walls to prevent thieves from stealing the laptops, especially overnight.

Solar Panels

In some laptops, solar panels are able to generate enough solar power for the laptop to operate. The One Laptop Per Child Initiative released the OLPC XO-1 laptop which was tested and successfully operated by use of solar panels. Presently, they are designing a OLPC XO-3 laptop with these features. The OLPC XO-3 can operate with 2 watts of electricity because its renewable energy resources generate a total of 4 watts. Samsung has also designed the NC215S solar–powered notebook that will be sold commercially in the U.S. market.

Advantages

A teacher using laptop as part of a workshop for school children

Wikipedia co-founder Jimmy Wales using a laptop on a park bench

Portability is usually the first feature mentioned in any comparison of laptops versus desktop PCs. Physical portability allows a laptop to be used in many places—not only at home and at the office, but also during commuting and flights, in coffee shops, in lecture halls and libraries, at clients' locations or at a meeting room, etc. Within a home, portability enables laptop users to move their device from the living room to the dining room to the family room. Portability offers several distinct advantages:

- Productivity: Using a laptop in places where a desktop PC cannot be used can help employees and students to increase their productivity on work or school tasks. For example, an office worker reading her work e-mails during an hour-long commute by train, or a student doing her homework at the university coffee shop during a break between lectures.

- Immediacy: Carrying an laptop means having instant access to information, including personal and work files. This allows better collaboration between coworkers or students, as a laptop can be flipped open to look at a report, document, spreadsheet, or presentation anytime and anywhere.

- Up-to-date information: If a person has more than one desktop PC, a problem of synchronization arises: changes made on one computer are not automatically propagated to the others. There are ways to resolve this problem, including physical transfer of updated files (using a USB flash memory stick or CD-ROMs) or using synchronization software over the Internet, such as cloud computing. However, transporting a single laptop to both locations avoids the problem entirely, as the files exist in a single location and are always up-to-date.

- Connectivity: In the 2010s, a proliferation of Wi-Fi wireless networks and cellular broadband data services (HSDPA, EVDO and others) in many urban centers, combined with near-ubiquitous Wi-Fi support by modern laptops meant that a laptop could now have easy Internet and local network connectivity while remaining mobile. Wi-Fi networks and laptop programs are especially widespread at university campuses.

Other advantages of laptops:

- Size: Laptops are smaller than desktop PCs. This is beneficial when space is at a premium, for example in small apartments and student dorms. When not in use, a laptop can be closed and put away in a desk drawer.

- Low power consumption: Laptops are several times more power-efficient than desktops. A typical laptop uses 20–120 W, compared to 100–800 W for desktops. This could be particularly beneficial for large businesses, which run hundreds of personal computers thus multiplying the potential savings, and homes where there is a computer running 24/7 (such as a home media server, print server, etc.).

- Quiet: Laptops are typically much quieter than desktops, due both to the components (quieter, slower 2.5-inch hard drives) and to less heat production leading to use of fewer and slower cooling fans.

- Battery: a charged laptop can continue to be used in case of a power outage and is not affected by short power interruptions and blackouts. A desktop PC needs an Uninterruptible power supply (UPS) to handle short interruptions, blackouts, and spikes; achieving on-battery time of more than 20–30 minutes for a desktop PC requires a large and expensive UPS.

- All-in-One: designed to be portable, most 2010-era laptops have all components integrated into the chassis (however, some small laptops may not have an internal CD/CDR/DVD drive, so an external drive needs to be used). For desktops (excluding all-in-ones) this is divided into the desktop "tower" (the unit with the CPU, hard drive, power supply, etc.), keyboard, mouse, display screen, and optional peripherals such as speakers.

Disadvantages

Compared to desktop PCs, laptops have disadvantages in the following areas:

Performance

While the performance of mainstream desktops and laptop is comparable, and the cost of laptops has fallen less rapidly than desktops, laptops remain more expensive than desktop PCs at the same performance level. The upper limits of performance of laptops remain much lower than the highest-end desktops (especially "workstation class" machines with two processor sockets), and "bleeding-edge" features usually appear first in desktops and only then, as the underlying technology matures, are adapted to laptops.

For Internet browsing and typical office applications, where the computer spends the majority of its time waiting for the next user input, even relatively low-end laptops (such as Netbooks) can be fast enough for some users. As of mid-2010, at the lowest end, the cheapest netbooks—between US$200–300—remain more expensive than the lowest-end desktop computers (around US$200), but only when those are priced without a screen/monitor. Once an inexpensive monitor is added, the prices are comparable.

Most higher-end laptops are sufficiently powerful for high-resolution movie playback, some 3D gaming and video editing and encoding. However, laptop processors can be disadvantaged when dealing with higher-end database, maths, engineering, financial software, virtualization, etc. This is because laptops use the mobile versions of processors to conserve power, and these lag behind desktop chips when it comes to performance. Also, the top-of-the-line mobile graphics processors (GPUs) are significantly behind the top-of-the-line desktop GPUs to a greater degree than the processors, which limits the utility of laptops for high-end 3D gaming and scientific visualization applications. Some manufacturers work around this performance problem by using desktop CPUs for laptops.

Upgradeability

Upgradeability of laptops is very limited compared to desktops, which are thoroughly standardized. In general, hard drives and memory can be upgraded easily. Optical drives and internal expansion cards may be upgraded if they follow an industry standard, but all other internal components, including the motherboard, CPU and graphics, are not always intended to be upgradeable. Intel, Asus, Compal, Quanta and some other laptop manufacturers have created the Common Building Block standard for laptop parts to address some of the inefficiencies caused by the lack of standards. The reasons for limited upgradeability are both technical and economic. There is no industry-wide standard form factor for laptops; each major laptop manufacturer pursues its own proprietary design and construction, with the result that laptops are difficult to upgrade and have high repair costs. Devices such as sound cards, network adapters, hard and optical drives, and numerous other peripherals are available, but these upgrades usually impair the laptop's portability, because they add cables and boxes to the setup and often have to be disconnected and reconnected when the laptop is on the move.

Ergonomics and Health Effects

Wrists

Prolonged use of laptops can cause repetitive strain injury because of their small, flat keyboard and trackpad pointing devices,. Usage of separate, external ergonomic keyboards and pointing devices is recommended to prevent injury when working for long periods of time; they can be connected to a laptop easily by USB or via a docking station. Some health standards require ergonomic keyboards at workplaces.

Laptop cooler (silver) under laptop (white), preventing heating of lap and improving laptop airflow

Neck and Spine

A laptop's integrated screen often requires users to lean over for a better view, which can cause neck and/or spinal injuries. A larger and higher-quality external screen can be connected to almost any laptop to alleviate this and to provide additional screen space for more productive work. Another solution is to use a computer stand.

Possible Effect on Fertility

A study by State University of New York researchers found that heat generated from laptops can increase the temperature of the lap of male users when balancing the computer on their lap, potentially putting sperm count at risk. The study, which included roughly two dozen men between the ages of 21 and 35, found that the sitting position required to balance a laptop can increase scrotum temperature by as much as 2.1 °C (4 °F). However, further research is needed to determine whether this directly affects male sterility. A later 2010 study of 29 males published in *Fertility and Sterility* found that men who kept their laptops on their laps experienced scrotal hyperthermia (overheating) in which their scrotal temperatures increased by up to 2.0 °C (4 °F). The resulting heat increase, which could not be offset by a laptop cushion, may increase male infertility.

A common practical solution to this problem is to place the laptop on a table or desk, or to use a book or pillow between the body and the laptop. Another solution is to obtain a cooling unit for the laptop. These are usually USB powered and consist of a hard thin plastic case housing one, two, or three cooling fans – with the entire assembly designed to sit under the laptop in question – which results in the laptop remaining cool to the touch, and greatly reduces laptop heat buildup.

Thighs

Heat generated from using a laptop on the lap can also cause skin discoloration on the thighs known as "toasted skin syndrome".

Durability

A clogged heat sink on a laptop after 2.5 years of use

Equipment Wear

Because of their portability, laptops are subject to more wear and physical damage than desktops. Components such as screen hinges, latches, power jacks, and power cords deteriorate gradually from ordinary use, and may have to be replaced. A liquid spill onto the keyboard, a rather minor mishap with a desktop system (given that a basic keyboard costs about US$20), can damage the internals of a laptop and result destroy the computer or result in a costly repair. One study found that a laptop is three times more likely to break during the first year of use than a desktop. To maintain a laptop, it is recommended to clean it every three months for dirt, debris, dust, and food particles. Most cleaning kits consist of a lint-free or Microfiber cloth for the LCD screen and keyboard, compressed air for getting dust out of the cooling fan, and cleaning solution. Harsh chemicals such as bleach should not be used to clean a laptop, as they can damage it.

Parts Replacement

Original external components are expensive, and usually proprietary and non-interchangeable; other parts are inexpensive—a power jack can cost a few dollars—but their replacement may require extensive disassembly and reassembly of the laptop by a technician. Other inexpensive but fragile parts often cannot be purchased separate from larger more expensive components. For example, the video display cable and the backlight power cable that pass through the lid hinges to connect the motherboard to the screen will eventually break from repeated opening and closing of the lid. These tiny cables usually cannot be purchased from the original manufacturer separate from the entire LCD panel, with the price of hundreds of dollars, although for popular models an aftermarket in pulled parts generally exists. The repair costs of a failed motherboard or LCD panel often exceeds the value of a used laptop. Parts can also be ordered from third party vendors.

Heating and Cooling

Laptops rely on extremely compact cooling systems involving a fan and heat sink that can fail from blockage caused by accumulated airborne dust and debris. Most laptops do not have any type of removable dust collection filter over the air intake for these cooling systems, resulting in a system that gradually conducts more heat and noise as the years pass. In some cases the laptop starts to overheat even at idle load levels. This dust is usually stuck inside where the fan and heat sink meet, where it can not be removed by a casual cleaning and vacuuming. Most of the time, compressed air can dislodge the dust and debris but may not entirely remove it. After the device is turned on, the loose debris is reaccumulated into the cooling system by the fans. A complete disassembly is usually required to clean the laptop entirely. However, preventative maintenance such as regular cleaning of the heat sink via compressed air can prevent dust build up on the heat sink. Many laptops are difficult to disassemble by the average user and contain components that are sensitive to electrostatic discharge (ESD).

Battery Life

Battery life is limited because the capacity drops with time, eventually requiring replacement after as little as a year. A new battery typically stores enough energy to run the laptop for three to five hours, depending on usage, configuration, and power management settings. Yet, as it ages, the battery's energy storage will dissipate progressively until it lasts only a few minutes. The battery is often easily replaceable and a higher capacity model may be obtained for longer charging and discharging time. Some laptops (specifically ultrabooks) do not have the usual removable battery and have to be brought to the service center of its manufacturer to have its battery replaced. Replacement batteries can also be expensive.

Security and Privacy

Because they are valuable, commonly used, portable, and easy to conceal in a backpack or other type of travel bag, laptops are prized targets for theft. Every day, over 1,600 laptops go missing from U.S. airports. The cost of stolen business or personal data, and of the resulting problems (identity theft, credit card fraud, breach of privacy), can be many times the value of the stolen laptop itself. Consequently, physical protection of laptops and the safeguarding of data contained on them are both of great importance. Most laptops have a Kensington security slot, which can be used to tether them to a desk or other immovable object with a security cable and lock. In addition, modern operating systems and third-party software offer disk encryption functionality, which renders the data on the laptop's hard drive unreadable without a key or a passphrase. As of 2015, some laptops also have additional security elements added, including eye recognition software and fingerprint scanning components.

Software such as LoJack for Laptops, Laptop Cop, and GadgetTrack have been engineered to help people locate and recover their stolen laptop in the event of theft. Setting one's laptop with a password on its firmware (protection against going to firmware setup or booting), internal HDD/SSD (protection against accessing it and loading an operating system on it afterwards), and every user account of the operating system are additional security measures that a user should do. Fewer than 5% of lost or stolen laptops are recovered by the companies that own them, however, that number may decrease due to a variety of companies and software solutions specializing in laptop recovery. In the 2010s, the common availability of webcams on laptops raised privacy concerns. In *Robbins v. Lower Merion School District* (Eastern District of Pennsylvania 2010), school-issued laptops loaded with special software enabled staff from two high schools to take secret webcam shots of students at home, via their students' laptops.

Major Brands and Manufacturers

There are a many laptop brands and manufacturers. Several major brands that offer notebooks in various classes are listed in the box to the right. The major brands usually offer good service and support, including well-executed documentation and driver downloads that remain available for many years after a particular laptop model is no longer produced. Capitalizing on service, support, and brand image, laptops from major brands are more expensive than laptops by smaller brands and ODMs. Some brands are specializing in a particular class of laptops, such as gaming laptops (Alienware), high-performance laptops (HP Envy), netbooks (EeePC) and laptops for children (OLPC).

Many brands, including the major ones, do not design and do not manufacture their laptops. Instead, a small number of Original Design Manufacturers (ODMs) design new models of laptops, and the brands choose the models to be included in their lineup. In 2006, 7 major ODMs manufactured 7 of every 10 laptops in the world, with the largest one (Quanta Computer) having 30% of world market share. Therefore, there often are identical models available both from a major label and from a low-profile ODM in-house brand.

Sales

Battery-powered portable computers had just 2% worldwide market share in 1986. However, laptops have become increasingly popular, both for business and personal use. Around 109 million notebook PCs shipped worldwide in 2007, a growth of 33% compared to 2006. In 2008 it was estimated that 145.9 million notebooks were sold, and that the number would grow in 2009 to 177.7 million. The third quarter of 2008 was the first time when worldwide notebook PC shipments exceeded desktops, with 38.6 million units versus 38.5 million units.

May 2005 was the first time notebooks outsold desktops in the US over the course of a full month; at the time notebooks sold for an average of $1,131 while desktops sold for an average of $696. When looking at operating systems, for Microsoft Windows laptops the average selling price (ASP) showed a decline in 2008/2009, possibly due to low-cost netbooks, drawing an average US$689 at U.S. retail stores in August 2008. In 2009, ASP had further fallen to $602 by January and to $560 in February. While Windows machines ASP fell $129 in these seven months, Apple (Mac) OS X laptop ASP declined just $12 from $1,524 to $1,512.

Extreme Environments

The ruggedized Grid Compass computer was used since the early days of the Space Shuttle program. The first commercial laptop used in space was a Macintosh portable in 1991 aboard Space Shuttle mission STS-43. Apple and other laptop computers continue to be flown aboard manned spaceflights, though the only long duration flight certified computer for the International Space Station is the ThinkPad. As of 2011, over 100 ThinkPads were aboard the ISS. Laptops used aboard the International Space Station and other spaceflights are generally the same ones that can be purchased by the general public but needed modifications are made to allow them to be used safely and effectively in a weightless environment such as updating the cooling systems to function without relying on hot air rising and accommodation for the lower cabin air pressure. Laptops operating in harsh usage environments and conditions, such as strong vibrations, extreme temperatures, and wet or dusty conditions differ from those used in space in that they are custom designed for the task and do not use commercial off-the-shelf hardware.

Accessories

A common accessory for laptops is a laptop sleeve, laptop skin, or laptop case, which provides a degree of protection from scratches. Sleeves, which are distinguished by being relatively thin and flexible, are most commonly made of neoprene, with sturdier ones made of low-resilience polyurethane. Some laptop sleeves are wrapped in ballistic nylon to provide some measure of waterproofing. Bulkier and sturdier cases can be made of metal with polyurethane padding inside, and may have locks for added security. Metal, padded cases also offer protection against impacts and drops. Another common accessory is a laptop cooler, a device which helps lower the internal temperature of the laptop either actively or passively. A common active method involves using electric fans to draw heat away from the laptop, while a passive method might involve propping the laptop up on some type of pad so it can receive more air flow. Some stores sell laptop pads which enable a reclining person on a bed to use a laptop.

Obsolete Features

Features that certain early models of laptops used to have that are not available in most 2016 laptops include:

- Reset ("cold restart") button in a hole (needed a thin metal tool to press)

- Instant power off button in a hole (needed a thin metal tool to press)

- Integrated charger or power adapter inside the laptop

- Floppy disk drive

- Serial port

- Parallel port

- Modem

- Shared PS/2 input device port

- VHS or 8mm VCR

- IrDA

- S-video port

Some 2016 laptops do not have an internal CD-ROM/DVD/CD drive.

Tablet Computer

A tablet computer, commonly shortened to tablet, is a thin, flat mobile computer with a touch-screen display, which in 2016 is usually color, processing circuitry, and a rechargeable battery in a single device. Tablets often come equipped with sensors, including digital cameras, a microphone, and an accelerometer. The touchscreen display uses the recognition of finger or stylus gestures to replace the mouse, trackpad and keyboard used in laptops. They usually feature on-screen, pop-up virtual keyboards for typing and inputting commands. Tablets may have physical buttons for basic features such as speaker volume and power, and ports for plugging in network communications, headphones and battery charging. Tablets are typically larger than smartphones or personal digital assistants with screens 7 inches (18 cm) or larger, measured diagonally. Tablets have Wi-Fi capability built in so that users can connect to the Internet and can have cellular network capabilities.

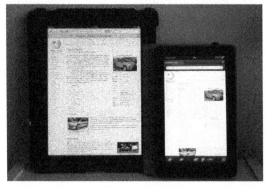

Apple's iPad (left) is the top-selling tablet with 170 million units sold by mid-October 2013, followed by Amazon's Kindle Fire (right) with an estimated 7 million sold as of May 2012.

Tablets can be classified according to the presence and physical appearance of keyboards. Slates and booklets do not have a physical keyboard and text input and other input is usually entered through the use of a virtual keyboard shown on a touchscreen-enabled display. Hybrids, convertibles and 2-in-1s do have physical keyboards (although these are usually concealable or detachable), yet they typically also make use of virtual keyboards. Most tablets can use separate keyboards connected using Bluetooth.

The format was conceptualized in the mid-20th century (Stanley Kubrick depicted fictional tablets in the 1968 film *2001: A Space Odyssey*) and prototyped and developed in the last two decades of that century. In April 2010, the iPad was released, which was the first mass-market tablet with finger-friendly multi-touch and a dedicated operating system. In the 2010s, tablets rapidly rose in popularity and ubiquity and became a large product category used for both personal and workplace applications.

History

Wireless tablet device portrayed in the movie *2001: A Space Odyssey* (1968)

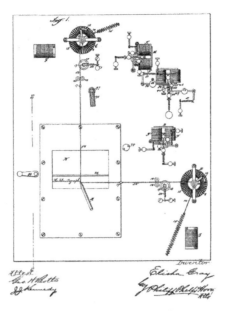

1888 telautograph patent schema

The tablet computer and its associated operating system began with the development of pen computing. Electrical devices with data input and output on a flat information display existed as early as 1888 with the telautograph, which used a sheet of paper as display and a pen attached to electromechanical actuators. Throughout the 20th century devices with these characteristics have been imagined and created whether as blueprints, prototypes, or commercial products. In addition to many academic and research systems, several companies released commercial products in the 1980s, with various input/output types tried out:

Fictional and prototype tablets

Tablet computers appeared in a number of works of science fiction in the second half of the 20th century; all helped to promote and disseminate the concept to a wider audience. Examples include:

- Isaac Asimov described a Calculator Pad in his novel *Foundation* (1951)

- Stanislaw Lem described the Opton in his novel *Return from the Stars* (1961)

- Numerous similar devices were depicted in Gene Roddenberry's 1966 *Star Trek: The Original Series*

- Arthur C. Clarke's NewsPad was depicted in Stanley Kubrick's film *2001: A Space Odyssey* (1968)

- Douglas Adams described a tablet computer in *The Hitchhiker's Guide to the Galaxy* and the associated comedy of the same name (1978)

- The sci-fi TV series *Star Trek The Next Generation* featured tablet computers which were designated as PADDs.

A device more powerful than today's tablets appeared briefly in Jerry Pournelle and Larry Niven's *The Mote in God's Eye* (1974).

Additionally, real-life projects either proposed or created tablet computers, such as:

- In 1968, computer scientist Alan Kay envisioned a KiddiComp, while a PhD candidate; he developed and described the concept as a Dynabook in his proposal, *A personal computer for children of all ages* (1972), which outlines the requirements for a conceptual portable educational device that would offer functionality similar to that supplied via a laptop computer, or (in some of its other incarnations) a tablet or slate computer, with the exception of the requirement for any Dynabook device offering near eternal battery life. Adults could also use a Dynabook, but the target audience was children.

- In 1992, Atari showed developers the Stylus, later renamed ST-Pad. The ST-Pad was based on the TOS/GEM Atari ST Platform and prototyped early handwriting recognition. Shiraz Shivji's company *Momentus* demonstrated in the same time a failed x86 MS-DOS based Pen Computer with its own GUI.

- In 1994, the European Union initiated the NewsPad project, inspired by Clarke and Kubrick's fictional work. Acorn Computers developed and delivered an ARM-based touch screen tablet computer for this program, branding it the "NewsPad"; the project ended in 1997.

Risc User: NewsPad Covered in the October 1996 edition

- During the November 2000 COMDEX, Microsoft used the term Tablet PC to describe a prototype handheld device they were demonstrating.

- In 2001, Ericsson Mobile Communications announced an experimental product named the DelphiPad, which was developed in cooperation with the Centre for Wireless Communications in Singapore, with a touch-sensitive screen, Netscape Navigator as a web browser, and Linux as its operating system.

Early Devices

Following their earlier tablet-computer products such as the Pencept PenPad and the CIC Handwriter, in September 1989, GRiD Systems released the first commercially available tablet-type portable computer, the GRiDPad. The GRiDPad was also manufactured by the Samsung Corporation after acquiring GRiD System. All three products were based on extended versions of the MS-DOS operating system. In 1992, IBM announced (in April) and shipped to developers (in October) the

2521 ThinkPad, which ran the GO Corporation's PenPoint OS. Also based on PenPoint was AT&T's EO Personal Communicator from 1993, which ran on AT&T's own hardware, including their own AT&T Hobbit CPU. Apple Computer launched the Apple Newton personal digital assistant in 1993. It utilised Apple's own new Newton OS, initially running on hardware manufactured by Motorola and incorporating an ARM CPU, that Apple had specifically co-developed with Acorn Computers. The operating system and platform design were later licensed to Sharp and Digital Ocean, who went on to manufacture their own variants.

Apple Newton MessagePad, the first tablet produced by Apple. It was released in 1993.

In 1996, Palm, Inc. released the first of the Palm OS based PalmPilot touch and stylus based PDA, the touch based devices initially incorporating a Motorola Dragonball (68000) CPU. Also in 1996 Fujitsu released the Stylistic 1000 tablet format PC, running Microsoft Windows 95, on a 100 MHz AMD486 DX4 CPU, with 8 MB RAM offering stylus input, with the option of connecting a conventional Keyboard and mouse. Intel announced a StrongARM processor-based touchscreen tablet computer in 1999, under the name WebPAD. It was later re-branded as the "Intel Web Tablet". In 2000, Norwegian company Screen Media AS and the German company Dosch & Amand Gmbh released the " FreePad". It was based on Linux and used the Opera browser. Internet access was provided by DECT DMAP, only available in Europe and provided up to 10Mbit/s. The device had 16 MB storage, 32 MB of RAM and x86 compatible 166 MHz "Geode"-Microcontroller by National Semiconductor. The screen was 10.4" or 12.1" and was touch sensitive. It had slots for SIM cards to enable support of television set-up box. FreePad were sold in Norway and the Middle East; but the company was dissolved in 2003.

In April 2000, Microsoft launched the Pocket PC 2000, utilizing their touch capable Windows CE 3.0 operating system. The devices were manufactured by several manufacturers, based on a mix of: x86, MIPS, ARM, and SuperH hardware. In 2002, Microsoft attempted to define the Microsoft Tablet PC as a mobile computer for field work in business, though their devices failed, mainly due to pricing and usability decisions that limited them to their original purpose - such as the existing devices being too heavy to be held with one hand for extended periods, and having legacy applications created for desktop interfaces and not well adapted to the slate format.

Nokia had plans for an Internet tablet since before 2000. An early model was test manufactured in 2001, the Nokia M510, which was running on EPOC and featuring an Opera browser, speakers and a 10-inch 800×600 screen, but it was not released because of fears that the market was not ready for it. In 2005, Nokia finally released the first of its Internet Tablet range, the Nokia 770. These tablets now ran a Debian based Linux OS called Maemo. Nokia used the term *internet tablet* to refer to a portable information appliance that focused on Internet use and media consumption, in the range between a personal digital assistant (PDA) and an Ultra-Mobile PC (UMPC). They made two mobile phones, the N900 that runs Maemo, and N9 that run Meego. Android was the first of the 2000s-era dominating platforms for tablet computers to reach the market. In 2008, the first plans for Android-based tablets appeared. The first products were released in 2009. Among them was the Archos 5, a pocket-sized model with a 5-inch touchscreen, that was first released with a proprietary operating system and later (in 2009) released with Android 1.4. The Camangi WebStation was released in Q2 2009. The first LTE Android tablet appeared late 2009 and was made by ICD for Verizon. This unit was called the Ultra, but a version called Vega was released around the same time. Ultra had a 7-inch display while Vega's was 15 inches. Many more products followed in 2010. Several manufacturers waited for Android Honeycomb, specifically adapted for use with tablets, which debuted in February 2011.

2010 and Afterwards

Apple is often credited for defining a new class of consumer device with the iPad, which shaped the commercial market for tablets in the following years, and was the most successful tablet at the time of its release. iPads and competing devices were tested by the US military in 2011 and cleared for secure use in 2013. Its debut in 2010 pushed tablets into the mainstream. Samsung's Galaxy Tab and others followed, continuing the trends towards the features listed above. In March 2012, *PC Magazine* reported that 31% of U.S. Internet users owned a tablet, used mainly for viewing published content such as video and news. The top-selling line of devices was Apple's iPad with 100 million sold between its release in April 2010 and mid-October 2012, but iPad market share (number of units) dropped to 36% in 2013 with Android tablets climbing to 62%. Android tablet sales volume was 121 million devices, plus 52 million, between 2012 and 2013 respectively. Individual brands of Android operating system devices or compatibles follow iPad with Amazon's Kindle Fire with 7 million, and Barnes & Noble's Nook with 5 million.

In 2013, Samsung announced a tablet running Android and Windows 8 operating systems concurrently; switching from one operating system to the other and vice versa does not require restarting the device, and data can be synchronized between the two operating systems. The device, named ATIV Q, was scheduled for release in late 2013 but its release has been indefinitely delayed. pre-sented its first tablet computer during its global press conference in New York on 23 November 2010. The family which is called Acer Iconia also includes a big screen smartphone called Iconia Smart. The Iconia series displays utilize Gorilla Glass. Meanwhile, Asus released its Transformer Book Trio, a tablet that is also capable of running the operating systems Windows 8 and Android. As of February 2014, 83% of mobile app developers were targeting tablets, but 93% of developers were targeting smartphones. By 2014 around 23% of B2B companies were said to have deployed tablets for sales-related activities, according to a survey report by Corporate Visions. As of November 2015, tablet use in the world is led by the iPad with a market share of 65.66% and Android tablets with a market share of 32.08%. The iPad holds majority use in North America, Western Europe, Japan, Australia, and most of the Americas. Android tablets are more popular in most of Asia (China and Russia an exception), Africa and Eastern Europe. In 2015 tablet sales did not increase. Apple remained the largest seller but its market share declined below 25% Samsung vice president Gary Riding said early in 2016 that tablets were only doing well among those using them for work. Newer models were more expensive and designed for a keyboard and stylus, which reflected the changing uses.

Touch Interface

Samsung Galaxy Tab demonstrating multi-touch

A key component among tablet computers is touch input on a touchscreen. This allows the user to navigate easily and type with a virtual keyboard on the screen or press other icons on the screen to open apps or files. The first tablet to do this was the GRiDPad by GRiD Systems Corporation; the tablet featured both a stylus, a pen-like tool to aid with precision in a touchscreen device as well as an on-screen keyboard. The system must respond to touches rather than clicks of a keyboard or mouse, which allows integrated hand-eye operation, a natural use of the somatosensory system. This is even more true of 2016-era multi-touch interface, which often emulates the way objects behave.

Handwriting Recognition

Chinese characters like this one meaning "person" can be written by handwriting recognition (人, Mandarin: *rén*, Korean: *in*, Japanese: *jin, nin*; *hito*, Cantonese: jan4). The character has two strokes, the first shown here in brown, and the second in red. The black area represents the starting position of the writing instrument.

Some ARM powered tablets, such as the Galaxy Note 10, support a stylus and support handwriting recognition. Wacom and N-trig digital pens provide approximately 2500 DPI resolution for handwriting, exceeding the resolution of capacitive touch screens by more than a factor of 10. These pens also support pressure sensitivity, allowing for "variable-width stroke-based" characters, such as Chinese/Japanese/Korean writing, due to their built-in capability of "pressure sensing". Pressure is also used in digital art applications such as Autodesk Sketchbook. Apps exist on both iOS and Android platforms for handwriting recognition and in 2015 Google introduced its own handwriting input with support for 82 languages.

Touchscreen Hardware

Touchscreens usually come in one of two forms:

- Resistive touchscreens are passive and respond to pressure on the screen. They allow a high level of precision, useful in emulating a pointer (as is common in tablet computers) but may require calibration. Because of the high resolution, a stylus or fingernail is often used. Stylus-oriented systems are less suited to multi-touch.

- Capacitive touchscreens tend to be less accurate, but more responsive than resistive devices. Because they require a conductive material, such as a finger tip, for input, they are not common among stylus-oriented devices, but are prominent on consumer devices. Most finger-driven capacitive screens do not currently support pressure input (except for the iPhone 6S), but some tablets use a pressure-sensitive stylus or active pen.

- Some tablets can recognize individual palms, while some professional-grade tablets use pressure-sensitive films, such as those on graphics tablets. Some capacitive touch-screens can detect the size of the touched area and the pressure used.

Features

As of 2016, most tablets use capacitive touchscreens with multi-touch, unlike earlier resistive touchscreen devices which users needed styluses to do input. After 2007, with access to capacitive screens and the success of the iPhone, other features became common, such as multi-touch features (in which the user can touch the screen in multiple places to trigger actions and other natural user interface features, as well as flash memory solid state storage and "instant on" warm-booting; external USB and Bluetooth keyboards defined tablets. Some tablets have 3G mobile telephony applications. Most tablets released since mid-2010 use a version of an ARM processor for longer battery life. The ARM Cortex family is powerful enough for tasks such as internet browsing, light production work and mobile games. As with smartphones, most mobile tablet apps are supplied through online distribution. These sources, known as "app stores", provide centralized catalogs of software and allow "one click" on-device software purchasing, installation and updates. The app store is often shared with smartphones that use the same operating system.

Hardware

- High-definition, anti-glare display

- Touchscreen

- Front- and/or back- facing camera(s) for photographs and video

- Lower weight and longer battery life than a comparably-sized laptop

- Wireless local area and internet connectivity (usually with Wi-Fi standard and optional mobile broadband)

- Bluetooth for connecting peripherals and communicating with local devices

- Ports for wired connections and charging, for example USB ports

- Early devices had IR support and could work as a TV remote controller.

- Docking station: Keyboard and additional connections

Special hardware: The tablets can be equipped with special hardware to provide functionality, such as camera, GPS and local data storage.

Software

- Mobile web browser

- E-book readers for digital books, periodicals and other content

- App store for adding apps such as games, education and utilities

- Portable media player function including video and music playback

- Email and social media

- Some have mobile phone functions (messaging, speakerphone, address book)

Data Storage

- On-board flash memory

- Ports for removable storage

- Various cloud storage services for backup and syncing data across devices

- Local storage on a LAN

Additional Inputs

Besides a touchscreen and keyboard, some tablets can also use these input methods:

- Accelerometer: Detects the physical movement and orientation of the tablet. This allows the touchscreen display to shift to either portrait or landscape mode. In addition, tilting the tablet may be used as an input (for instance to steer in a driving game)

- Ambient light and proximity sensors, to detect if the device is close to something, in particular, to your ear, etc., which help to distinguish between intentional and unintentional touches.

- Speech recognition Google introduced voice input in Android 2.1 in 2009 and voice actions in 2.2 in 2010, with up to five languages (now around 40). Siri was introduced as a system-wide personal assistant on the iPhone 4S in 2011 and now supports nearly 20 languages. In both cases the voice input is sent to central servers to perform general speech recognition and therefore requires a network connection for more than simple commands.

- Gesture recognition

- Character recognition to write text on the tablet, that can be stored as any other text in the intended storage, instead of using a keyboard.

- Near field communication with other compatible devices including ISO/IEC 14443 RFID tags.

Types

Crossover tablet device types from 2014: Microsoft Surface Pro 3 laplet, and Sony Xperia Z Ultra phablet, shown next to a generic blue-colored lighter to indicate their size.

Comparison of several mini tablet computers: Amazon Kindle Fire (left), iPad Mini (center) and Google Nexus 7 (right)

Samsung's Galaxy Note series were the first commercially successful phablet devices

Tablets can be loosely grouped into several categories, by physical size, operating system installed, input/output technology and usage.

Slate

A slate's size may vary, starting from 6 inches (approximately 15 cm). Some models in the larger than 10-inch category include the Samsung Galaxy Tab Pro 12.2 at 12.2 inches, the Toshiba Excite at 13.3 inches and the Dell XPS 18 at 18.4 inches. As of March 2013, the thinnest tablet on the market was the Sony Xperia Tablet Z at only 0.27 inches (6.9 mm) thick. On 9 September 2015, Apple released the iPad Pro with a 12.9 inches (33 cm) screen size, larger than the regular iPad.

Mini Tablet

Mini tablets are smaller and lighter than standard slates, with a typical screen size between 7–8 inches (18–20 cm). The first successful ones were introduced by Amazon (Kindle Fire), Barnes & Noble (Nook Tablet), and Samsung (7-inch Galaxy Tab) in 2011, and by Google (the Nexus 7) in 2012. They work the same as larger tablets, however with lower specifications when compared to the larger tablets. On September 14, 2012, Amazon released an upgraded version of the Kindle Fire, called the Kindle Fire HD, with higher resolution and more features compared with the original Kindle Fire, though it remained 7 inches. In October 2012, Apple released the iPad Mini with a 7.9 inch screen size, about 2 inches smaller than the regular iPad, but less powerful than the then current iPad 3. On July 24, 2013, Google released an upgraded version of the Nexus 7, with FHD display, dual cameras, stereo speakers, more color accuracy, performance improvement, built-in wireless charging, and a variant with 4G LTE support for AT&T, T-Mobile, and Verizon. In September 2013, Amazon further updated the Fire tablet with the Kindle Fire HDX. In November 2013, Apple released the iPad Mini 2, which remained at 7.9 inches and nearly matched the hardware of the iPad Air.

Phablet

Since 2010, crossover touch-screen mobile phones with screens larger than 5-inches have been released. That size is generally considered larger than a traditional smartphone, creating a hybrid category called a *phablet* by *Engadget* and *Forbes*. Phablet is a portmanteau of phone and tablet. Examples of phablets are the Dell Streak, LG Optimus Vu, and Samsung Galaxy Note. Samsung announced they had shipped a million units of the Galaxy Note within two months of introducing it.

Convertible, Hybrid, 2-in-1

Convertibles and hybrids are crossover devices, featuring traits of both tablets and laptops. *Convertibles* have a chassis design allowing to conceal the keyboard, for example folding it behind the chassis. *Hybrids'* keyboards can be completely detached even when the device is running. *2-in-1s* can have both the convertible or hybrid form, dubbed *2-in-1 convertibles* and *2-in-1 detachables* respectively, but distinct by a support of desktop operating system, such as Windows 10. When traditional tablets are primarily used as a media consumption devices, 2-in-1s capable of both that and a content creation, and due to this fact they are often dubbed as a *laptop or desktop replacements*. 2-in-1s have a number of typical laptop I/O-ports, such as USB 3 and DisplayPort, run desktop operating system, like Windows 10, and can connect to a number of traditional PC peripheral devices and external displays. Asus Transformer Pad-series devices, which run variants of Android OS, are example of hybrids. The latest addition to the Apple iPad series, iPad Pro with an optional detachable keyboard and a stylus is a prominent example of a modern hybrid. Microsoft's Surface Pro-series devices and Surface Book exemplify 2-in-1 detachables, whereas Lenovo Yoga-series computers are notable 2-in-1 convertibles.

Gaming Tablet

Some tablets are modified by adding physical gamepad buttons such as D-pad and thumb sticks for better gaming experience combined with the touchscreen and all other features of a typical tablet computer. Most of these tablets are targeted to run native OS games and emulator games. Nvidia's Shield Tablet, with a 8 inches (200 mm) display, and running Android, is an example. It runs Android games purchased from Google Play store. PC games can also be streamed to the tablet from computers with some models of Nvidia-powered video cards.

Booklet

Booklets are dual-touchscreen tablet computers with a clamshell design that can fold like a laptop. Examples include the Microsoft Courier, which was discontinued in 2010, the Sony Tablet P (which was considered a flop), and the Toshiba Libretto W100.

Customized Business Tablet

In contrast to consumer-grade tablet computers, customized business tablets are built specifically for a business entity to achieve customized functionality from a hardware and software perspective, and delivered in a business-to-business transaction. For example, in hardware, a transportation company may find that the consumer-grade GPS module in an off-the-shelf tablet provides insufficient accuracy, so a tablet can be customized and embedded with a professional-grade antenna to provide a better GPS signal. For a software example, the same transportation company might remove certain software functions in the Android system, such as the internet browser, to reduce costs from unnecessary cellular network data consumption of an employee.

System Architecture

Two major architectures dominate the tablet market, ARM Holdings' ARM architecture and Intel's and AMD's x86. Intel's x86, including x86-64 has powered the "IBM compatible" PC since 1981 and Apple's Macintosh computers since 2006. The CPUs have been incorporated into tablet PCs over the years and generally offer greater performance along with the ability to run full versions of Microsoft Windows, along with Windows desktop and enterprise applications. Non-Windows based x86 tablets include the JooJoo. Intel announced plans to enter the tablet market with its Atom in 2010. In October 2013, Intel's foundry operation announced plans to build FPGA-based quad cores for ARM and x86 processors.

ARM has been the CPU architecture of choice for manufacturers of smartphones (95% ARM), PDAs, digital cameras (80% ARM), set-top boxes, DSL routers, smart televisions (70% ARM), storage devices and tablet computers (95% ARM). This dominance began with the release of the mobile-focused and comparatively power-efficient 32-bit ARM610 processor originally designed for the Apple Newton in 1993 and ARM3-using Acorn A4 laptop in 1992. The chip was adopted by Psion, Palm and Nokia for PDAs and later smartphones, camera phones, cameras, etc. ARM's licensing model supported this success by allowing device manufacturers to license, alter and fabricate custom SoC derivatives tailored to their own products. This has helped manufacturers extend battery life and shrink component count along with the size of devices.

The multiple licensees ensured that multiple fabricators could supply near-identical products, while encouraging price competition. This forced unit prices down to a fraction of their x86 equivalents. The architecture has historically had limited support from Microsoft, with only Windows CE available, but with the 2012 release of Windows 8, Microsoft announced additional support for the architecture, shipping their own ARM-based tablet computer, branded the Microsoft Surface, as well as an x86-64 Intel Core i5 variant branded as Microsoft Surface Pro. Intel tablet chip sales were 1 million units in 2012, and 12 million units in 2013. Intel chairman Andy Bryant has stated that its 2014 goal is to quadruple its tablet chip sales to 40 million units by the end of that year, as an investment for 2015.

Operating System

Tablets, like conventional PCs, run on multiple operating systems (though dual-booting on tablets is relatively rare). These operating systems come in two classes, desktop-based and mobile-based ("phone-like") OS. Desktop OS-based tablets are currently thicker and heavier, require more storage, more cooling and give less battery life, but can run processor-intensive applications such as Adobe Photoshop in addition to mobile apps and have more ports, while mobile-based tablets are the reverse, only run mobile apps. Those that focus more so on mobile apps use battery life conservatively because the processor is significantly smaller. This allows the battery to last much longer than the common laptop. At the end of Q1 2013, GlobalWebIndex noted that in two years tablet usage increased by 282 percent, with 156 million Android tablet users and 122 million iPad users making up 75 percent. By year-end 2013, Gartner found that 121 million (plus 53M in 2012) Android tablets, 70 million (plus 61M in 2012) iOS tablets, and 4 million (plus 1M in 2012) Windows tablets had been sold to end-users (2013 and 2012 results).

Android

Android is a Linux-based operating system that Google offers as open source under the Apache license. It is designed primarily for mobile devices such as smartphones and tablet computers. Android supports low-cost ARM systems and others. Many such systems were announced in 2010. Vendors such as Motorola and Lenovo delayed deployment of their tablets until after 2011, when Android was reworked to include more tablet features. Android 3.0 (Honeycomb) and later versions support larger screen sizes, mainly tablets, and have access to the Google Play service. Android includes operating system, middleware and key applications. Other vendors sell customized Android tablets, such as Kindle Fire and Nook, which are used to consume mobile content and provide their own app store, rather than using the larger Google Play system, thereby fragmenting the Android market. Hardware makers that have shipped Android tablets include Acer, Asus, Samsung, Sony, and Toshiba. Additionally, Google introduced the Nexus 7 and Nexus 10 tablets in 2012.

iOS

The iPad runs on iOS, which was created for the iPhone and iPod Touch. Although built on the same underlying Unix implementation as MacOS, its user interface is radically different. iOS is designed for fingers and has none of the features that required a stylus on earlier tablets. Apple introduced multi-touch gestures, such as moving two fingers apart or together to zoom in or out, also known as "pinch to zoom". iOS is built for the ARM architecture.

Modbook

Previous to the iPad, Axiotron introduced an aftermarket, heavily modified Apple MacBook called Modbook, a Mac OS X-based tablet personal computer. The Modbook uses Apple's Inkwell for handwriting and gesture recognition, and uses digitization hardware from Wacom. To get Mac OS X to talk to the digitizer on the integrated tablet, the Modbook is supplied with a third-party driver called TabletMagic; Wacom does not provide driver support for this device. Another predecessor to the iPad was the Apple MessagePad introduced in 1993.

Windows

Windows 3.1 to 7

Following Windows for Pen Computing for Windows 3.1 in 1991, Microsoft supported tablets running Windows XP under the Microsoft Tablet PC name. According to Microsoft in 2001, "Microsoft Tablet PCs" are pen-based, fully functional x86 PCs with handwriting and voice recognition functionality. Tablet PCs used the same hardware as laptops but added support for pen input. Windows XP Tablet PC Edition provided pen support. Tablet support was added to both Home and Business versions of Windows Vista and Windows 7. Tablets running Windows could use the touchscreen for mouse input, hand writing recognition and gesture support. Following Tablet PC, Microsoft announced the Ultra-mobile PC initiative in 2006 which brought Windows tablets to a smaller, touch-centric form factor. In 2008, Microsoft showed a prototype of a two-screen tablet called Microsoft Courier, but cancelled the project. A model of the Asus Eee Pad shown in 2010 was to use Windows CE but switched to Android.

Windows 8

In October 2012, Microsoft released Windows 8, which features significant changes to various aspects of the operating system's user interface and platform which are designed for touch-based devices such as tablets. The operating system also introduced an application store and a new style of application optimized primarily for use on tablets. Microsoft also introduced Windows RT, an edition of Windows 8 for use on ARM-based devices. The launch of Windows 8 and RT was accompanied by the release of devices with the two operating systems by various manufacturers (including Microsoft themselves, with the release of Surface), such as slate tablets, hybrids, and convertibles. Windows RT is likely to be discontinued. In the first half of 2014, Windows tablets have grown 33%.

Windows 10

Released in July 2015, Windows 10 introduces what Microsoft described as "universal apps"; expanding on Metro-style apps, these apps can be designed to run across multiple Microsoft product families with nearly identical code—including PCs, tablets, smartphones, embedded systems, Xbox One, Surface Hub and Windows Holographic. The Windows user interface was revised to handle transitions between a mouse-oriented interface and a touchscreen-optimized interface based on available input devices—particularly on 2-in-1 PCs; both interfaces include an updated Start menu which incorporates elements of Windows 7's traditional Start menu with the tiles of Windows 8.

Firefox OS

Firefox OS is an open-source operating system based on Linux and the Firefox web browser, targeting low-end smartphones, tablet computers and smart TV devices. In 2013, the Mozilla Foundation started a prototype tablet model with Foxconn.

Linux

The ProGear by FrontPath was an early implementation of a Linux tablet that used a Transmeta chip and a resistive digitizer. The ProGear initially came with a version of Slackware Linux, and later with Windows 98. They can run many operating systems. However, the device is no longer for sale and FrontPath has ceased operations. Many touch screen sub-notebook computers can run any of several Linux distributions with little customization. X.org now supports screen rotation and tablet input through Wacom drivers, and handwriting recognition software from both the Qt-based Qtopia and GTK+-based Internet Tablet OS provide open source systems. KDE's Plasma Active is a graphical environment for tablet. Linux open source note taking software includes Xournal (which supports PDF file annotation), Gournal (a Gnome-based note taking application), and the Java-based Jarnal (which supports handwriting recognition as a built-in function). A standalone handwriting recognition program, CellWriter, requires users to write letters separately in a grid.

Many desktop distributions include tablet-friendly interfaces smaller devices. These open source libraries are freely available and can be run or ported to devices that conform to the tablet PC design. Maemo (rebranded MeeGo in 2010), a Debian Linux based user environment, was developed for the Nokia Internet Tablet devices (770, N800, N810 & N900). It is currently in generation 5, and has many applications. Ubuntu uses the Unity UI, and many other distributions (such as Fedora) use the Gnome shell (which also supports Ubuntu). Canonical hinted that Ubuntu would be available on tablets by 2014. In February 2016 there was a commercial release of an Ubuntu tablet. TabletKiosk was the first to offer a hybrid digitizer / touch device running openSUSE Linux.

Nokia's Use

The Nokia N800

Nokia entered the tablet space in May 2005 with the Nokia 770 running Maemo, a Debian-based Linux distribution custom-made for their Internet tablet line. The product line continued with the N900, with phone capabilities. The user interface and application framework layer, named Hildon, was an early instance of a software platform for generic computing in a tablet device intended for internet consumption. But Nokia didn't commit to it as their only platform for their future mobile devices and the project competed against other in-house platforms and later replaced it with the Series 60. Following the launch of the Ultra-mobile PC, Intel started the Mobile Internet Device initiative, which took the same hardware and combined it with a tabletized Linux configuration. Intel co-developed the lightweight Moblin (mobile Linux) operating system following the successful launch of the Atom CPU series on netbooks.

Tizen

MeeGo was a Linux-based operating system developed by Intel and Nokia that supports netbooks, smartphones and tablet PCs. In 2010, Nokia and Intel combined the Maemo and Moblin projects to form MeeGo. The first tablet using MeeGo is the Neofonie WeTab launched September 2010 in Germany. The WeTab uses an extended version of the MeeGo operating system called WeTab OS. WeTab OS adds runtimes for Android and Adobe AIR and provides a proprietary user interface optimized for the WeTab device. On September 27, 2011 the Linux Foundation announced that MeeGo would be replaced in 2012 by Tizen.

Hybrid OS Operation

Several hardware companies have built hybrid devices with the possibility to work with both the Windows 8 and Android operating systems. In mid-2014, Asus planned to release a hybrid touch-screen Windows tablet/laptop with a detachable Android smartphone. When docked to the back of the tablet/laptop display, the Android phone is displayed within the Windows 8 screen, which is switchable to Android tablet and Android laptop. However this device was never released and the only hybrid which was sold was the Asus Transformer Book Trio.

Discontinued Tablets

Blackberry OS

The BlackBerry PlayBook is a tablet computer announced in September 2010 that runs the Black-Berry Tablet OS. The OS is based on the QNX system that Research in Motion acquired in early 2010. Delivery to developers and enterprise customers was expected in October 2010. The Black-Berry PlayBook was officially released to US and Canadian consumers on April 19, 2011. As of 2014, Playbook is not available on sale on any Blackberry websites. The OS though continues on its smartphones.

WebOS

Hewlett Packard announced that the TouchPad, running WebOS 3.0 on a 1.2 GHz Qualcomm Snapdragon CPU, would be released in June 2011. On August 18, 2011, HP announced the discontinuation of the TouchPad, due to sluggish sales. In February 2013, HP announced they had sold WebOS to LG Electronics.

Commercialization

Application markets and software *walled gardens*

Mobile device suppliers typically adopt a "walled garden" approach, wherein the supplier controls what software applications ("apps") are available. Software development kits are restricted to approved software developers. This can be used to reduce the impact of malware, provide software with an approved content rating, control application quality and exclude competing vendors. Apple, Google, Amazon, Microsoft and Barnes & Noble all adopted the strategy. B&N originally allowed arbitrary apps to be installed, but, in December 2011, excluded third parties. Apple and IBM have agreed to cooperate in cross-selling IBM-developed applications for iPads and iPhones in enterprise-level accounts. Proponents of open source software say that it violates the spirit of personal control that traditional personal computers have always provided.

Market Share

As of October 2012, display screen shipments for tablets began surpassing shipments for laptop display screens. According to a survey conducted by the Online Publishers Association (OPA) now called Digital Content Next (DCN) in March 2012, 31% percent of Internet users in the United States owned a tablet, up from 12% in 2011. The survey also found that 72% of tablet owners had an iPad, while 32% had an Android tablet. By 2012, Android tablet adoption had increased. 52% of tablet owners owned an iPad, while 51% owned an Android-powered tablet (percentages do not add up to 100% because some tablet owners own more than one type). By end of 2013, Android's market share rose to 61.9%, followed by iOS at 36%. By late 2014, Android's market share rose to 72%, followed by iOS at 22.3% and Windows at 5.7%.

Note: Others consists of small vendors with market share about one percent or mostly less. In one year Apple market share dropped significantly and, on the other side, Android vendors' market share increased with Samsung dominating.

Sales

Research firms Gartner and IDC both predict that tablet sales will exceed traditional personal computer (desktops, notebooks) sales in 2015. As per the report from ABI Research in 2014 December, globally the average selling price of Ultrabooks and tablets declined 7.8 percent in 2014.

Usage

Around 2010, tablet use by businesses jumped, as business have started to use them for conferences, events, and trade shows. In 2012, Intel reported that their tablet program improved productivity for about 19,000 of their employees by an average of 57 minutes a day. In the US and Canada, it is estimated that 60% of online consumers will own a tablet by 2017 and in Europe, 42% of online consumers will own one. As of the beginning of 2013, 29% of US online consumers owned tablet computers, a significant jump from 5% in 2011. As of the beginning of 2014, 44% of US online consumers own tablets. Tablet use has also become increasingly common amongst children. A 2014 survey found that touch screens were the most frequently used object for play amongst American children under the age of 12. Touch screen devices were used more often in play than game consoles, board games, puzzles, play vehicles, blocks and dolls/action figures. Despite this, the majority of parents said that a touch screen device was "never" or only "sometimes" a toy. As of 2014, nearly two-thirds of American 2- to 10-year-olds have access to a tablet or e-reader. The large use of tablets by adults is as a personal internet-connected TV. A recent study has found that a third of children under five have their own tablet device. While Android tablets sell more units than iPad, the web browser usage share of iPads is about 65% as of the middle of 2015.

Effects on Sleep

The blue wavelength of light from back-lit tablets may impact one's ability to fall asleep when reading at night, through the suppression of melatonin. Experts at Harvard Medical School suggest limiting tablets for reading use in the evening. Those who have a delayed body clock, such as teenagers, which makes them prone to stay up late in the evening and sleep later in the morning, may be at particular risk for increases in sleep deficiencies. PC apps such as and F.lux and Android apps such as CF.lumen and Twilight attempt to decrease the impact on sleep by filtering blue wavelengths from the display. iOS 9.3 has "Night Shift" built-in that shifts the colors of the device's display to be warmer.

Smart card

A smart card, chip card, or integrated circuit card (ICC) is any pocket-sized card that has embedded integrated circuits. Smart cards are made of plastic, generally polyvinyl chloride, but sometimes polyethylene terephthalate based polyesters, acrylonitrile butadiene styrene or polycarbonate. Since April 20

One of the first smart card prototype created by its inventor Roland Moreno circa 1975. The chip is still to be miniaturized

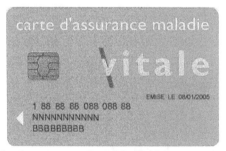

Carte Vitale, the smart card used for health insurance in France

09, a Japanese company has manufactured reusable financial smart cards made from paper.

Smart cards can be either contact or contactless smart card. Smart cards can provide personal identification, authentication, data storage, and application processing. Smart cards may provide strong security authentication for single sign-on (SSO) within large organizations.

History

Invention

In 1968 and 1969 Helmut Gröttrup and Jürgen Dethloff jointly filed patents for the automated chip card. Roland Moreno patented the memory card concept in 1974. An important patent for smart cards with a microprocessor and memory as used today was filed by Jürgen Dethloff in 1976 and granted as USP 4105156 in 1978. In 1977, Michel Ugon from Honeywell Bull invented the first microprocessor smart card with two chips: one microprocessor and one memory, and in 1978, he has patented the self-programmable one-chip microcomputer (SPOM) that defines the necessary architecture to program the chip. Three years later, Motorola used this patent in its "CP8". At that time, Bull had 1,200 patents related to smart cards. In 2001, Bull sold its CP8 division together with its patents to Schlumberger, who subsequently combined its own internal smart card department and CP8 to create Axalto. In 2006, Axalto and Gemplus, at the time the world's top two smart card manufacturers, merged and became Gemalto. In 2008 Dexa Systems spun off from Schlumberger and acquired Enterprise Security Services business, which included the smart card solutions division responsible for deploying the first large scale public key infrastructure (PKI) based smart card management systems.

The first mass use of the cards was as a telephone card for payment in French pay phones, starting in 1983.

Carte Bleue

After the Télécarte, microchips were integrated into all French *Carte Bleue* debit cards in 1992. Customers inserted the card into the merchant's point of sale (POS) terminal, then typed the personal identification number (PIN), before the transaction was accepted. Only very limited transactions (such as paying small highway tolls) are processed without a PIN.

Smart-card-based "electronic purse" systems store funds on the card so that readers do not need network connectivity. They entered European service in the mid-1990s. They have been common in Germany (Geldkarte), Austria (Quick Wertkarte), Belgium (Proton), France (Moneo), the Netherlands (Chipknip Chipper (decommissioned in 2001)), Switzerland ("Cash"), Norway ("Mondex"), Sweden ("Cash", decommissioned in 2004), Finland ("Avant"), UK ("Mondex"), Denmark ("Danmønt") and Portugal ("Porta-moedas Multibanco").

Since the 1990s, smart-cards have been the Subscriber Identity Modules (SIMs) used in European GSM mobile phone equipment. Mobile phones are widely used in Europe, so smart cards have become very common.

EMV

Europay MasterCard Visa (EMV)-compliant cards and equipment are widespread. The United States started using the EMV technology in 2014. Typically, a country's national payment association, in coordination with MasterCard International, Visa International, American Express and Japan Credit Bureau (JCB), jointly plan and implement EMV systems.

Historically, in 1993 several international payment companies agreed to develop smart-card specifications for debit and credit cards. The original brands were MasterCard, Visa, and Europay. The first version of the EMV system was released in 1994. In 1998 the specifications became stable.

EMVCo maintains these specifications. EMVco's purpose is to assure the various financial institutions and retailers that the specifications retain backward compatibility with the 1998 version. EMVco upgraded the specifications in 2000 and 2004.

EMV compliant cards were accepted into the United States in 2014. MasterCard was the first company that has been allowed to use the technology in the United States. The United States has felt pushed to use the technology because of the increase in identity theft. The credit card information stolen from Target in late 2013 was one of the largest indicators that American credit card information is not safe. Target has made the decision on April 30, 2014 that they are going to try and implement the smart chip technology in order to protect themselves from future credit card identity theft.

Before 2014, the consensus in America was that there was enough security measures to avoid credit card theft and that the smart chip was not necessary. The cost of the smart chip technology was significant, which was why most of the corporations did not want to pay for it in the United States. The debate came when online credit theft was insecure enough for the United States to invest in the technology. The adaptation of EMV's increased significantly in 2015 when the liability shifts occurred in October by the credit card companies.

Development of Contactless Systems

Contactless smart cards do not require physical contact between a card and reader. They are becoming more popular for payment and ticketing. Typical uses include mass transit and motorway tolls. Visa and MasterCard implemented a version deployed in 2004–2006 in the U.S. Most contactless fare collection systems are incompatible, though the MIFARE Standard card from NXP Semiconductors has a considerable market share in the US and Europe.

Smart cards are also being introduced for identification and entitlement by regional, national, and international organizations. These uses include citizen cards, drivers' licenses, and patient cards. In Malaysia, the compulsory national ID MyKad enables eight applications and has 18 million users. Contactless smart cards are part of ICAO biometric passports to enhance security for international travel.

Design

A smart card may have the following generic characteristics:

- Dimensions similar to those of a credit card. ID-1 of the ISO/IEC 7810 standard defines cards as nominally 85.60 by 53.98 millimetres (3.370 in × 2.125 in). Another popular size is ID-000 which is nominally 25 by 15 millimetres (0.984 in × 0.591 in) (commonly used in SIM cards). Both are 0.76 millimetres (0.030 in) thick.

- Contains a tamper-resistant security system (for example a secure cryptoprocessor and a secure file system) and provides security services (e.g., protects in-memory information).

- Managed by an administration system which securely interchanges information and configuration settings with the card, controlling card blacklisting and application-data updates.

- Communicates with external services via card-reading devices, such as ticket readers, ATMs, DIP reader, etc.

Contact Smart Cards

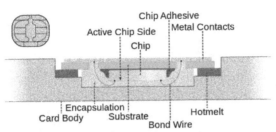

Illustration of smart card structure and packaging

Smart card reader on a laptop

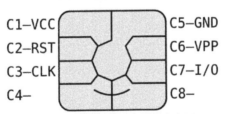

A smart card pinout. VCC: Power supply. RST: Reset signal, used to reset the card's communications. CLK: Provides the card with a clock signal, from which data communications timing is derived. GND: Ground (reference voltage). VPP: ISO/IEC 7816-3:1997 designated this as a programming voltage: an input for a higher voltage to program persistent memory (e.g., EEPROM). ISO/IEC 7816-3:2006 designates it SPU, for either standard or proprietary use, as input and/or output. I/O: Serial input and output (half-duplex). C4, C8: The two remaining contacts are AUX1 and AUX2 respectively, and used for USB interfaces and other uses. However, the usage defined in ISO/IEC 7816-2:1999/ Amd 1:2004 may have been superseded by ISO/IEC 7816-2:2007.

Contact smart cards have a contact area of approximately 1 square centimetre (0.16 sq in), comprising several gold-plated contact pads. These pads provide electrical connectivity when inserted into a reader, which is used as a communications medium between the smart card and a host (e.g., a computer, a point of sale terminal) or a mobile telephone. Cards do not contain batteries; power is supplied by the card reader.

The ISO/IEC 7810 and ISO/IEC 7816 series of standards define:

- physical shape and characteristics

- electrical connector positions and shapes

- electrical characteristics

- communications protocols, including commands sent to and responses from the card

- basic functionality

Because the chips in financial cards are the same as those used in subscriber identity modules (SIMs) in mobile phones, programmed differently and embedded in a different piece of PVC, chip manufacturers are building to the more demanding GSM/3G standards. So, for example, although the EMV standard allows a chip card to draw 50 mA from its terminal, cards are normally well below the telephone industry's 6 mA limit. This allows for smaller and cheaper financial card terminals.

Communication protocols for contact smart cards include T=0 (character-level transmission protocol, defined in ISO/IEC 7816-3) and T=1 (block-level transmission protocol, defined in ISO/IEC 7816-3).

Contactless Smart Cards

A second card type is the *contactless smart card*, in which the card communicates with and is powered by the reader through RF induction technology (at data rates of 106–848 kbit/s). These cards require only proximity to an antenna to communicate. Like smart cards with contacts, contactless cards do not have an internal power source. Instead, they use an inductor to capture some of the incident radio-frequency interrogation signal, rectify it, and use it to power the card's electronics.

APDU transmission via a contactless interface is defined in ISO/IEC 14443-4.

Hybrids

A hybrid smart card which clearly shows the antenna connected to the main chip

Hybrid cards implement contactless and contact interfaces on a single card with dedicated modules/storage and processing.

Dual Interface

Dual-interface cards implement contactless and contact interfaces on a single card with some shared storage and processing. An example is Porto's multi-application transport card, called Andante, which uses a chip with both contact and contactless (ISO/IEC 14443 Type B) interfaces.

USB

The CCID (Chip Card Interface Device) is a USB protocol that allows a smartcard to be connected to a Computer, using a standard USB interface. This allows the smartcard to be used as a security token for authentication and data encryption such as Bitlocker. CCID devices typically look like a standard USB dongle and may contain a SIM card inside the USB dongle.

Applications

Financial

Smart cards serve as credit or ATM cards, fuel cards, mobile phone SIMs, authorization cards for pay television, household utility pre-payment cards, high-security identification and access-control cards, and public transport and public phone payment cards.

Smart cards may also be used as electronic wallets. The smart card chip can be "loaded" with funds to pay parking meters, vending machines or merchants. Cryptographic protocols protect the exchange of money between the smart card and the machine. No connection to a bank is needed. The holder of the card may use it even if not the owner. Examples are Proton, Geldkarte, Chipknip and Moneo. The German Geldkarte is also used to validate customer age at vending machines for cigarettes.

These are the best known payment cards (classic plastic card):

- Visa: Visa Contactless, Quick VSDC, "qVSDC", Visa Wave, MSD, payWave

- MasterCard: PayPass Magstripe, PayPass MChip

- American Express: ExpressPay

- Discover: Zip

Roll-outs started in 2005 in the U.S. Asia and Europe followed in 2006. Contactless (non-PIN) transactions cover a payment range of ~$5–50. There is an ISO/IEC 14443 PayPass implementation. Some, but not all PayPass implementations conform to EMV.

Non-EMV cards work like magnetic stripe cards. This is common in the U.S. (PayPass Magstripe and Visa MSD). The cards do not hold or maintain the account balance. All payment passes without a PIN, usually in off-line mode. The security of such a transaction is no greater than with a magnetic stripe card transaction.

EMV cards can have either contact or contactless interfaces. They work as if they were a normal EMV card with a contact interface. Via the contactless interface they work somewhat differently, in that the card commands enabled improved features such as lower power and shorter transaction times.

SIM

The subscriber identity modules used in mobile-phone systems are reduced-size smart cards, using otherwise identical technologies.

Identification

Smart-cards can authenticate identity. Sometimes they employ a public key infrastructure (PKI). The card stores an encrypted digital certificate issued from the PKI provider along with other relevant information. Examples include the U.S. Department of Defense (DoD) Common Access Card (CAC), and other cards used by other governments for their citizens. If they include biometric identification data, cards can provide superior two- or three-factor authentication.

Smart cards are not always privacy-enhancing, because the subject may carry incriminating information on the card. Contactless smart cards that can be read from within a wallet or even a garment simplify authentication; however, criminals may access data from these cards.

Cryptographic smart cards are often used for single sign-on. Most advanced smart cards include specialized cryptographic hardware that uses algorithms such as RSA and Digital Signature Algorithm (DSA). Today's cryptographic smart cards generate key pairs on board, to avoid the risk from having more than one copy of the key (since by design there usually isn't a way to extract private keys from a smart card). Such smart cards are mainly used for digital signatures and secure identification.

The most common way to access cryptographic smart card functions on a computer is to use a vendor-provided PKCS#11 library. On Microsoft Windows the Cryptographic Service Provider (CSP) API is also supported.

The most widely used cryptographic algorithms in smart cards (excluding the GSM so-called "crypto algorithm") are Triple DES and RSA. The key set is usually loaded (DES) or generated (RSA) on the card at the personalization stage.

Some of these smart cards are also made to support the National Institute of Standards and Technology (NIST) standard for Personal Identity Verification, FIPS 201.

Turkey implemented the first smart card driver's license system in 1987. Turkey had a high level of road accidents and decided to develop and use digital tachograph devices on heavy vehicles, instead of the existing mechanical ones, to reduce speed violations. Since 1987, the professional driver's licenses in Turkey have been issued as smart cards. A professional driver is required to insert his driver's license into a digital tachograph before starting to drive. The tachograph unit records speed violations for each driver and gives a printed report. The driving hours for each driver are also being monitored and reported. In 1990 the European Union conducted a feasibility study through BEVAC Consulting Engineers, titled "Feasibility study with respect to a European electronic drivers license (based on a smart-card) on behalf of Directorate General VII". In this study, chapter seven describes Turkey's experience.

Argentina's Mendoza province began using smart card driver's licenses in 1995. Mendoza also had a high level of road accidents, driving offenses, and a poor record of recovering fines. Smart licenses hold up-to-date records of driving offenses and unpaid fines. They also store personal information, license type and number, and a photograph. Emergency medical information such as blood type, allergies, and biometrics (fingerprints) can be stored on the chip if the card holder wishes. The Argentina government anticipates that this system will help to collect more than $10 million per year in fines.

In 1999 Gujarat was the first Indian state to introduce a smart card license system. As of 2005, it has issued 5 million smart card driving licenses to its people.

In 2002, the Estonian government started to issue smart cards named ID Kaart as primary identification for citizens to replace the usual passport in domestic and EU use. As of 2010 about 1 million smart cards have been issued (total population is about 1.3 million) and they are widely used in internet banking, buying public transport tickets, authorization on various websites etc.

By the start of 2009 the entire population of Spain and Belgium became eID cards that are used for identification. These cards contain two certificates: one for authentication and one for signature. This signature is legally enforceable. More and more services in these countries use eID for authorization.

On August 14, 2012, the ID cards in Pakistan were replaced. The Smart Card is a third generation chip-based identity document that is produced according to international standards and requirements. The card has over 36 physical security features and has the latest encryption codes. This smart card replaced the NICOP (the ID card for overseas Pakistani).

Smart cards may identify emergency responders and their skills. Cards like these allow first responders to bypass organizational paperwork and focus more time on the emergency resolution. In 2004, The Smart Card Alliance expressed the needs: "to enhance security, increase government efficiency, reduce identity fraud, and protect personal privacy by establishing a mandatory, Government-wide standard for secure and reliable forms of identification". emergency response personnel can carry these cards to be positively identified in emergency situations. WidePoint Corporation, a smart card provider to FEMA, produces cards that contain additional personal information, such as medical records and skill sets.

In 2007, the Open Mobile Alliance (OMA) proposed a new standard defining V1.0 of the Smart Card Web Server (SCWS), an HTTP server embedded in a SIM card intended for a smartphone user. The non-profit trade association SIMalliance has been promoting the development and adoption of SCWS. SIMalliance states that SCWS offers end-users a familiar, OS-independent, browser-based interface to secure, personal SIM data. As of mid-2010, SIMalliance had not reported widespread industry acceptance of SCWS. The OMA has been maintaining the standard, approving V1.1 of the standard in May 2009, and V1.2 is expected was approved in October 2012.

Public Transit

Smart cards and integrated ticketing are used by many public transit operators. Card users may also make small purchases using the cards. Some operators offer points for usage, exchanged at retailers or for other benefits. Examples include Singapore's CEPAS, Toronto's Presto card, Hong Kong's Octopus Card, London's Oyster Card, Dublin's Leap card, Brussels' MoBIB, Québec's OPUS card, San Francisco's Clipper card, Auckland's AT Hop, Brisbane's go card, Perth's SmartRider and Sydney's Opal card. However, these present a privacy risk because they allow the mass transit operator (and the government) to track an individual's movement. In Finland, for example, the Data Protection Ombudsman prohibited the transport operator Helsinki Metropolitan Area Council (YTV) from collecting such information, despite YTV's argument that the card owner has the right to a list of trips paid with the card. Earlier, such information was used in the investigation of the Myyrmanni bombing.

The UK's Department for Transport mandated smart cards to administer travel entitlements for elderly and disabled residents. These schemes let residents use the cards for more than just bus passes. They can also be used for taxi and other concessionary transport. One example is the "Smartcare go" scheme provided by Ecebs. The UK systems use the ITSO Ltd specification.

Computer Security

Smart cards can be used as a security token.

The Mozilla Firefox web browser can use smart cards to store certificates for use in secure web browsing.

Some disk encryption systems, such as Microsoft's BitLocker, can use smart cards to securely hold encryption keys, and also to add another layer of encryption to critical parts of the secured disk.

GnuPG, the well known encryption suite, also supports storing keys in a smart card.

Smart cards are also used for single sign-on to log on to computers.

Schools

Smart cards are being provided to students at some schools and colleges. Uses include:

- Tracking student attendance

- As an electronic purse, to pay for items at canteens, vending machines, laundry facilities, etc.

- Tracking and monitoring food choices at the canteen, to help the student maintain a healthy diet

- Tracking loans from the school library

- Access control for admittance to restricted buildings, dormitories, and other facilities. This requirement may be enforced at all times (such as for a laboratory containing valuable equipment), or just during after-hours periods (such as for an academic building that is open during class times, but restricted to authorized personnel at night), depending on security needs.

- Access to transportation services

Healthcare

Smart health cards can improve the security and privacy of patient information, provide a secure carrier for portable medical records, reduce health care fraud, support new processes for portable medical records, provide secure access to emergency medical information, enable compliance with government initiatives (e.g., organ donation) and mandates, and provide the platform to implement other applications as needed by the health care organization.

Other Uses

Smart cards are widely used to protect digital television streams. VideoGuard is a specific example of how smart card security worked.

Multiple-use Systems

The Malaysian government promotes MyKad as a single system for all smart-card applications. MyKad started as identity cards carried by all citizens and resident non-citizens. Available applications now include identity, travel documents, drivers license, health information, an electronic wallet, ATM bank-card, public toll-road and transit payments, and public key encryption infrastructure. The personal information inside the MYKAD card can be read using special APDU commands.

Security

Smart cards have been advertised as suitable for personal identification tasks, because they are engineered to be tamper resistant. The chip usually implements some cryptographic algorithm. There are, however, several methods for recovering some of the algorithm's internal state.

Differential power analysis involves measuring the precise time and electric current required for certain encryption or decryption operations. This can deduce the on-chip private key used by public key algorithms such as RSA. Some implementations of symmetric ciphers can be vulnerable to timing or power attacks as well.

Smart cards can be physically disassembled by using acid, abrasives, solvents, or some other technique to obtain unrestricted access to the on-board microprocessor. Although such techniques may involve a risk of permanent damage to the chip, they permit much more detailed information (e.g., photomicrographs of encryption hardware) to be extracted.

Benefits

The benefits of smart cards are directly related to the volume of information and applications that are programmed for use on a card. A single contact/contactless smart card can be programmed with multiple banking credentials, medical entitlement, driver's license/public transport entitlement, loyalty programs and club memberships to name just a few. Multi-factor and proximity authentication can and has been embedded into smart cards to increase the security of all services on the card. For example, a smart card can be programmed to only allow a contactless transaction if it is also within range of another device like a uniquely paired mobile phone. This can significantly increase the security of the smart card.

Governments and regional authorities save money because of improved security, better data and reduced processing costs. These savings help reduce public budgets or enhance public services. There are many examples in the UK, many using a common open LASSeO specification.

Individuals have better security and more convenience with using smart cards that perform multiple services. For example, they only need to replace one card if their wallet is lost or stolen. The data storage on a card can reduce duplication, and even provide emergency medical information.

Advantages

The first main advantage of smart cards is their flexibility. Smart cards have multiple functions which simultaneously can be an ID, a credit card, a stored-value cash card, and a repository of personal information such as telephone numbers or medical history. The card can be easily replaced if lost, and, the requirement for a PIN (or other form of security) provides additional security from unauthorised access to information by others. At the first attempt to use it illegally, the card would be deactivated by the card reader itself.

The second main advantage is security. Smart cards can be electronic key rings, giving the bearer ability to access information and physical places without need for online connections. They are encryption devices, so that the user can encrypt and decrypt information without relying on unknown, and therefore potentially untrustworthy, appliances such as ATMs. Smart cards are very flexible in providing authentication at different level of the bearer and the counterpart. Finally, with the information about the user that smart cards can provide to the other parties, they are useful devices for customizing products and services.

Other general benefits of smart cards are:

- Portability

- Increasing data storage capacity

- Reliability that is virtually unaffected by electrical and magnetic fields.

Smart Cards and Electronic Commerce

Smart cards can be used in electronic commerce, over the Internet, though the business model used in current electronic commerce applications still cannot use the full potential of the electronic medium. An advantage of smart cards for electronic commerce is their use customize services. For example, in order for the service supplier to deliver the customized service, the user may need to provide each supplier with their profile, a boring and time-consuming activity. A smart card can contain a non-encrypted profile of the bearer, so that the user can get customized services even without previous contacts with the supplier.

Disadvantages

The plastic card in which the chip is embedded is fairly flexible. The larger the chip, the higher the probability that normal use could damage it. Cards are often carried in wallets or pockets, a harsh environment for a chip. However, for large banking systems, failure-management costs can be more than offset by fraud reduction.

If the account holder's computer hosts malware, the smart card security model may be broken. Malware can override the communication (both input via keyboard and output via application screen) between the user and the application. Man-in-the-browser malware (e.g., the Trojan Silentbanker) could modify a transaction, unnoticed by the user. Banks like Fortis and Belfius in Belgium and Rabobank ("random reader") in the Netherlands combine a smart card with an unconnected card reader to avoid this problem. The customer enters a challenge received from the bank's website, a PIN and the transaction amount into the reader, The reader returns an 8-digit signature. This signature is manually entered into the personal computer and verified by the bank, preventing point-of-sale-malware from changing the transaction amount.

Smart cards have also been the targets of security attacks. These attacks range from physical invasion of the card's electronics, to non-invasive attacks that exploit weaknesses in the card's software or hardware. The usual goal is to expose private encryption keys and then read and manipulate secure data such as funds. Once an attacker develops a non-invasive attack for a particular smart card model, he is typically able to perform the attack on other cards of that model in seconds, often using equipment that can be disguised as a normal smart card reader. While manufacturers may develop new card models with additional security, it may be costly or inconvenient for users to upgrade vulnerable systems. Tamper-evident and audit features in a smart card system help manage the risks of compromised cards.

Another problem is the lack of standards for functionality and security. To address this problem, the Berlin Group launched the ERIDANE Project to propose "a new functional and security framework for smart-card based Point of Interaction (POI) equipment".

Personal Digital Assistant

A personal digital assistant (PDA), also known as a handheld PC, or personal data assistant, is a mobile device that functions as a personal information manager. The term evolved from Personal Desktop Assistant, a software term for an application that prompts or prods the user of a computer with suggestions or provides quick reference to contacts and other lists. PDAs were largely discontinued in the early 2010s after the widespread adoption of highly capable smartphones, in particular those based on iOS and Android.

The Palm TX

Nearly all PDAs have the ability to connect to the Internet. A PDA has an electronic visual display, enabling it to include a web browser, all models also have audio capabilities enabling use as a portable media player, and also enabling most of them to be used as mobile phones. Most PDAs can access the Internet, intranets or extranets via Wi-Fi or Wireless Wide Area Networks. Most PDAs employ touchscreen technology.

The first PDA was released in 1984 by Psion, the Organizer. Followed by Psion's Series 3, in 1991, which began to resemble the more familiar PDA style. It also had a full keyboard. The term *PDA* was first used on January 7, 1992 by Apple Computer CEO John Sculley at the Consumer Electronics Show in Las Vegas, Nevada, referring to the Apple Newton. In 1994, IBM introduced the first PDA with full mobile phone functionality, the IBM Simon, which can also be considered the first smartphone. Then in 1996, Nokia introduced a PDA with full mobile phone functionality, the 9000 Communicator, which became the world's best-selling PDA. The Communicator spawned a new category of PDAs: the "PDA phone", now called "smartphone". Another early entrant in this market was Palm, with a line of PDA products which began in March 1996. The terms "personal digital assistant" and "PDA" apply to smartphones but are not used in marketing, media, or general conversation to refer to devices such as the BlackBerry, iPad, iPhone or Android devices.

Typical Features

A typical PDA has a touchscreen for entering data, a memory card slot for data storage, and IrDA, Bluetooth and/or Wi-Fi. However, some PDAs may not have a touchscreen, using softkeys, a directional pad, and a numeric keypad or a thumb keyboard for input; this is typically seen on telephones that are also PDAs.To have the functions expected of a PDA, a device's software typically includes an appointment calendar, a to-do list, an address book for contacts, a calculator, and some sort of memo (or "note") program. PDAs with wireless data connections also typically include an email client and a Web browser.

Touchscreen

Many of the original PDAs, such as the Apple Newton and Palm Pilot, featured a touchscreen for user interaction, having only a few buttons—usually reserved for shortcuts to often-used programs. Some touchscreen PDAs, including Windows Mobile devices, had a detachable stylus to facilitate making selections. The user interacts with the device by tapping the screen to select buttons or issue commands, or by dragging a finger (or the stylus) on the screen to make selections or scroll.

Typical methods of entering text on touchscreen PDAs include:

- A virtual keyboard, where a keyboard is shown on the touchscreen. Text is entered by tapping the on-screen keyboard with a finger or stylus.

- An external keyboard connected via USB, Infrared port, or Bluetooth. Some users may choose a chorded keyboard for one-handed use.

- Handwriting recognition, where letters or words are written on the touchscreen, often with a stylus, and the PDA converts the input to text. Recognition and computation of handwritten horizontal and vertical formulas, such as "1 + 2 =", may also be a feature.

- Stroke recognition allows the user to make a predefined set of strokes on the touchscreen, sometimes in a special input area, representing the various characters to be input. The strokes are often simplified character shapes, making them easier for the device to recognize. One widely known stroke recognition system is Palm's Graffiti.

Despite research and development projects, end-users experience mixed results with handwriting recognition systems. Some find it frustrating and inaccurate, while others are satisfied with the quality of the recognition.

Touchscreen PDAs intended for business use, such as the BlackBerry and Palm Treo, usually also offer full keyboards and scroll wheels or thumbwheels to facilitate data entry and navigation. Many touchscreen PDAs support some form of external keyboard as well. Specialized folding keyboards, which offer a full-sized keyboard but collapse into a compact size for transport, are available for many models. External keyboards may attach to the PDA directly, using a cable, or may use wireless technology such as infrared or Bluetooth to connect to the PDA. Newer PDAs, such as the HTC HD2, Apple iPhone, Apple iPod Touch, and Palm Pre, Palm Pre Plus, Palm Pixi, Palm Pixi Plus, Google Android (operating system) include more advanced forms of touchscreen that can register multiple touches simultaneously. These "multi-touch" displays allow for more sophisticated interfaces using various gestures entered with one or more fingers.

Memory Cards

Although many early PDAs did not have memory card slots, now most have either some form of Secure Digital (SD) slot, a CompactFlash slot or a combination of the two. Although designed for memory, Secure Digital Input/Output (SDIO) and CompactFlash cards are available that provide accessories like Wi-Fi or digital cameras, if the device can support them. Some PDAs also have a USB port, mainly for USB flash drives.[dubious – discuss] Some PDAs use microSD cards, which are electronically compatible with SD cards, but have a much smaller physical size.

Wired Connectivity

While early PDAs connected to a user's personal computer via serial ports or another proprietary connection, many today connect via a USB cable. Older PDAs were unable to connect to each other via USB, as their implementations of USB didn't support acting as the "host". Some early PDAs were able to connect to the Internet indirectly by means of an external modem connected via the PDA's serial port or "sync" connector, or directly by using an expansion card that provided an Ethernet port.

Wireless Connectivity

Most modern PDAs have Bluetooth, a popular wireless protocol for mobile devices. Bluetooth can be used to connect keyboards, headsets, GPS receivers, and other nearby accessories. It's also possible to transfer files between PDAs that have Bluetooth. Many modern PDAs have Wi-Fi wireless network connectivity and can connect to Wi-Fi hotspots. All smartphones, and some other modern PDAs, can connect to Wireless Wide Area Networks, such as those provided by cellular telecommunications companies. Older PDAs from the 1990s to 2006 typically had an IrDA (infrared) port allowing short-range, line-of-sight wireless communication. Few current models use this technology, as it has been supplanted by Bluetooth and Wi-Fi. IrDA allows communication between two PDAs, or between a PDA and any device with an IrDA port or adapter. Some printers have IrDA receivers, allowing IrDA-equipped PDAs to print to them, if the PDA's operating system supports it. Universal PDA keyboards designed for these older PDAs use infrared technology. Infrared technology is low-cost and has the advantage of being allowed aboard.

Synchronization

Most PDAs can synchronize their data with applications on a user's computer. This allows the user to update contact, schedule, or other information on their computer, using software such as Microsoft Outlook or ACT!, and have that same data transferred to PDA—or transfer updated information from the PDA back to the computer. This eliminates the need for the user to update their data in two places. Synchronization also prevents the loss of information stored on the device if it is lost, stolen, or destroyed. When the PDA is repaired or replaced, it can be "re-synced" with the computer, restoring the user's data. Some users find that data input is quicker on their computer than on their PDA, since text input via a touchscreen or small-scale keyboard is slower than a full-size keyboard. Transferring data to a PDA via the computer is therefore a lot quicker than having to manually input all data on the handheld device.

Most PDAs come with the ability to synchronize to a computer. This is done through *synchronization software* provided with the handheld, or sometime with the computer's operating system. Examples of synchronization software include:

- HotSync Manager, for Palm OS PDAs

- 'Microsoft ActiveSync, used by Windows XP and older Windows operating systems to synchronize with Windows Mobile, Pocket PC, and Windows CE PDAs, as well as PDAs running iOS, Palm OS, and Symbian

- Microsoft Windows Mobile Device Center for Windows Vista, which supports Microsoft Windows Mobile and Pocket PC devices.

- Apple iTunes, used on Mac OS X and Microsoft Windows to sync iOS devices (such as the iPhone and iPod touch)

- iSync, included with Mac OS X, can synchronize many SyncML-enabled PDAs

- BlackBerry Desktop Software, used to sync BlackBerry devices.

These programs allow the PDA to be synchronized with a personal information manager, which may be part of the computer's operating system, provided with the PDA, or sold separately by a third party. For example, the RIM BlackBerry comes with RIM's *Desktop Manager* program, which can synchronize to both Microsoft Outlook and ACT!. Other PDAs come only with their own proprietary software. For example, some early Palm OS PDAs came only with Palm Desktop, while later Palm PDAs—such as the Treo 650—have the ability to sync to Palm Desktop or Microsoft Outlook. Microsoft's ActiveSync and Windows Mobile Device Center only synchronize with Microsoft Outlook or a Microsoft Exchange server. Third-party synchronization software is also available for some PDAs from companies like CommonTime and CompanionLink. Third-party software can be used to synchronize PDAs to other personal information managers that are not supported by the PDA manufacturers (for example, GoldMine and IBM Lotus Notes).

Wireless Synchronization

Some PDAs can synchronize some or all of their data using their wireless networking capabilities, rather than having to be directly connected to a personal computer via a cable. Apple iOS devices, like the iPhone, iPod Touch, and iPad can use Apple's iCloud service (formerly MobileMe) to synchronize calendar, address book, mail account, Internet bookmark, and other data with one or more Macintosh or Windows computers using Wi-Fi or cellular data connections. Devices running Palm's webOS or Google's Android operating system primarily sync with the cloud. For example, if Gmail is used, information in contacts, email, and calendar can be synchronized between the phone and Google's servers. RIM sells BlackBerry Enterprise Server to corporations so that corporate BlackBerry users can wirelessly synchronize their PDAs with the company's Microsoft Exchange Server, IBM Lotus Domino, or Novell GroupWise servers. Email, calendar entries, contacts, tasks, and memos kept on the company's server are automatically synchronized with the BlackBerry.

Operating Systems of PDAs

The most common operating systems pre-installed on PDAs are:-

- Palm OS

- Microsoft Windows Mobile (Pocket PC) with a Windows CE kernel.

- WebOS

Other, rarely used operating systems:

- EPOC, then Symbian OS (in mobile phone + PDA combos)

- Linux (e.g. VR3, iPAQ, Sharp Zaurus PDA, Opie, GPE, Familiar Linux etc.)

- Newton

- QNX (also on iPAQ)

Automobile Navigation

Some PDAs include Global Positioning System (GPS) receivers; this is particularly true of smartphones. Other PDAs are compatible with external GPS-receiver add-ons that use the PDA's processor and screen to display location information. PDAs with GPS functionality can be used for automotive navigation. PDAs are increasingly being fitted as standard on new cars. PDA-based GPS can also display traffic conditions, perform dynamic routing, and show known locations of roadside mobile radar guns. TomTom, Garmin, and iGO offer GPS navigation software for PDAs.

Ruggedized

Some businesses and government organizations rely upon rugged PDAs, sometimes known as enterprise digital assistants (EDAs) or mobile computers, for mobile data applications. These PDAs have features that make them more robust and able to handle inclement weather, bumps and moisture. EDAs often have extra features for data capture, such as barcode readers, radio-frequency identification (RFID) readers, magnetic stripe card readers, or smart card readers. These features are designed to facilitate the use of these devices to scan in product or item codes.

Typical applications include:

- Military (U.S. Army)

- Rangers

- Wildlife biologists

- Supply chain management in warehouses

- Package delivery

- Route accounting

- Medical treatment and recordkeeping in hospitals

- Facilities maintenance and management

- Parking enforcement

- Access control and security

- Capital asset maintenance

- Meter reading by utilities

- Waiter and waitress applications in restaurants and hospitality venues

- Infection control audit and surveillance within healthcare environments

- Taxicab allocation and routing

Medical and Scientific Uses

Many companies have developed PDA products aimed at the medical profession's, such as PDAs loaded with drug databases, treatment information, and medical information. Services such as AvantGo translate medical journals into PDA-readable formats. WardWatch organizes medical records, providing reminders of information such as the treatment regimens of patients to doctors making ward rounds. Pendragon and Syware provide tools for conducting research with, allowing the user to enter data into a centralized database using their PDA. Microsoft Visual Studio and Sun Java also provide programming tools for developing survey instruments on the handheld. These development tools allow for integration with SQL databases that are stored on the handheld and can be synchronized with a desktop- or server-based database. PDAs have been used by doctors to aid diagnosis and drug selection and some studies have concluded that when patients can use PDAs to record their symptoms, they communicate more effectively with hospital staff during follow-up visits. The development of Sensor Web technology may lead to wearable bodily sensors to monitor ongoing conditions, like diabetes or epilepsy, which would alert patients and doctors when treatment is required using wireless communication and PDAs.

Educational Uses

PDAs and handheld devices are allowed in many classrooms for digital note-taking. Students can spell-check, modify, and amend their class notes on a PDA. Some educators distribute course material through the Internet or infrared file-sharing functions of the PDA. Textbook publishers have begun to release e-books, which can be uploaded directly to a PDA, reducing the number of textbooks students must carry. Brighton and SUSSEX Medical School in the UK was the first medical school to provide wide scale use of PDAs to its undergraduate students. The learning opportunities provided by having PDAs complete with a suite of key medical texts was studied with results showing that learning occurred in context with timely access to key facts and through consolidation of knowledge via repetition. The PDA was an important addition to the learning ecology rather than a replacement. Software companies have developed PDA programs to meet the instructional needs of educational institutions, such as dictionaries, thesauri, word processing software, encyclopedias, webinar and digital lesson planners.

Recreational Uses

PDAs may be used by music enthusiasts to play a variety of music file formats. Many PDAs include the functionality of an MP3 player. Road rally enthusiasts can use PDAs to calculate distance, speed, and time. This information may be used for navigation, or the PDA's GPS functions can be used for navigation. Underwater divers can use PDAs to plan breathing gas mixtures and decompression schedules using software such as "V-Planner".

Models

Consumer

- Acer N Series
- AlphaSmart
- Apple Newton
- Atari Portfolio
- Dell Axim
- E-TEN
- Fujitsu Siemens Computers Pocket LOOX
- Handspring (company)
- HP iPAQ
- HTC (Dopod, Qtek)'s series of Windows Mobile PDA/phones
- Huawei series
- I-mate
- iPhone
- HP Jornada Pocket PC

- LifeDrive

- NEC MobilePro

- Osaris running EPOC OS distributed by Oregon Scientific

- Palm (PDA) (Tungsten E2, TX, Treo, Zire Handheld)

- Palm, Inc. smartphones under Palm OS and under the successor WebOS (Pre, and Pixi).

- Philips Nino

- Casio Pocket Viewer

- PocketMail (email PDA with built-in acoustic coupler)

- Psion

- Sharp Wizard and Sharp Zaurus

- Tapwave Zodiac

- Toshiba e310

- Abacus PDA Watch

- Amida Simputer

- GMate Yopy

- Roland PMA-5 (Personal Music Assistant)

- Royal (ezVue 7, etc.)

- Sony CLIÉ

- Sony Magic Link with the Magic Cap operating system

Ruggedized

- Getac

- Handheld Group

- Honeywell (Hand Held Products)

- Intermec

- Motorola (Symbol Technologies)

- Psion Teklogix

- Trimble Navigation

- American Industrial Systems (Mil-Spec, IP67)

- Catchwell

- Datalogic Mobile

- ecom instruments

- M3 Mobile

- Pidion (Bluebird Soft Inc.)

- Skeye (Hoeft & Wessel AG)

- Two Technologies, Inc. (Ultra Rugged Handheld Computers)

- Unitech

References

- Linzmayer, Owen W. (2004). Apple confidential 2.0 : the definitive history of the world's most colorful company ([Rev. 2. ed.]. ed.). San Francisco, Calif.: No Starch Press. ISBN 1-59327-010-0.

- Andrew Smith, Faithe Wempen (2011). CompTIA Strata Study Guide. John Wiley & Sons. p. 140. ISBN 978-0-470-97742-2. Retrieved July 5, 2012.

- Thorp-Lancaster, Dan (Apr 15, 2015). "Google releases new Handwriting Input keyboard with support for 82 languages". Android Central. Retrieved Dec 30, 2015.

- "IDC Forecasts Worldwide Tablet Shipments to Surpass Portable PC Shipments in 2013, Total PC Shipments in 2015". International Data Corporation. May 28, 2013. Archived from the original on June 10, 2014. Retrieved May 22, 2014.

- "Gartner Says Worldwide Traditional PC, Tablet, Ultramobile and Mobile Phone Shipments Are On Pace to Grow 6.9 Percent in 2014". Gartner. March 27, 2014. Retrieved May 23, 2014.

- Strickland, Jonathan. "What's the difference between notebooks, netbooks and ultra-mobile PCs?". HowStuffWorks.com. Retrieved 23 September 2014.

- "U.S. Commercial Channel Computing Device Sales Set to End 2013 with Double-Digit Growth, According to NPD". NPD Group. Retrieved 23 September 2014.

- "HP EliteBook 6930p Notebook PC specifications – HP Products and Services Products". H10010.www1.hp.com. 25 May 2009. Retrieved 17 June 2013.

- Poeter, Damon (December 27, 2012). "Non-Apple Tablets Making Small Gains on iPad | News & Opinion". PCMag.com. Retrieved July 8, 2013.

- "Gartner Says Worldwide Tablet Sales Grew 68 Percent in 2013, With Android Capturing 62 Percent of the Market". Gartner. March 3, 2014. Retrieved April 17, 2014.

- BeHardware reported lower retailer return rates for SSDs than HDDs between April and October 2010. Prieur, Marc (6 May 2011). "Components returns rates". BeHardware. Retrieved 10 February 2012.

- A 2011 study by Intel on the use of 45,000 SSDs reported an annualized failure rate of 0.61% for SSDs, compared with 4.85% for HDDs. "Validating the Reliability of Intel® Solid-State Drives". Intel. July 2011. Retrieved 10 February 2012.

- "Samsung's Solar Powered Laptop Will Be First Sun Powered Laptop Sold in US | Inhabitat – Sustainable Design Innovation, Eco Architecture, Green Building". Inhabitat. Retrieved 23 October 2012.

- "This Week in Apple History – August 22–31: "Welcome, IBM. Seriously", Too Late to License". The Mac Observer. 31 October 2004. Retrieved 23 October 2012.

- Moscaritolo, Angela (June 18, 2012). "Survey: 31 Percent of U.S. Internet Users Own Tablets". PC Magazine. Retrieved October 20, 2012.

- Chen, Brian X. (October 23, 2012). "Apple, Facing Competition, Introduces a Smaller iPad of no significant change". The New York Times. Retrieved October 24, 2012.

- Franklin, Eric (April 10, 2012). "The Toshiba Excite 13 sports the largest tablet screen yet | Android Atlas - CNET Reviews". Reviews.cnet.com. Retrieved June 14, 2013.

Various Networks in Mobile Computing

Cellular networks are networks where the last link of the network uses wireless technology. Code division multiple access is the network used by radio technologies. Alternatively, the networks also used in mobile computing are code division multiple access, public switched telephone network, 4G, 5G and general packet radio service. The major networks of mobile computing are discussed in this chapter.

Cellular Network

A cellular network or mobile network is a communication network where the last link is wireless. The network is distributed over land areas called cells, each served by at least one fixed-location transceiver, known as a cell site or base station. This base station provides the cell with the network coverage which can be used for transmission of voice, data and others. A cell might use a different set of frequencies from neighboring cells, to avoid interference and provide guaranteed service quality within each cell.

Top of a cellular radio tower

When joined together these cells provide radio coverage over a wide geographic area. This enables a large number of portable transceivers (e.g., mobile phones, pagers, etc.) to communicate with each other and with fixed transceivers and telephones anywhere in the network, via base stations, even if some of the transceivers are moving through more than one cell during transmission.

Cellular networks offer a number of desirable features:

- More capacity than a single large transmitter, since the same frequency can be used for multiple links as long as they are in different cells

- Mobile devices use less power than with a single transmitter or satellite since the cell towers are closer

- Larger coverage area than a single terrestrial transmitter, since additional cell towers can be added indefinitely and are not limited by the horizon

Major telecommunications providers have deployed voice and data cellular networks over most of the inhabited land area of the Earth. This allows mobile phones and mobile computing devices to be connected to the public switched telephone network and public Internet. Private cellular networks can be used for research or for large organizations and fleets, such as dispatch for local public safety agencies or a taxicab company.

Concept

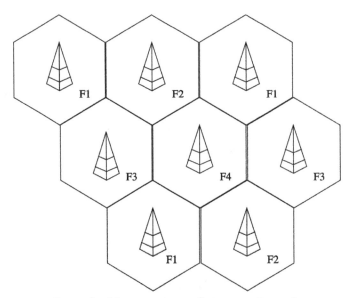

Example of frequency reuse factor or pattern 1/4

In a cellular radio system, a land area to be supplied with radio service is divided into regular shaped cells, which can be hexagonal, square, circular or some other regular shapes, although hexagonal cells are conventional. Each of these cells is assigned with multiple frequencies $(f_1 - f_6)$ which have corresponding radio base stations. The group of frequencies can be reused in other cells, provided that the same frequencies are not reused in adjacent neighboring cells as that would cause co-channel interference.

The increased capacity in a cellular network, compared with a network with a single transmitter, comes from the mobile communication switching system developed by Amos Joel of Bell Labs that permitted multiple callers in the same area to use the same frequency by switching calls made using the same frequency to the nearest available cellular tower having that frequency available and from the fact that the same radio frequency can be reused in a different area for a completely different transmission. If there is a single plain transmitter, only one transmission can be used on any given frequency. Unfortunately, there is inevitably some level of interference from the signal from the other cells which use the same frequency. This means that, in a standard FDMA system, there must be at least a one cell gap between cells which reuse the same frequency.

In the simple case of the taxi company, each radio had a manually operated channel selector knob to tune to different frequencies. As the drivers moved around, they would change from channel to channel. The drivers knew which frequency covered approximately what area. When they did not receive a signal from the transmitter, they would try other channels until they found one that worked. The taxi drivers would only speak one at a time, when invited by the base station operator (this is, in a sense, time division multiple access (TDMA)).

Cell Signal Encoding

To distinguish signals from several different transmitters, time division multiple access (TDMA), frequency division multiple access (FDMA), code division multiple access (CDMA), and orthogonal frequency division multiple access (OFDMA) were developed.

With TDMA, the transmitting and receiving time slots used by different users in each cell are different from each other.

With FDMA, the transmitting and receiving frequencies used by different users in each cell are different from each other. In a simple taxi system, the taxi driver manually tuned to a frequency of a chosen cell to obtain a strong signal and to avoid interference from signals from other cells.

The principle of CDMA is more complex, but achieves the same result; the distributed transceivers can select one cell and listen to it.

Other available methods of multiplexing such as polarization division multiple access (PDMA) cannot be used to separate signals from one cell to the next since the effects of both vary with position and this would make signal separation practically impossible. Time division multiple access is used in combination with either FDMA or CDMA in a number of systems to give multiple channels within the coverage area of a single cell.

Frequency Reuse

The key characteristic of a cellular network is the ability to re-use frequencies to increase both coverage and capacity. As described above, adjacent cells must use different frequencies, however there is no problem with two cells sufficiently far apart operating on the same frequency, provided the masts and cellular network users' equipment do not transmit with too much power.

The elements that determine frequency reuse are the reuse distance and the reuse factor. The reuse distance, D is calculated as

$$D = R\sqrt{3N},$$

where R is the cell radius and N is the number of cells per cluster. Cells may vary in radius from 1 to 30 kilometres (0.62 to 18.64 mi). The boundaries of the cells can also overlap between adjacent cells and large cells can be divided into smaller cells.

The frequency reuse factor is the rate at which the same frequency can be used in the network. It is $1/K$ (or K according to some books) where K is the number of cells which cannot use the same frequencies for transmission. Common values for the frequency reuse factor are 1/3, 1/4, 1/7, 1/9 and 1/12 (or 3, 4, 7, 9 and 12 depending on notation).

In case of N sector antennas on the same base station site, each with different direction, the base station site can serve N different sectors. N is typically 3. A reuse pattern of N/K denotes a further division in frequency among N sector antennas per site. Some current and historical reuse patterns are 3/7 (North American AMPS), 6/4 (Motorola NAMPS), and 3/4 (GSM).

If the total available bandwidth is B, each cell can only use a number of frequency channels corresponding to a bandwidth of B/K, and each sector can use a bandwidth of B/NK.

Code division multiple access-based systems use a wider frequency band to achieve the same rate of transmission as FDMA, but this is compensated for by the ability to use a frequency reuse factor of 1, for example using a reuse pattern of 1/1. In other words, adjacent base station sites use the same frequencies, and the different base stations and users are separated by codes rather than frequencies. While N is shown as 1 in this example, that does not mean the CDMA cell has only one sector, but rather that the entire cell bandwidth is also available to each sector individually.

Depending on the size of the city, a taxi system may not have any frequency-reuse in its own city, but certainly in other nearby cities, the same frequency can be used. In a large city, on the other hand, frequency-reuse could certainly be in use.

Recently also orthogonal frequency-division multiple access based systems such as LTE are being deployed with a frequency reuse of 1. Since such systems do not spread the signal across the frequency band, inter-cell radio resource management is important to coordinate resource allocation between different cell sites and to limit the inter-cell interference. There are various means of Inter-Cell Interference Coordination (ICIC) already defined in the standard. Coordinated scheduling, multi-site MIMO or multi-site beam forming are other examples for inter-cell radio resource management that might be standardized in the future.

Directional Antennas

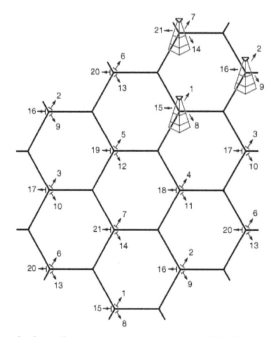

Cellular telephone frequency reuse pattern. See U.S. Patent 4,144,411

Cell towers frequently use a directional signal to improve reception in higher-traffic areas. In the United States, the FCC limits omnidirectional cell tower signals to 100 watts of power. If the tower has directional antennas, the FCC allows the cell operator to broadcast up to 500 watts of effective radiated power (ERP).

Cell phone companies use this directional signal to improve reception along highways and inside buildings like stadiums and arenas. As a result, a cell phone user may be standing in sight of a cell tower, but still have trouble getting a good signal because the directional antennas point in a different direction.

Although the original cell towers created an even, omnidirectional signal, were at the centers of the cells and were omnidirectional, a cellular map can be redrawn with the cellular telephone towers located at the corners of the hexagons where three cells converge. Each tower has three sets of directional antennas aimed in three different directions with 120 degrees for each cell (totaling 360 degrees) and receiving/transmitting into three different cells at different frequencies. This provides a minimum of three channels, and three towers for each cell and greatly increases the chances of receiving a usable signal from at least one direction.

The numbers in the illustration are channel numbers, which repeat every 3 cells. Large cells can be subdivided into smaller cells for high volume areas.

Broadcast Messages and Paging

Practically every cellular system has some kind of broadcast mechanism. This can be used directly for distributing information to multiple mobiles. Commonly, for example in mobile telephony systems, the most important use of broadcast information is to set up channels for one to one communication between the mobile transceiver and the base station. This is called paging. The three different paging procedures generally adopted are sequential, parallel and selective paging.

The details of the process of paging vary somewhat from network to network, but normally we know a limited number of cells where the phone is located (this group of cells is called a Location Area in the GSM or UMTS system, or Routing Area if a data packet session is involved; in LTE, cells are grouped into Tracking Areas). Paging takes place by sending the broadcast message to all of those cells. Paging messages can be used for information transfer. This happens in pagers, in CDMA systems for sending SMS messages, and in the UMTS system where it allows for low downlink latency in packet-based connections.

Movement from Cell to Cell and Handing Over

In a primitive taxi system, when the taxi moved away from a first tower and closer to a second tower, the taxi driver manually switched from one frequency to another as needed. If a communication was interrupted due to a loss of a signal, the taxi driver asked the base station operator to repeat the message on a different frequency.

In a cellular system, as the distributed mobile transceivers move from cell to cell during an ongoing continuous communication, switching from one cell frequency to a different cell frequency is done electronically without interruption and without a base station operator or manual switching. This is called the handover or handoff. Typically, a new channel is automatically selected for the mobile unit on the new base station which will serve it. The mobile unit then automatically switches from the current channel to the new channel and communication continues.

The exact details of the mobile system's move from one base station to the other varies considerably from system to system.

Mobile Phone Network

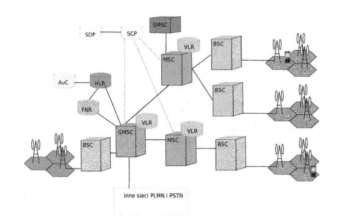

GSM network architecture

The most common example of a cellular network is a mobile phone (cell phone) network. A mobile phone is a portable telephone which receives or makes calls through a cell site (base station), or transmitting tower. Radio waves are used to transfer signals to and from the cell phone.

Modern mobile phone networks use cells because radio frequencies are a limited, shared resource. Cell-sites and handsets change frequency under computer control and use low power transmitters so that the usually limited number of radio frequencies can be simultaneously used by many callers with less interference.

A cellular network is used by the mobile phone operator to achieve both coverage and capacity for their subscribers. Large geographic areas are split into smaller cells to avoid line-of-sight signal loss and to support a large number of active phones in that area. All of the cell sites are connected to telephone exchanges (or switches), which in turn connect to the public telephone network.

In cities, each cell site may have a range of up to approximately $\frac{1}{2}$ mile (0.80 km), while in rural areas, the range could be as much as 5 miles (8.0 km). It is possible that in clear open areas, a user may receive signals from a cell site 25 miles (40 km) away.

Since almost all mobile phones use cellular technology, including GSM, CDMA, and AMPS (analog), the term "cell phone" is in some regions, notably the US, used interchangeably with "mobile phone". However, satellite phones are mobile phones that do not communicate directly with a ground-based cellular tower, but may do so indirectly by way of a satellite.

There are a number of different digital cellular technologies, including: Global System for Mobile Communications (GSM), General Packet Radio Service (GPRS), cdmaOne, CDMA2000, Evolution-Data Optimized (EV-DO), Enhanced Data Rates for GSM Evolution (EDGE), Universal

Mobile Telecommunications System (UMTS), Digital Enhanced Cordless Telecommunications (DECT), Digital AMPS (IS-136/TDMA), and Integrated Digital Enhanced Network (iDEN). The transition from existing analog to the digital standard followed a very different path in Europe and the US. As a consequence multiple digital standard surfaced in the US, while Europe and many countries converged towards the GSM standard.

Structure of the Mobile Phone Cellular Network

A simple view of the cellular mobile-radio network consists of the following:

- A network of radio base stations forming the base station subsystem.

- The core circuit switched network for handling voice calls and text

- A packet switched network for handling mobile data

- The public switched telephone network to connect subscribers to the wider telephony network

This network is the foundation of the GSM system network. There are many functions that are performed by this network in order to make sure customers get the desired service including mobility management, registration, call set-up, and handover.

Any phone connects to the network via an RBS (Radio Base Station) at a corner of the corresponding cell which in turn connects to the Mobile switching center (MSC). The MSC provides a connection to the public switched telephone network (PSTN). The link from a phone to the RBS is called an *uplink* while the other way is termed *downlink*.

Radio channels effectively use the transmission medium through the use of the following multiplexing and access schemes: frequency division multiple access (FDMA), time division multiple access (TDMA), code division multiple access (CDMA), and space division multiple access (SDMA).

Small Cells

Small cells, which have a smaller coverage area than base stations, are categorised as follows:

- Microcell, less than 2 kilometres

- Picocell, less than 200 metres

- Femtocell, around 10 metres

Cellular Handover in Mobile Phone Networks

As the phone user moves from one cell area to another cell while a call is in progress, the mobile station will search for a new channel to attach to in order not to drop the call. Once a new channel is found, the network will command the mobile unit to switch to the new channel and at the same time switch the call onto the new channel.

With CDMA, multiple CDMA handsets share a specific radio channel. The signals are separated by using a pseudonoise code (PN code) specific to each phone. As the user moves from one cell to another, the handset sets up radio links with multiple cell sites (or sectors of the same site) simultaneously. This is known as "soft handoff" because, unlike with traditional cellular technology, there is no one defined point where the phone switches to the new cell.

In IS-95 inter-frequency handovers and older analog systems such as NMT it will typically be impossible to test the target channel directly while communicating. In this case other techniques have to be used such as pilot beacons in IS-95. This means that there is almost always a brief break in the communication while searching for the new channel followed by the risk of an unexpected return to the old channel.

If there is no ongoing communication or the communication can be interrupted, it is possible for the mobile unit to spontaneously move from one cell to another and then notify the base station with the strongest signal.

Cellular Frequency Choice in Mobile Phone Networks

The effect of frequency on cell coverage means that different frequencies serve better for different uses. Low frequencies, such as 450 MHz NMT, serve very well for countryside coverage. GSM 900 (900 MHz) is a suitable solution for light urban coverage. GSM 1800 (1.8 GHz) starts to be limited by structural walls. UMTS, at 2.1 GHz is quite similar in coverage to GSM 1800.

Higher frequencies are a disadvantage when it comes to coverage, but it is a decided advantage when it comes to capacity. Pico cells, covering e.g. one floor of a building, become possible, and the same frequency can be used for cells which are practically neighbours.

Cell service area may also vary due to interference from transmitting systems, both within and around that cell. This is true especially in CDMA based systems. The receiver requires a certain signal-to-noise ratio, and the transmitter should not send with too high transmission power in view to not cause interference with other transmitters. As the receiver moves away from the transmitter, the power received decreases, so the power control algorithm of the transmitter increases the power it transmits to restore the level of received power. As the interference (noise) rises above the received power from the transmitter, and the power of the transmitter cannot be increased any more, the signal becomes corrupted and eventually unusable. In CDMA-based systems, the effect of interference from other mobile transmitters in the same cell on coverage area is very marked and has a special name, *cell breathing*.

One can see examples of cell coverage by studying some of the coverage maps provided by real operators on their web sites or by looking at independently crowdsourced maps such as OpenSignal. In certain cases they may mark the site of the transmitter, in others it can be calculated by working out the point of strongest coverage.

A cellular repeater is used to extend cell coverage into larger areas. They range from wideband repeaters for consumer use in homes and offices to smart or digital repeaters for industrial needs.

Coverage Comparison of Different Frequencies

The following table shows the dependency of the coverage area of one cell on the frequency of a CDMA2000 network:

GSM

GSM (Global System for Mobile Communications, originally *Groupe SpécialMobile*), is a standard developed by the European Telecommunications Standards Institute (ETSI) to describe the protocols for second-generation (2G) digital cellular networks used by mobile phones, first deployed in Finland in July 1991. As of 2014 it has become the de facto global standard for mobile communications - with over 90% market share, operating in over 219 countries and territories.

The GSM logo is used to identify compatible handsets and equipment. The dots symbolize three clients in the home network and one roaming client.

2G networks developed as a replacement for first generation (1G) analog cellular networks, and the GSM standard originally described a digital, circuit-switched network optimized for full duplex voice telephony. This expanded over time to include data communications, first by circuit-switched transport, then by packet data transport via GPRS (General Packet Radio Services) and EDGE (Enhanced Data rates for GSM Evolution or EGPRS).

Subsequently, the 3GPP developed third-generation (3G) UMTS standards followed by fourth-generation (4G) LTE Advanced standards, which do not form part of the ETSI GSM standard.

"GSM" is a trademark owned by the GSM Association. It may also refer to the (initially) most common voice codec used, Full Rate.

History

In 1982 work began to develop a European standard for digital cellular voice telecommunications when the European Conference of Postal and Telecommunications Administrations (CEPT) set up the Groupe Spécial Mobile committee and later provided a permanent technical-support group based in Paris. Five years later, in 1987, 15 representatives from 13 European countries signed a memorandum of understanding in Copenhagen to develop and deploy a common cellular telephone system across Europe, and EU rules were passed to make GSM a mandatory standard. The decision to develop a continental standard eventually resulted in a unified, open, standard-based network which was larger than that in the United States.

In February 1987 Europe produced the very first agreed GSM Technical Specification. Ministers from the four big EU countries cemented their political support for GSM with the Bonn Declaration on Global Information Networks in May and the GSM MoU was tabled for signature in September. The MoU drew-in mobile operators from across Europe to pledge to invest in new GSM networks to an ambitious common date. It got GSM up-and-running fast.

In this short 38-week period the whole of Europe (countries and industries) had been brought behind GSM in a rare unity and speed guided by four public officials: Armin Silberhorn (Germany), Stephen Temple (UK), Philippe Dupuis (France), and Renzo Failli (Italy). In 1989, the Groupe Spécial Mobile committee was transferred from CEPT to the European Telecommunications Standards Institute (ETSI).

In parallel, France and Germany signed a joint development agreement in 1984 and were joined by Italy and the UK in 1986. In 1986 the European Commission proposed reserving the 900 MHz spectrum band for GSM. The former Finnish prime minister Harri Holkeri made the world's first GSM call on July 1, 1991, calling Kaarina Suonio (mayor of the city of Tampere) using a network built by Telenokia and Siemens and operated by Radiolinja. In the following year, 1992, saw the sending of the first short messaging service (SMS or "text message") message, and Vodafone UK and Telecom Finland signed the first international roaming agreement.

Work began in 1991 to expand the GSM standard to the 1800 MHz frequency band and the first 1800 MHz network became operational in the UK by 1993. Also that year, Telecom Australia became the first network operator to deploy a GSM network outside Europe and the first practical hand-held GSM mobile phone became available.

In 1995, fax, data and SMS messaging services were launched commercially, the first 1900 MHz GSM network became operational in the United States and GSM subscribers worldwide exceeded 10 million. In the same year, the GSM Association formed. Pre-paid GSM SIM cards were launched in 1996 and worldwide GSM subscribers passed 100 million in 1998.

In 2000 the first commercial GPRS services were launched and the first GPRS-compatible handsets became available for sale. In 2001 the first UMTS (W-CDMA) network was launched, a 3G technology that is not part of GSM. Worldwide GSM subscribers exceeded 500 million. In 2002 the first Multimedia Messaging Service (MMS) were introduced and the first GSM network in the 800 MHz frequency band became operational. EDGE services first became operational in a network in 2003 and the number of worldwide GSM subscribers exceeded 1 billion in 2004.

By 2005, GSM networks accounted for more than 75% of the worldwide cellular network market, serving 1.5 billion subscribers. In 2005 the first HSDPA-capable network also became operational. The first HSUPA network launched in 2007. (High-Speed Packet Access (HSPA) and its uplink and downlink versions are 3G technologies, not part of GSM.) Worldwide GSM subscribers exceeded three billion in 2008.

The GSM Association estimated in 2010 that technologies defined in the GSM standard serve 80% of the global mobile market, encompassing more than 5 billion people across more than 212 countries and territories, making GSM the most ubiquitous of the many standards for cellular networks.

Note that GSM is a second-generation (2G) standard employing Time-Division Multiple-Access (TDMA) spectrum-sharing, issued by the European Telecommunications Standards Institute (ETSI). The GSM standard does not include the 3G UMTS CDMA-based technology nor the 4G LTE OFDMA-based technology standards issued by the 3GPP.

Telstra in Australia plans to shut down its 2G GSM network on December 1, 2016, which will make it the first mobile network operator to decommission a GSM network. The second mobile provider planning to shut down its GSM network (on January 1, 2017) is AT&T Mobility from the United States. Singapore will phase out 2G services by April 2017.

Technical Details

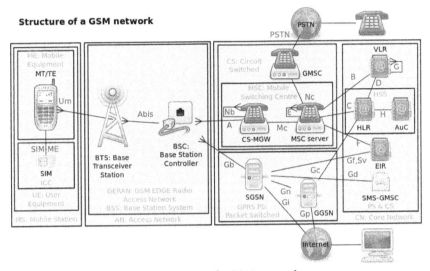

The structure of a GSM network

Network Structure

The network is structured into a number of discrete sections:

- Base Station Subsystem – the base stations and their controllers explained

- Network and Switching Subsystem – the part of the network most similar to a fixed network, sometimes just called the "core network"

- GPRS Core Network – the optional part which allows packet-based Internet connections

- Operations support system (OSS) – network maintenance

Base Station Subsystem

GSM is a cellular network, which means that cell phones connect to it by searching for cells in the immediate vicinity. There are five different cell sizes in a GSM network—macro, micro, pico, femto, and umbrella cells. The coverage area of each cell varies according to the implementation environment. Macro cells can be regarded as cells where the base station antenna is installed on a mast or a building above average rooftop level. Micro cells are cells whose antenna height is under average rooftop level; they are typically used in urban areas. Picocells are small cells whose coverage diameter is a few dozen metres; they are mainly used indoors. Femtocells are cells designed for use in residential or small business environments and connect to the service provider's network via a broadband internet connection. Umbrella cells are used to cover shadowed regions of smaller cells and fill in gaps in coverage between those cells.

GSM cell site antennas in the Deutsches Museum, Munich, Germany

Cell horizontal radius varies depending on antenna height, antenna gain, and propagation conditions from a couple of hundred meters to several tens of kilometres. The longest distance the GSM specification supports in practical use is 35 kilometres (22 mi). There are also several implementations of the concept of an extended cell, where the cell radius could be double or even more, depending on the antenna system, the type of terrain, and the timing advance.

Indoor coverage is also supported by GSM and may be achieved by using an indoor picocell base station, or an indoor repeater with distributed indoor antennas fed through power splitters, to deliver the radio signals from an antenna outdoors to the separate indoor distributed antenna system. These are typically deployed when significant call capacity is needed indoors, like in shopping centers or airports. However, this is not a prerequisite, since indoor coverage is also provided by in-building penetration of the radio signals from any nearby cell.

GSM Carrier Frequencies

GSM networks operate in a number of different carrier frequency ranges (separated into GSM frequency ranges for 2G and UMTS frequency bands for 3G), with most 2G GSM networks operating in the 900 MHz or 1800 MHz bands. Where these bands were already allocated, the 850 MHz and 1900 MHz bands were used instead (for example in Canada and the United States). In rare cases the 400 and 450 MHz frequency bands are assigned in some countries because they were previously used for first-generation systems.

Most 3G networks in Europe operate in the 2100 MHz frequency band. For more information on worldwide GSM frequency usage, see GSM frequency bands.

Regardless of the frequency selected by an operator, it is divided into timeslots for individual phones. This allows eight full-rate or sixteen half-rate speech channels per radio frequency. These eight radio timeslots (or burst periods) are grouped into a TDMA frame. Half-rate channels use alternate frames in the same timeslot. The channel data rate for all 8 channels is 270.833 kbit/s, and the frame duration is 4.615 ms.

The transmission power in the handset is limited to a maximum of 2 watts in GSM 850/900 and 1 watt in GSM 1800/1900.

Voice Codecs

GSM has used a variety of voice codecs to squeeze 3.1 kHz audio into between 6.5 and 13 kbit/s. Originally, two codecs, named after the types of data channel they were allocated, were used, called Half Rate (6.5 kbit/s) and Full Rate (13 kbit/s). These used a system based on linear predictive coding (LPC). In addition to being efficient with bitrates, these codecs also made it easier to identify more important parts of the audio, allowing the air interface layer to prioritize and better protect these parts of the signal.

As GSM was further enhanced in 1997 with the Enhanced Full Rate (EFR) codec, a 12.2 kbit/s codec that uses a full-rate channel. Finally, with the development of UMTS, EFR was refactored into a variable-rate codec called AMR-Narrowband, which is high quality and robust against interference when used on full-rate channels, or less robust but still relatively high quality when used in good radio conditions on half-rate channel.

Subscriber Identity Module (SIM)

One of the key features of GSM is the Subscriber Identity Module, commonly known as a SIM card. The SIM is a detachable smart card containing the user's subscription information and phone book. This allows the user to retain his or her information after switching handsets. Alternatively, the user can also change operators while retaining the handset simply by changing the SIM. Some operators will block this by allowing the phone to use only a single SIM, or only a SIM issued by them; this practice is known as SIM locking.

Phone Locking

Sometimes mobile network operators restrict handsets that they sell for use with their own network. This is called *locking* and is implemented by a software feature of the phone. A subscriber may usually contact the provider to remove the lock for a fee, utilize private services to remove the lock, or use software and websites to unlock the handset themselves. It is possible to illegally hack past a phone locked by a network operator.

In some countries (e.g., Bangladesh, Belgium, Brazil, Chile, Germany, Hong Kong, India, Iran, Lebanon, Malaysia, Nepal, Pakistan, Poland, Singapore, South Africa, Thailand) all phones are sold unlocked.

GSM Security

GSM was intended to be a secure wireless system. It has considered the user authentication using a pre-shared key and challenge-response, and over-the-air encryption. However, GSM is vulnerable to different types of attack, each of them aimed at a different part of the network.

The development of UMTS introduces an optional Universal Subscriber Identity Module (USIM), that uses a longer authentication key to give greater security, as well as mutually authenticating the network and the user, whereas GSM only authenticates the user to the network (and not vice versa). The security model therefore offers confidentiality and authentication, but limited authorization capabilities, and no non-repudiation.

GSM uses several cryptographic algorithms for security. The A5/1, A5/2, and A5/3 stream ciphers are used for ensuring over-the-air voice privacy. A5/1 was developed first and is a stronger algorithm used within Europe and the United States; A5/2 is weaker and used in other countries. Serious weaknesses have been found in both algorithms: it is possible to break A5/2 in real-time with a ciphertext-only attack, and in January 2007, The Hacker's Choice started the A5/1 cracking project with plans to use FPGAs that allow A5/1 to be broken with a rainbow table attack. The system supports multiple algorithms so operators may replace that cipher with a stronger one.

Since 2000, different efforts have been done in order to crack the A5 encryption algorithms. Both A5/1 and A5/2 algorithms are broken, and their cryptanalysis has been considered in the literature. As an example, Karsten Nohl (de) developed a number of rainbow tables (static values which reduce the time needed to carry out an attack) and have found new sources for known plaintext attacks. He said that it is possible to build "a full GSM interceptor...from open-source components" but that they had not done so because of legal concerns. Nohl claimed that he was able to intercept voice and text conversations by impersonating another user to listen to voicemail, make calls, or send text messages using a seven-year-old Motorola cellphone and decryption software available for free online.

New attacks have been observed that take advantage of poor security implementations, architecture, and development for smartphone applications. Some wiretapping and eavesdropping techniques hijack the audio input and output providing an opportunity for a third party to listen in to the conversation.

GSM uses General Packet Radio Service (GPRS) for data transmissions like browsing the web. The most commonly deployed GPRS ciphers were publicly broken in 2011.

The researchers revealed flaws in the commonly used GEA/1 and GEA/2 ciphers and published the open-source "gprsdecode" software for sniffing GPRS networks. They also noted that some carriers do not encrypt the data (i.e., using GEA/0) in order to detect the use of traffic or protocols they do not like (e.g., Skype), leaving customers unprotected. GEA/3 seems to remain relatively hard to break and is said to be in use on some more modern networks. If used with USIM to prevent connections to fake base stations and downgrade attacks, users will be protected in the medium term, though migration to 128-bit GEA/4 is still recommended.

Standards Information

The GSM systems and services are described in a set of standards governed by ETSI, where a full list is maintained.

GSM open-source software

Several open-source software projects exist that provide certain GSM features:

- gsmd daemon by Openmoko

- OpenBTS develops a Base transceiver station

- *The GSM Software Project* aims to build a GSM analyzer for less than $1,000

- *OsmocomBB* developers intend to replace the proprietary baseband GSM stack with a free software implementation

- YateBTS develops a Base transceiver station

Issues with Patents and Open Source

Patents remain a problem for any open-source GSM implementation, because it is not possible for GNU or any other free software distributor to guarantee immunity from all lawsuits by the patent holders against the users. Furthermore, new features are being added to the standard all the time which means they have patent protection for a number of years.

The original GSM implementations from 1991 may now be entirely free of patent encumbrances, however patent freedom is not certain due to the United States' "first to invent" system that was in place until 2012. The "first to invent" system, coupled with "patent term adjustment" can extend the life of a U.S. patent far beyond 20 years from its priority date. It is unclear at this time whether OpenBTS will be able to implement features of that initial specification without limit. As patents subsequently expire, however, those features can be added into the open-source version. As of 2011, there have been no lawsuits against users of OpenBTS over GSM use.

Code Division Multiple Access

Code division multiple access (CDMA) is a channel access method used by various radio communication technologies.

CDMA is an example of multiple access, where several transmitters can send information simultaneously over a single communication channel. This allows several users to share a band of frequencies. To permit this without undue interference between the users, CDMA employs spread-spectrum technology and a special coding scheme (where each transmitter is as-signed a code).

CDMA is used as the access method in many mobile phone standards. IS-95, also called "cdma-One", and its 3G evolution CDMA2000, are often simply referred to as "CDMA"', but UMTS, the 3G standard used by GSM carriers, also uses "wideband CDMA", or W-CDMA, as well as TD-CD-MA and TD-SCDMA, as its radio technologies.

History

The technology of code division multiple access channels has long been known. In the Soviet Union (USSR), the first work devoted to this subject was published in 1935 by professor Dmitriy V. Ageev. It was shown that through the use of linear methods, there are three types of signal separation: frequency, time and compensatory. The technology of CDMA was used in 1957, when the young military radio engineer Leonid Kupriyanovich in Moscow, made an experimental model of a wearable automatic mobile phone, called LK-1 by him, with a base station. LK-1 has a weight of

3 kg, 20–30 km operating distance, and 20–30 hours of battery life. The base station, as described by the author, could serve several customers. In 1958, Kupriyanovich made the new experimental "pocket" model of mobile phone. This phone weighed 0.5 kg. To serve more customers, Kupriyanovich proposed the device, named by him as correllator. In 1958, the USSR also started the development of the "Altai" national civil mobile phone service for cars, based on the Soviet MRT-1327 standard. The phone system weighed 11 kg (24 lb). It was placed in the trunk of the vehicles of high-ranking officials and used a standard handset in the passenger compartment. The main developers of the Altai system were VNIIS (Voronezh Science Research Institute of Communications) and GSPI (State Specialized Project Institute). In 1963 this service started in Moscow and in 1970 Altai service was used in 30 USSR cities.

Uses

A CDMA2000 mobile phone

- One of the early applications for code division multiplexing is in the Global Positioning System (GPS). This predates and is distinct from its use in mobile phones.

- The Qualcomm standard IS-95, marketed as cdmaOne.

- The Qualcomm standard IS-2000, known as CDMA2000, is used by several mobile phone companies, including the Globalstar satellite phone network.

- The UMTS 3G mobile phone standard, which uses W-CDMA.

- CDMA has been used in the OmniTRACS satellite system for transportation logistics.

Steps in CDMA Modulation

CDMA is a spread-spectrum multiple access technique. A spread spectrum technique spreads the bandwidth of the data uniformly for the same transmitted power. A spreading code is a pseudo-random code that has a narrow ambiguity function, unlike other narrow pulse codes. In CDMA a locally generated code runs at a much higher rate than the data to be transmitted. Data for transmission is combined via bitwise XOR (exclusive OR) with the faster code. The figure shows how a spread spectrum signal is generated. The data signal with pulse duration of T_b (symbol period) is XOR'ed with the code signal with pulse duration of T_c (chip period). (Note: bandwidth is proportional to $1/T$, where T = bit time.) Therefore, the bandwidth of the data signal is $1/T_c$ and the bandwidth of the spread spectrum signal is $1/T_c$. Since T_c is much smaller than T_b, the bandwidth of the spread spectrum signal is much larger than the bandwidth of the original signal. The ratio $T_b >$ is called the spreading factor or processing gain and determines to a certain extent the upper limit of the total number of users supported simultaneously by a base station.

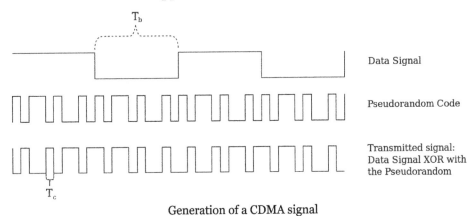

Generation of a CDMA signal

Each user in a CDMA system uses a different code to modulate their signal. Choosing the codes used to modulate the signal is very important in the performance of CDMA systems. The best performance will occur when there is good separation between the signal of a desired user and the signals of other users. The separation of the signals is made by correlating the received signal with the locally generated code of the desired user. If the signal matches the desired user's code then the correlation function will be high and the system can extract that signal. If the desired user's code has nothing in common with the signal the correlation should be as close to zero as possible (thus eliminating the signal); this is referred to as cross-correlation. If the code is correlated with the signal at any time offset other than zero, the correlation should be as close to zero as possible. This is referred to as auto-correlation and is used to reject multi-path interference.

An analogy to the problem of multiple access is a room (channel) in which people wish to talk to each other simultaneously. To avoid confusion, people could take turns speaking (time division), speak at different pitches (frequency division), or speak in different languages (code division). CDMA is analogous to the last example where people speaking the same language can understand each other, but other languages are perceived as noise and rejected. Similarly, in radio CDMA, each group of users is given a shared code. Many codes occupy the same channel, but only users associated with a particular code can communicate.

In general, CDMA belongs to two basic categories: synchronous (orthogonal codes) and asynchronous (pseudorandom codes).

Code Division Multiplexing (Synchronous CDMA)

The digital modulation method is analogous to those used in simple radio transceivers. In the analog case, a low frequency data signal is time multiplied with a high frequency pure sine wave carrier, and transmitted. This is effectively a frequency convolution (Wiener–Khinchin theorem) of the two signals, resulting in a carrier with narrow sidebands. In the digital case, the sinusoidal carrier is replaced by Walsh functions. These are binary square waves that form a complete orthonormal set. The data signal is also binary and the time multiplication is achieved with a simple XOR function. This is usually a Gilbert cell mixer in the circuitry.

Synchronous CDMA exploits mathematical properties of orthogonality between vectors representing the data strings. For example, binary string *1011* is represented by the vector (1, 0, 1, 1). Vectors can be multiplied by taking their dot product, by summing the products of their respective components (for example, if u = (a, b) and v = (c, d), then their dot product u·v = ac + bd). If the dot product is zero, the two vectors are said to be *orthogonal* to each other. Some properties of the dot product aid understanding of how W-CDMA works. If vectors *a* and *b* are orthogonal, then $\mathbf{a \cdot b} = 0$ and:

$$\mathbf{a \cdot (a+b)} = \|\mathbf{a}\|^2 \quad \text{since} \quad \mathbf{a \cdot a + a \cdot b} = \|\mathbf{a}\|^2 + 0$$

$$\mathbf{a \cdot (-a+b)} = -\|\mathbf{a}\|^2 \quad \text{since} \quad -\mathbf{a \cdot a + a \cdot b} = -\|\mathbf{a}\|^2 + 0$$

$$\mathbf{b \cdot (a+b)} = \|\mathbf{b}\|^2 \quad \text{since} \quad \mathbf{b \cdot a + b \cdot b} = 0 + \|\mathbf{b}\|^2$$

$$\mathbf{b \cdot (a-b)} = -\|\mathbf{b}\|^2 \quad \text{since} \quad \mathbf{b \cdot a - b \cdot b} = 0 - \|\mathbf{b}\|^2$$

Each user in synchronous CDMA uses a code orthogonal to the others' codes to modulate their signal. An example of four mutually orthogonal digital signals is shown in the figure. Orthogonal codes have a cross-correlation equal to zero; in other words, they do not interfere with each other. In the case of IS-95 64 bit Walsh codes are used to encode the signal to separate different users. Since each of the 64 Walsh codes are orthogonal to one another, the signals are channelized into 64 orthogonal signals. The following example demonstrates how each user's signal can be encoded and decoded.

Example

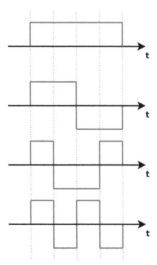

An example of four mutually orthogonal digital signals.

Start with a set of vectors that are mutually orthogonal. (Although mutual orthogonality is the only condition, these vectors are usually constructed for ease of decoding, for example columns or rows from Walsh matrices.) An example of orthogonal functions is shown in the picture on the right. These vectors will be assigned to individual users and are called the *code*, *chip code*, or *chipping code*. In the interest of brevity, the rest of this example uses codes, v, with only two bits.

Each user is associated with a different code, say v. A 1 bit is represented by transmitting a positive code, v, and a 0 bit is represented by a negative code, −v. For example, if $v = (v_0, v_1) = (1, −1)$ and the data that the user wishes to transmit is (1, 0, 1, 1), then the transmitted symbols would be $(v, −v, v, v) = (v_0, v_1, −v_0, −v_1, v_0, v_1, v_0, v_1) = (1, −1, −1, 1, 1, −1, 1, −1)$. For the purposes of this article, we call this constructed vector the *transmitted vector*.

Each sender has a different, unique vector v chosen from that set, but the construction method of the transmitted vector is identical.

Now, due to physical properties of interference, if two signals at a point are in phase, they add to give twice the amplitude of each signal, but if they are out of phase, they subtract and give a signal that is the difference of the amplitudes. Digitally, this behaviour can be modelled by the addition of the transmission vectors, component by component.

If sender0 has code (1, −1) and data (1, 0, 1, 1), and sender1 has code (1, 1) and data (0, 0, 1, 1), and both senders transmit simultaneously, then this table describes the coding steps:

Because signal0 and signal1 are transmitted at the same time into the air, they add to produce the raw signal:

$$(1, −1, −1, 1, 1, −1, 1, −1) + (−1, −1, −1, −1, 1, 1, 1, 1) = (0, −2, −2, 0, 2, 0, 2, 0)$$

This raw signal is called an interference pattern. The receiver then extracts an intelligible signal for any known sender by combining the sender's code with the interference pattern. The following table explains how this works, and shows that the signals do not interfere with one another:

Further, after decoding, all values greater than 0 are interpreted as 1 while all values less than zero are interpreted as 0. For example, after decoding, data0 is (2, −2, 2, 2), but the receiver interprets this as (1, 0, 1, 1). Values of exactly 0 means that the sender did not transmit any data, as in the following example:

Assume signal0 = (1, −1, −1, 1, 1, −1, 1, −1) is transmitted alone. The following table shows the decode at the receiver:

When the receiver attempts to decode the signal using sender1's code, the data is all zeros, therefore the cross correlation is equal to zero and it is clear that sender1 did not transmit any data.

Asynchronous CDMA

When mobile-to-base links cannot be precisely coordinated, particularly due to the mobility of the handsets, a different approach is required. Since it is not mathematically possible to create signature sequences that are both orthogonal for arbitrarily random starting points and which make full use of the code space, unique "pseudo-random" or "pseudo-noise" (PN) sequences are used in *asynchronous* CDMA systems. A PN code is a binary sequence that appears random but can be reproduced in a deterministic manner by intended receivers. These PN codes are used to encode and decode a user's signal in Asynchronous CDMA in the same manner as the orthogonal codes in synchronous CDMA (shown in the example above). These PN sequences are statistically uncorrelated, and the sum of a large number of PN sequences results in *multiple access interference* (MAI) that is approximated by a Gaussian noise process (following the central limit theorem in statistics). Gold codes are an example of a PN suitable for this purpose, as there is low correlation between the codes. If all of the users are received with the same power level, then the variance (e.g., the noise power) of the MAI increases in direct proportion to the number of users. In other words, unlike synchronous CDMA, the signals of other users will appear as noise to the signal of interest and interfere slightly with the desired signal in proportion to number of users.

All forms of CDMA use spread spectrum process gain to allow receivers to partially discriminate against unwanted signals. Signals encoded with the specified PN sequence (code) are received, while signals with different codes (or the same code but a different timing offset) appear as wideband noise reduced by the process gain.

Since each user generates MAI, controlling the signal strength is an important issue with CDMA transmitters. A CDM (synchronous CDMA), TDMA, or FDMA receiver can in theory completely reject arbitrarily strong signals using different codes, time slots or frequency channels due to the orthogonality of these systems. This is not true for Asynchronous CDMA; rejection of unwanted signals is only partial. If any or all of the unwanted signals are much stronger than the desired signal, they will overwhelm it. This leads to a general requirement in any asynchronous CDMA system to approximately match the various signal power levels as seen at the receiver. In CDMA cellular, the base station uses a fast closed-loop power control scheme to tightly control each mobile's transmit power.

Advantages of Asynchronous CDMA Over Other Techniques

Efficient Practical Utilization of the Fixed Frequency Spectrum

In theory CDMA, TDMA and FDMA have exactly the same spectral efficiency but practically, each has its own challenges – power control in the case of CDMA, timing in the case of TDMA, and frequency generation/filtering in the case of FDMA.

TDMA systems must carefully synchronize the transmission times of all the users to ensure that they are received in the correct time slot and do not cause interference. Since this cannot be perfectly controlled in a mobile environment, each time slot must have a guard-time, which reduces the probability that users will interfere, but decreases the spectral efficiency. Similarly, FDMA systems must use a guard-band between adjacent channels, due to the unpredictable doppler shift of the signal spectrum because of user mobility. The guard-bands will reduce the probability that adjacent channels will interfere, but decrease the utilization of the spectrum.

Flexible Allocation of Resources

Asynchronous CDMA offers a key advantage in the flexible allocation of resources i.e. allocation of a PN codes to active users. In the case of CDM (synchronous CDMA), TDMA, and FDMA the number of simultaneous orthogonal codes, time slots and frequency slots respectively are fixed hence the capacity in terms of number of simultaneous users is limited. There are a fixed number of orthogonal codes, time slots or frequency bands that can be allocated for CDM, TDMA, and FDMA systems, which remain underutilized due to the bursty nature of telephony and packetized data transmissions. There is no strict limit to the number of users that can be supported in an asynchronous CDMA system, only a practical limit governed by the desired bit error probability, since the SIR (Signal to Interference Ratio) varies inversely with the number of users. In a bursty traffic environment like mobile telephony, the advantage afforded by asynchronous CDMA is that the performance (bit error rate) is allowed to fluctuate randomly, with an average value determined by the number of users times the percentage of utilization. Suppose there are 2N users that only talk half of the time, then 2N users can be accommodated with the same *average* bit error probability as N users that talk all of the time. The key difference here is that the bit error probability for N users talking all of the time is constant, whereas it is a *random* quantity (with the same mean) for 2N users talking half of the time.

In other words, asynchronous CDMA is ideally suited to a mobile network where large numbers of transmitters each generate a relatively small amount of traffic at irregular intervals. CDM (synchronous CDMA), TDMA, and FDMA systems cannot recover the underutilized resources inherent to bursty traffic due to the fixed number of orthogonal codes, time slots or frequency channels that can be assigned to individual transmitters. For instance, if there are N time slots in a TDMA system and 2N users that talk half of the time, then half of the time there will be more than N users needing to use more than N time slots. Furthermore, it would require significant overhead to continually allocate and deallocate the orthogonal code, time slot or frequency channel resources. By comparison, asynchronous CDMA transmitters simply send when they have something to say, and go off the air when they don't, keeping the same PN signature sequence as long as they are connected to the system.

Spread-spectrum Characteristics of CDMA

Most modulation schemes try to minimize the bandwidth of this signal since bandwidth is a limited resource. However, spread spectrum techniques use a transmission bandwidth that is several orders of magnitude greater than the minimum required signal bandwidth. One of the initial reasons for doing this was military applications including guidance and communication systems. These systems were designed using spread spectrum because of its security and resistance to jamming. Asynchronous CDMA has some level of privacy built in because the signal is spread using a pseudo-random code; this code makes the spread spectrum signals appear random or have noise-like properties. A receiver cannot demodulate this transmission without knowledge of the pseudo-random sequence used to encode the data. CDMA is also resistant to jamming. A jamming signal only has a finite amount of power available to jam the signal. The jammer can either spread its energy over the entire bandwidth of the signal or jam only part of the entire signal.

CDMA can also effectively reject narrow band interference. Since narrow band interference affects only a small portion of the spread spectrum signal, it can easily be removed through notch filtering without much loss of information. Convolution encoding and interleaving can be used to assist in recovering this lost data. CDMA signals are also resistant to multipath fading. Since the spread spectrum signal occupies a large bandwidth only a small portion of this will undergo fading due to multipath at any given time. Like the narrow band interference this will result in only a small loss of data and can be overcome.

Another reason CDMA is resistant to multipath interference is because the delayed versions of the transmitted pseudo-random codes will have poor correlation with the original pseudo-random code, and will thus appear as another user, which is ignored at the receiver. In other words, as long as the multipath channel induces at least one chip of delay, the multipath signals will arrive at the receiver such that they are shifted in time by at least one chip from the intended signal. The correlation properties of the pseudo-random codes are such that this slight delay causes the multipath to appear uncorrelated with the intended signal, and it is thus ignored.

Some CDMA devices use a rake receiver, which exploits multipath delay components to improve the performance of the system. A rake receiver combines the information from several correlators, each one tuned to a different path delay, producing a stronger version of the signal than a simple receiver with a single correlation tuned to the path delay of the strongest signal.

Frequency reuse is the ability to reuse the same radio channel frequency at other cell sites within a cellular system. In the FDMA and TDMA systems frequency planning is an important consideration. The frequencies used in different cells must be planned carefully to ensure signals from different cells do not interfere with each other. In a CDMA system, the same frequency can be used in every cell, because channelization is done using the pseudo-random codes. Reusing the same frequency in every cell eliminates the need for frequency planning in a CDMA system; however, planning of the different pseudo-random sequences must be done to ensure that the received signal from one cell does not correlate with the signal from a nearby cell.

Since adjacent cells use the same frequencies, CDMA systems have the ability to perform soft hand offs. Soft hand offs allow the mobile telephone to communicate simultaneously with two or more cells. The best signal quality is selected until the hand off is complete. This is different from hard hand offs utilized in other cellular systems. In a hard hand off situation, as the mobile telephone approaches a hand off, signal strength may vary abruptly. In contrast, CDMA systems use the soft hand off, which is undetectable and provides a more reliable and higher quality signal.

Collaborative CDMA

In a recent study, a novel collaborative multi-user transmission and detection scheme called Collaborative CDMA has been investigated for the uplink that exploits the differences between users' fading channel signatures to increase the user capacity well beyond the spreading length in multiple access interference (MAI) limited environment. The authors show that it is possible to achieve this increase at a low complexity and high bit error rate performance in flat fading channels, which is a major research challenge for overloaded CDMA systems. In this approach, instead of using one sequence per user as in conventional CDMA, the authors group a small number of users to share the same spreading sequence and enable group spreading and despreading operations. The new collaborative multi-user receiver consists of two stages: group multi-user detection (MUD) stage to suppress the MAI between the groups and a low complexity maximum-likelihood detection stage to recover jointly the co-spread users' data using minimum Euclidean distance measure and users' channel gain coefficients. In CDMA, signal security is high.

Public Switched Telephone Network

The public switched telephone network (PSTN) is the aggregate of the world's circuit-switched telephone networks that are operated by national, regional, or local telephony operators, providing infrastructure and services for public telecommunication. The PSTN consists of telephone lines, fiber optic cables, microwave transmission links, cellular networks, communications satellites, and undersea telephone cables, all interconnected by switching centers, thus allowing most telephones to communicate with each other. Originally a network of fixed-line analog telephone systems, the PSTN is now almost entirely digital in its core network and includes mobile and other networks, as well as fixed telephones.

The technical operation of the PSTN adheres to the standards created by the ITU-T. These standards allow different networks in different countries to interconnect seamlessly. The E.163 and E.164 standards provide a single global address space for telephone numbers. The combination of the interconnected networks and the single numbering plan allow telephones around the world to dial each other.

History (America)

The first telephones had no network but were in private use, wired together in pairs. Users who wanted to talk to different people had as many telephones as necessary for the purpose. A user who wished to speak whistled loudly into the transmitter until the other party heard.

However, a bell was added soon for signaling, so an attendant no longer need wait for the whistle, and then a switch hook. Later telephones took advantage of the exchange principle already employed in telegraph networks. Each telephone was wired to a local telephone exchange, and the exchanges were wired together with trunks. Networks were connected in a hierarchical manner until they spanned cities, countries, continents and oceans. This was the beginning of the PSTN[when?], though the term was not used for many decades.

Automation introduced pulse dialing between the phone and the exchange, and then among exchanges, followed by more sophisticated address signaling including multi-frequency, culminating in the SS7 network that connected most exchanges by the end of the 20th century.

The growth of the PSTN meant that teletraffic engineering techniques needed to be deployed to deliver quality of service (QoS) guarantees for the users. The work of A. K. Erlang established the mathematical foundations of methods required to determine the capacity requirements and configuration of equipment and the number of personnel required to deliver a specific level of service.

In the 1970s the telecommunications industry began implementing packet switched network data services using the X.25 protocol transported over much of the end-to-end equipment as was already in use in the PSTN.

In the 1980s the industry began planning for digital services assuming they would follow much the same pattern as voice services, and conceived a vision of end-to-end circuit switched services, known as the Broadband Integrated Services Digital Network (B-ISDN). The B-ISDN vision has been overtaken by the disruptive technology of the Internet.

At the turn of the 21st century, the oldest parts of the telephone network still use analog technology for the last mile loop to the end user. However, digital technologies such as DSL, ISDN, FTTx, and cable modems have become more common in this portion of the network.

Several large private telephone networks are not linked to the PSTN, usually for military purposes. There are also private networks run by large companies which are linked to the PSTN only through limited gateways, such as a large private branch exchange (PBX).

Operators

The task of building the networks and selling services to customers fell to the network operators. The first company to be incorporated to provide PSTN services was the Bell Telephone Company in the United States.

In some countries, however, the job of providing telephone networks fell to government as the investment required was very large and the provision of telephone service was increasingly becoming an essential public utility. For example, the General Post Office in the United Kingdom brought together a number of private companies to form a single nationalized company. In recent decades however, these state monopolies were broken up or sold off through privatization.

Regulation

In most countries, the central has a regulator dedicated to monitoring the provision of PSTN services in that country. Their tasks may be for example to ensure that end customers are not overcharged for services where monopolies may exist. They may also regulate the prices charged between the operators to carry each other's traffic.

Technology

Network Topology

The PSTN network architecture had to evolve over the years to support increasing numbers of subscribers, calls, connections to other countries, direct dialing and so on. The model developed by the United States and Canada was adopted by other nations, with adaptations for local markets.

The original concept was that the telephone exchanges are arranged into hierarchies, so that if a call cannot be handled in a local cluster, it is passed to one higher up for onward routing. This reduced the number of connecting trunks required between operators over long distances and also kept local traffic separate.

However, in modern networks the cost of transmission and equipment is lower and, although hierarchies still exist, they are much flatter, with perhaps only two layers.

Digital Channels

As described above, most automated telephone exchanges now use digital switching rather than mechanical or analog switching. The trunks connecting the exchanges are also digital, called circuits or channels. However analog two-wire circuits are still used to connect the last mile from the exchange to the telephone in the home (also called the local loop). To carry a typical phone call from a calling party to a called party, the analog audio signal is digitized at an 8 kHz sample rate with 8-bit resolution using a special type of nonlinear pulse code modulation known as G.711. The call is then transmitted from one end to another via telephone exchanges. The call is switched using a call set up protocol (usually ISUP) between the telephone exchanges under an overall routing strategy.

The call is carried over the PSTN using a 64 kbit/s channel, originally designed by Bell Labs. The name given to this channel is Digital Signal 0 (DS0). The DS0 circuit is the basic granularity of circuit switching in a telephone exchange. A DS0 is also known as a timeslot because DS0s are aggregated in time-division multiplexing (TDM) equipment to form higher capacity communication links.

A Digital Signal 1 (DS1) circuit carries 24 DS0s on a North American or Japanese T-carrier (T1) line, or 32 DS0s (30 for calls plus two for framing and signaling) on an E-carrier (E1) line used in most other countries. In modern networks, the multiplexing function is moved as close to the end user as possible, usually into cabinets at the roadside in residential areas, or into large business premises.

These aggregated circuits are conveyed from the initial multiplexer to the exchange over a set of equipment collectively known as the access network. The access network and inter-exchange transport use synchronous optical transmission, for example, SONET and Synchronous Digital Hierarchy (SDH) technologies, although some parts still use the older PDH technology.

Within the access network, there are a number of reference points defined. Most of these are of interest mainly to ISDN but one – the V reference point – is of more general interest. This is the reference point between a primary multiplexer and an exchange. The protocols at this reference point were standardized in ETSI areas as the V5 interface.

Impact on IP Standards

Voice quality over PSTN networks was used as the benchmark for the development of the Telecommunications Industry Association's TIA-TSB-116 standard on voice-quality recommendations for IP telephony, to determine acceptable levels of audio delay and echo.

4G

4G is the fourth generation of wireless mobile telecommunications technology, succeeding 3G. A 4G system must provide capabilities defined by ITU in IMT Advanced. Potential and current applications include amended mobile web access, IP telephony, gaming services, high-definition mobile TV, video conferencing, 3D television.

Two 4G candidate systems are commercially deployed: the Mobile WiMAX standard (first used in South Korea in 2007), and the first-release Long Term Evolution (LTE) standard (in Oslo, Norway, and Stockholm, Sweden since 2009). It has, however, been debated whether these first-release versions should be considered 4G, as discussed in the technical-definition section below.

In the United States, Sprint (previously Clearwire) has deployed Mobile WiMAX networks since 2008, while MetroPCS became the first operator to offer LTE service in 2010. USB wireless modems were among the first devices able to access these networks, with WiMAX smartphones becoming available during 2010, and LTE smartphones arriving in 2011. 3G and 4G equipment made for other continents are not always compatible because of different frequency bands. Mobile Wi-MAX is not available for the European market as of April 2012.

Technical Understanding

In March 2008, the International Telecommunications Union-Radio communications sector (ITU-R) specified a set of requirements for 4G standards, named the International Mobile Telecommunications Advanced (IMT-Advanced) specification, setting peak speed requirements for 4G service at 100 megabits per second (Mbit/s) for high mobility communication (such as from trains and cars) and 1 gigabit per second (Gbit/s) for low mobility communication (such as pedestrians and stationary users).

Since the first-release versions of Mobile WiMAX and LTE support much less than 1 Gbit/s peak bit rate, they are not fully IMT-Advanced compliant, but are often branded 4G by service providers. According to operators, a generation of the network refers to the deployment of a new non-backward-compatible technology. On December 6, 2010, ITU-R recognized that these two technologies, as well as other beyond-3G technologies that do not fulfill the IMT-Advanced requirements, could nevertheless be considered "4G", provided they represent forerunners to IMT-Advanced compliant versions and "a substantial level of improvement in performance and capabilities with respect to the initial third generation systems now deployed".

Mobile WiMAX Release 2 (also known as *WirelessMAN-Advanced* or *IEEE 802.16m'*) and LTE Advanced (LTE-A) are IMT-Advanced compliant backwards compatible versions of the above two systems, standardized during the spring 2011, and promising speeds in the order of 1 Gbit/s. Services were expected in 2013.

As opposed to earlier generations, a 4G system does not support traditional circuit-switched telephony service, but all-Internet Protocol (IP) based communication such as IP telephony. As seen below, the spread spectrum radio technology used in 3G systems, is abandoned in all 4G candidate systems and replaced by OFDMA multi-carrier transmission and other frequency-domain equalization (FDE) schemes, making it possible to transfer very high bit rates despite extensive multipath radio propagation (echoes). The peak bit rate is further improved by smart antenna arrays for multiple-input multiple-output (MIMO) communications.

Background

The nomenclature of the generations generally refers to a change in the fundamental nature of the service, non-backwards-compatible transmission technology, higher peak bit rates, new frequency bands, wider channel frequency bandwidth in Hertz, and higher capacity for many simultaneous data transfers (higher system spectral efficiency in bit/second/Hertz/site).

New mobile generations have appeared about every ten years since the first move from 1981 analog (1G) to digital (2G) transmission in 1992. This was followed, in 2001, by 3G multi-media support, spread spectrum transmission and, at least, 200 kbit/s peak bit rate, in 2011/2012 to be followed by "real" 4G, which refers to all-Internet Protocol (IP) packet-switched networks giving mobile ultra-broadband (gigabit speed) access.

While the ITU has adopted recommendations for technologies that would be used for future global communications, they do not actually perform the standardization or development work themselves, instead relying on the work of other standard bodies such as IEEE, The WiMAX Forum, and 3GPP.

In the mid-1990s, the ITU-R standardization organization released the IMT-2000 requirements as a framework for what standards should be considered 3G systems, requiring 200 kbit/s peak bit rate. In 2008, ITU-R specified the IMT-Advanced (International Mobile Telecommunications Advanced) requirements for 4G systems.

The fastest 3G-based standard in the UMTS family is the HSPA+ standard, which is commercially available since 2009 and offers 28 Mbit/s downstream (22 Mbit/s upstream) without MIMO, i.e. only with one antenna, and in 2011 accelerated up to 42 Mbit/s peak bit rate downstream using either DC-HSPA+ (simultaneous use of two 5 MHz UMTS carriers) or 2x2 MIMO. In theory speeds up to 672 Mbit/s are possible, but have not been deployed yet. The fastest 3G-based standard in the CDMA2000 family is the EV-DO Rev. B, which is available since 2010 and offers 15.67 Mbit/s downstream.

IMT-Advanced Requirements

This article refers to 4G using IMT-Advanced (*International Mobile Telecommunications Advanced*), as defined by ITU-R. An IMT-Advanced cellular system must fulfill the following requirements:

- Be based on an all-IP packet switched network.

- Have peak data rates of up to approximately 100 Mbit/s for high mobility such as mobile access and up to approximately 1 Gbit/s for low mobility such as nomadic/local wireless access.

- Be able to dynamically share and use the network resources to support more simultaneous users per cell.

- Use scale-able channel bandwidths of 5–20 MHz, optionally up to 40 MHz.*Rumney, Moray (September 2008). "IMT-Advanced: 4G Wireless Takes Shape in an Olympic Year" (PDF). Agilent Measurement Journal.*

- Have peak link spectral efficiency of 15-bit/s/Hz in the downlink, and 6.75-bit/s/Hz in the uplink (meaning that 1 Gbit/s in the downlink should be possible over less than 67 MHz bandwidth).

- System spectral efficiency is, in indoor cases, 3-bit/s/Hz/cell for downlink and 2.25-bit/s/Hz/cell for uplink.

- Smooth handovers across heterogeneous networks.

- The ability to offer high quality of service for next generation multimedia support.

In September 2009, the technology proposals were submitted to the International Telecommunication Union (ITU) as 4G candidates. Basically all proposals are based on two technologies.:

- LTE Advanced standardized by the 3GPP

- 802.16m standardized by the IEEE (i.e. WiMAX)

Implementations of Mobile WiMAX and first-release LTE are largely considered a stopgap solution that will offer a considerable boost until WiMAX 2 (based on the 802.16m spec) and LTE Advanced are deployed. The latter's standard versions were ratified in spring 2011, but are still far from being implemented.

The first set of 3GPP requirements on LTE Advanced was approved in June 2008. LTE Advanced was to be standardized in 2010 as part of Release 10 of the 3GPP specification. LTE Advanced will be based on the existing LTE specification Release 10 and will not be defined as a new specification series. A summary of the technologies that have been studied as the basis for LTE Advanced is included in a technical report.

Some sources consider first-release LTE and Mobile WiMAX implementations as pre-4G or near-4G, as they do not fully comply with the planned requirements of 1 Gbit/s for stationary reception and 100 Mbit/s for mobile.

Confusion has been caused by some mobile carriers who have launched products advertised as 4G but which according to some sources are pre-4G versions, commonly referred to as '3.9G', which do not follow the ITU-R defined principles for 4G standards, but today can be called 4G according to ITU-R. Vodafone NL for example, advertised LTE as '4G', while advertising now LTE Advanced as their '4G+' service which actually is (True) 4G. A common argument for branding 3.9G systems as new-generation is that they use different frequency bands from 3G technologies ; that they are based on a new radio-interface paradigm ; and that the standards are not backwards compatible with 3G, whilst some of the standards are forwards compatible with IMT-2000 compliant versions of the same standards.

System Standards

IMT-2000 Compliant 4G Standards

As of October 2010, ITU-R Working Party 5D approved two industry-developed technologies (LTE Advanced and WirelessMAN-Advanced) for inclusion in the ITU's International Mobile Telecommunications Advanced program (IMT-Advanced program), which is focused on global communication systems that will be available several years from now.

LTE Advanced

LTE Advanced (Long Term Evolution Advanced) is a candidate for IMT-Advanced standard, formally submitted by the 3GPP organization to ITU-T in the fall 2009, and expected to be released in 2013. The target of 3GPP LTE Advanced is to reach and surpass the ITU requirements. LTE Advanced is essentially an enhancement to LTE. It is not a new technology, but rather an improvement on the existing LTE network. This upgrade path makes it more cost effective for vendors to offer LTE and then upgrade to LTE Advanced which is similar to the upgrade from WCDMA to HSPA. LTE and LTE Advanced will also make use of additional spectrums and multiplexing to allow it to achieve higher data speeds. Coordinated Multi-point Transmission will also allow more system capacity to help handle the enhanced data speeds. Release 10 of LTE is expected to achieve the IMT Advanced speeds. Release 8 currently supports up to 300 Mbit/s of download speeds which is still short of the IMT-Advanced standards.

Data speeds of LTE-Advanced	
	LTE Advanced
Peak download	1 Gbps
Peak upload	500 Mbps

IEEE 802.16m or WirelessMAN-Advanced

The IEEE 802.16m or WirelessMAN-Advanced evolution of 802.16e is under development, with the objective to fulfill the IMT-Advanced criteria of 1 Gbit/s for stationary reception and 100 Mbit/s for mobile reception.

Forerunner Versions

3GPP Long Term Evolution (LTE)

Telia-branded Samsung LTE modem

The pre-4G 3GPP Long Term Evolution (LTE) technology is often branded "4G-LTE", but the first LTE release does not fully comply with the IMT-Advanced requirements. LTE has a theoretical net bit rate capacity of up to 100 Mbit/s in the downlink and 50 Mbit/s in the uplink if a 20 MHz channel is used — and more if multiple-input multiple-output (MIMO), i.e. antenna arrays, are used.

The physical radio interface was at an early stage named *High Speed OFDM Packet Access* (HSO-

PA), now named Evolved UMTS Terrestrial Radio Access (E-UTRA). The first LTE USB dongles do not support any other radio interface.

The world's first publicly available LTE service was opened in the two Scandinavian capitals, Stockholm (Ericsson and Nokia Siemens Networks systems) and Oslo (a Huawei system) on December 14, 2009, and branded 4G. The user terminals were manufactured by Samsung. As of November 2012, the five publicly available LTE services in the United States are provided by MetroPCS, Verizon Wireless, AT&T Mobility, U.S. Cellular, Sprint, and T-Mobile US.

T-Mobile Hungary launched a public beta test (called *friendly user test*) on 7 October 2011, and has offered commercial 4G LTE services since 1 January 2012.

In South Korea, SK Telecom and LG U+ have enabled access to LTE service since 1 July 2011 for data devices, slated to go nationwide by 2012. KT Telecom closed its 2G service by March 2012, and complete the nationwide LTE service in the same frequency around 1.8 GHz by June 2012.

In the United Kingdom, LTE services were launched by EE in October 2012, and by O2 and Vodafone in August 2013.

Data speeds of LTE	
	LTE
Peak download	100 Mbit/s
Peak upload	50 Mbit/s

Mobile WiMAX (IEEE 802.16e)

The Mobile WiMAX (IEEE 802.16e-2005) mobile wireless broadband access (MWBA) standard (also known as WiBro in South Korea) is sometimes branded 4G, and offers peak data rates of 128 Mbit/s downlink and 56 Mbit/s uplink over 20 MHz wide channels.

In June 2006, the world's first commercial mobile WiMAX service was opened by KT in Seoul, South Korea.

Sprint has begun using Mobile WiMAX, as of 29 September 2008, branding it as a "4G" network even though the current version does not fulfill the IMT Advanced requirements on 4G systems.

In Russia, Belarus and Nicaragua WiMax broadband internet access is offered by a Russian company Scartel, and is also branded 4G, Yota.

Data speeds of WiMAX	
	WiMAX
Peak download	128 Mbit/s
Peak upload	56 Mbit/s

TD-LTE for China market

Just as Long-Term Evolution (LTE) and WiMAX are being vigorously promoted in the global telecommunications industry, the former (LTE) is also the most powerful 4G mobile communications leading technology and has quickly occupied the Chinese market. TD-LTE, one of the two variants of the LTE air interface technologies, is not yet mature, but many domestic and international wireless carriers are, one after the other turning to TD-LTE.

IBM's data shows that 67% of the operators are considering LTE because this is the main source of their future market. The above news also confirms IBM's statement that while only 8% of the operators are considering the use of WiMAX, WiMAX can provide the fastest network transmission to its customers on the market and could challenge LTE.

TD-LTE is not the first 4G wireless mobile broadband network data standard, but it is China's 4G standard that was amended and published by China's largest telecom operator – China Mobile. After a series of field trials, is expected to be released into the commercial phase in the next two years. Ulf Ewaldsson, Ericsson's vice president said: "the Chinese Ministry of Industry and China Mobile in the fourth quarter of this year will hold a large-scale field test, by then, Ericsson will help the hand." But viewing from the current development trend, whether this standard advocated by China Mobile will be widely recognized by the international market is still debatable.

Discontinued Candidate Systems

UMB (Formerly EV-DO Rev. C)

UMB (Ultra Mobile Broadband) was the brand name for a discontinued 4G project within the 3GPP2 standardization group to improve the CDMA2000 mobile phone standard for next generation applications and requirements. In November 2008, Qualcomm, UMB's lead sponsor, announced it was ending development of the technology, favouring LTE instead. The objective was to achieve data speeds over 275 Mbit/s downstream and over 75 Mbit/s upstream.

Flash-OFDM

At an early stage the Flash-OFDM system was expected to be further developed into a 4G standard.

iBurst and MBWA (IEEE 802.20) Systems

The iBurst system (or HC-SDMA, High Capacity Spatial Division Multiple Access) was at an early stage considered to be a 4G predecessor. It was later further developed into the Mobile Broadband Wireless Access (MBWA) system, also known as IEEE 802.20.

Principal Technologies in All Candidate Systems

Key Features

The following key features can be observed in all suggested 4G technologies:

- Physical layer transmission techniques are as follows:

    o MIMO: To attain ultra high spectral efficiency by means of spatial processing including multi-antenna and multi-user MIMO

    o Frequency-domain-equalization, for example *multi-carrier modulation* (OFDM) in the downlink or *single-carrier frequency-domain-equalization* (SC-FDE) in the uplink: To exploit the frequency selective channel property without complex equalization

   o Frequency-domain statistical multiplexing, for example (OFDMA) or (single-carrier FDMA) (SC-FDMA, a.k.a. linearly precoded OFDMA, LP-OFDMA) in the uplink: Variable bit rate by assigning different sub-channels to different users based on the channel conditions

   o Turbo principle error-correcting codes: To minimize the required SNR at the reception side

- Channel-dependent scheduling: To use the time-varying channel

- Link adaptation: Adaptive modulation and error-correcting codes

- Mobile IP utilized for mobility

- IP-based femtocells (home nodes connected to fixed Internet broadband infrastructure)

As opposed to earlier generations, 4G systems do not support circuit switched telephony. IEEE 802.20, UMB and OFDM standards lack soft-handover support, also known as cooperative relaying.

Multiplexing and Access Schemes

Recently, new access schemes like Orthogonal FDMA (OFDMA), Single Carrier FDMA (SC-FDMA), Interleaved FDMA, and Multi-carrier CDMA (MC-CDMA) are gaining more importance for the next generation systems. These are based on efficient FFT algorithms and frequency domain equalization, resulting in a lower number of multiplications per second. They also make it possible to control the bandwidth and form the spectrum in a flexible way. However, they require advanced dynamic channel allocation and adaptive traffic scheduling.

WiMax is using OFDMA in the downlink and in the uplink. For the LTE (telecommunication), OFDMA is used for the downlink; by contrast, Single-carrier FDMA is used for the uplink since OFDMA contributes more to the PAPR related issues and results in nonlinear operation of amplifiers. IFDMA provides less power fluctuation and thus requires energy-inefficient linear amplifiers. Similarly, MC-CDMA is in the proposal for the IEEE 802.20 standard. These access schemes offer the same efficiencies as older technologies like CDMA. Apart from this, scalability and higher data rates can be achieved.

The other important advantage of the above-mentioned access techniques is that they require less complexity for equalization at the receiver. This is an added advantage especially in the MIMO environments since the spatial multiplexing transmission of MIMO systems inherently require high complexity equalization at the receiver.

In addition to improvements in these multiplexing systems, improved modulation techniques are being used. Whereas earlier standards largely used Phase-shift keying, more efficient systems such as 64QAM are being proposed for use with the 3GPP Long Term Evolution standards.

IPv6 Support

Unlike 3G, which is based on two parallel infrastructures consisting of circuit switched and packet switched network nodes, 4G will be based on packet switching *only*. This will require low-latency data transmission.

By the time that 4G was deployed, the process of IPv4 address exhaustion was expected to be in its final stages. Therefore, in the context of 4G, IPv6 is essential to support a large number of wireless-enabled devices. By increasing the number of IP addresses available, IPv6 removes the need for network address translation (NAT), a method of sharing a limited number of addresses among a larger group of devices, although NAT will still be required to communicate with devices that are on existing IPv4 networks.

As of June 2009, Verizon has posted specifications that require any 4G devices on its network to support IPv6.

Advanced Antenna Systems

The performance of radio communications depends on an antenna system, termed smart or intelligent antenna. Recently, multiple antenna technologies are emerging to achieve the goal of 4G systems such as high rate, high reliability, and long range communications. In the early 1990s, to cater for the growing data rate needs of data communication, many transmission schemes were proposed. One technology, spatial multiplexing, gained importance for its bandwidth conservation and power efficiency. Spatial multiplexing involves deploying multiple antennas at the transmitter and at the receiver. Independent streams can then be transmitted simultaneously from all the antennas. This technology, called MIMO (as a branch of intelligent antenna), multiplies the base data rate by (the smaller of) the number of transmit antennas or the number of receive antennas. Apart from this, the reliability in transmitting high speed data in the fading channel can be improved by using more antennas at the transmitter or at the receiver. This is called *transmit* or *receive diversity*. Both transmit/receive diversity and transmit spatial multiplexing are categorized into the space-time coding techniques, which does not necessarily require the channel knowledge at the transmitter. The other category is closed-loop multiple antenna technologies, which require channel knowledge at the transmitter.

Open-wireless Architecture and Software-defined Radio (SDR)

One of the key technologies for 4G and beyond is called Open Wireless Architecture (OWA), supporting multiple wireless air interfaces in an open architecture platform.

SDR is one form of open wireless architecture (OWA). Since 4G is a collection of wireless standards, the final form of a 4G device will constitute various standards. This can be efficiently realized using SDR technology, which is categorized to the area of the radio convergence.

History of 4G and Pre-4G Technologies

The 4G system was originally envisioned by the Defense Advanced Research Projects Agency (DARPA). The DARPA selected the distributed architecture and end-to-end Internet protocol (IP), and believed at an early stage in peer-to-peer networking in which every mobile device would be both a transceiver and a router for other devices in the network, eliminating the spoke-and-hub weakness of 2G and 3G cellular systems. Since the 2.5G GPRS system, cellular systems have provided dual infrastructures: packet switched nodes for data services, and circuit switched nodes for voice calls. In 4G systems, the circuit-switched infrastructure is abandoned and only a packet-switched network is provided, while 2.5G and 3G systems require both packet-switched and circuit-switched network nodes, i.e. two infrastructures in parallel. This means that in 4G, traditional voice calls are replaced by IP telephony.

- In 2002, the strategic vision for 4G — which ITU designated as IMT Advanced— was laid out.

- In 2005, OFDMA transmission technology is chosen as candidate for the HSOPA downlink, later renamed 3GPP Long Term Evolution (LTE) air interface E-UTRA.

- In November 2005, KT demonstrated mobile WiMAX service in Busan, South Korea.

- In April 2006, KT started the world's first commercial mobile WiMAX service in Seoul, South Korea.

- In mid-2006, Sprint announced that it would invest about US$5 billion in a WiMAX technology buildout over the next few years ($5.88 billion in real terms). Since that time Sprint has faced many setbacks that have resulted in steep quarterly losses. On 7 May 2008, Sprint, Imagine, Google, Intel, Comcast, Bright House, and Time Warner announced a pooling of an average of 120 MHz of spectrum; Sprint merged its Xohm WiMAX division with Clearwire to form a company which will take the name "Clear".

- In February 2007, the Japanese company NTT DoCoMo tested a 4G communication system prototype with 4×4 MIMO called VSF-OFCDM at 100 Mbit/s while moving, and 1 Gbit/s while stationary. NTT DoCoMo completed a trial in which they reached a maximum packet transmission rate of approximately 5 Gbit/s in the downlink with 12×12 MIMO using a 100 MHz frequency bandwidth while moving at 10 km/h, and is planning on releasing the first commercial network in 2010.

- In September 2007, NTT Docomo demonstrated e-UTRA data rates of 200 Mbit/s with power consumption below 100 mW during the test.

- In January 2008, a U.S. Federal Communications Commission (FCC) spectrum auction for the 700 MHz former analog TV frequencies began. As a result, the biggest share of the spectrum went to Verizon Wireless and the next biggest to AT&T. Both of these companies have stated their intention of supporting LTE.

- In January 2008, EU commissioner Viviane Reding suggested re-allocation of 500–800 MHz spectrum for wireless communication, including WiMAX.

- On 15 February 2008, Skyworks Solutions released a front-end module for e-UTRAN.

- In November 2008, ITU-R established the detailed performance requirements of IMT-Advanced, by issuing a Circular Letter calling for candidate Radio Access Technologies (RATs) for IMT-Advanced.

- In April 2008, just after receiving the circular letter, the 3GPP organized a workshop on IMT-Advanced where it was decided that LTE Advanced, an evolution of current LTE standard, will meet or even exceed IMT-Advanced requirements following the ITU-R agenda.

- In April 2008, LG and Nortel demonstrated e-UTRA data rates of 50 Mbit/s while travelling at 110 km/h.

- On 12 November 2008, HTC announced the first WiMAX-enabled mobile phone, the Max 4G

- On 15 December 2008, San Miguel Corporation, the largest food and beverage conglomerate in southeast Asia, has signed a memorandum of understanding with Qatar Telecom QSC (Qtel) to build wireless broadband and mobile communications projects in the Philippines. The joint-venture formed wi-tribe Philippines, which offers 4G in the country. Around the same time Globe Telecom rolled out the first WiMAX service in the Philippines.

- On 3 March 2009, Lithuania's LRTC announcing the first operational "4G" mobile WiMAX network in Baltic states.

- In December 2009, Sprint began advertising "4G" service in selected cities in the United States, despite average download speeds of only 3–6 Mbit/s with peak speeds of 10 Mbit/s (not available in all markets).

- On 14 December 2009, the first commercial LTE deployment was in the Scandinavian capitals Stockholm and Oslo by the Swedish-Finnish network operator TeliaSonera and its Norwegian brandname NetCom (Norway). TeliaSonera branded the network "4G". The modem devices on offer were manufactured by Samsung (dongle GT-B3710), and the network infrastructure created by Huawei (in Oslo) and Ericsson (in Stockholm). TeliaSonera plans to roll out nationwide LTE across Sweden, Norway and Finland. TeliaSonera used spectral bandwidth of 10 MHz, and single-in-single-out, which should provide physical layer net bitrates of up to 50 Mbit/s downlink and 25 Mbit/s in the uplink. Introductory tests showed a TCP throughput of 42.8 Mbit/s downlink and 5.3 Mbit/s uplink in Stockholm.

- On 4 June 2010, Sprint released the first WiMAX smartphone in the US, the HTC Evo 4G.

- On November 4, 2010, the Samsung Galaxy Craft offered by MetroPCS is the first commercially available LTE smartphone

- On 6 December 2010, at the ITU World Radiocommunication Seminar 2010, the ITU stated that LTE, WiMax and similar "evolved 3G technologies" could be considered "4G".

- In 2011, Argentina's Claro launched a pre-4G HSPA+ network in the country.

- In 2011, Thailand's Truemove-H launched a pre-4G HSPA+ network with nationwide availability.

- On March 17, 2011, the HTC Thunderbolt offered by Verizon in the U.S. was the second LTE smartphone to be sold commercially.

- In February 2012, Ericsson demonstrated mobile-TV over LTE, utilizing the new eMBMS service (enhanced Multimedia Broadcast Multicast Service).

Since 2009 the LTE-Standard has strongly evolved over the years, resulting in many deployments by various operators across the globe. For an overview of commercial LTE networks and their respective historic development : list of LTE networks. Among the vast range of deployments many operators are considering the deployment and operation of LTE networks. A compilation of planned LTE deployments can be found at: List of planned LTE networks.

Beyond 4G Research

A major issue in 4G systems is to make the high bit rates available in a larger portion of the cell, especially to users in an exposed position in between several base stations. In current research, this issue is addressed by macro-diversity techniques, also known as group cooperative relay, and also by Beam-Division Multiple Access (BDMA).

Pervasive networks are an amorphous and at present entirely hypothetical concept where the user can be simultaneously connected to several wireless access technologies and can seamlessly move between them. These access technologies can be Wi-Fi, UMTS, EDGE, or any other future access technology. Included in this concept is also smart-radio (also known as cognitive radio) technology to efficiently manage spectrum use and transmission power as well as the use of mesh routing protocols to create a pervasive network.

5G

5G (5th generation mobile networks or 5th generation wireless systems) denotes the proposed next major phase of mobile telecommunications standards beyond the current 4G/IMT-Advanced standards. Rather than faster peak Internet connection speeds, 5G planning aims at higher capacity than current 4G, allowing higher number of mobile broadband users per area unit, and allowing consumption of higher or unlimited data quantities in gigabyte per month and user. This would make it feasible for a large portion of the population to consume high-quality streaming media many hours per day in their mobile devices, also when out of reach of wifi hotspots. 5G research and development also aims at improved support of machine to machine communication, also known as the Internet of things, aiming at lower cost, lower battery consumption and lower latency than 4G equipment.

There is currently no standard for 5G deployments. The Next Generation Mobile Networks Alliance defines the following requirements that a 5G standard should fulfill:

- Data rates of tens of megabits per second for tens of thousands of users

- 1 Gb per second simultaneously to many workers on the same office floor

- Several hundreds of thousands of simultaneous connections for massive wireless sensor network

- Spectral efficiency significantly enhanced compared to 4G

- Coverage improved

- Signalling efficiency enhanced

- Latency reduced significantly compared to LTE.

The Next Generation Mobile Networks Alliance feels that 5G should be rolled out by 2020 to meet business and consumer demands. In addition to providing simply faster speeds, they predict that 5G networks also will need to meet new use cases, such as the Internet of Things (internet connected devices) as well as broadcast-like services and lifeline communication in times of natural disaster. Carriers, chipmakers, OEMS and OSATs, such as Advanced Semiconductor Engineering (ASE), have been gearing up for this next-generation (5G) wireless standard, as mobile systems and base stations will require new and faster application processors, basebands and RF devices.

Although updated standards that define capabilities beyond those defined in the current 4G standards are under consideration, those new capabilities have been grouped under the current ITU-T 4G standards. The U.S. Federal Communications Commission (FCC) approved the spectrum for 5G, including the 28 Gigahertz, 37 GHz and 39 GHz bands, on July 14, 2016.

Background

A new mobile generation has appeared approximately every 10 years since the first 1G system, Nordic Mobile Telephone, was introduced in 1982. The first '2G' system was commercially deployed in 1992, and the 3G system appeared in 2001. 4G systems fully compliant with IMT Advanced were first standardized in 2012. The development of the 2G (GSM) and 3G (IMT-2000 and UMTS) standards took about 10 years from the official start of the R&D projects, and development of 4G systems began in 2001 or 2002. Predecessor technologies have been on the market a few years before the new mobile generation, for example the pre-3G system CdmaOne/IS95 in the US in 1995, and the pre-4G systems Mobile WiMAX in South-Korea 2006, and first release-LTE in Scandinavia 2009. In April 2008, NASA partnered with Machine-to-Machine Intelligence (M2Mi) Corp to develop 5G communication technology

Mobile generations typically refer to non–backward-compatible cellular standards following requirements stated by ITU-R, such as IMT-2000 for 3G and IMT-Advanced for 4G. In parallel with the development of the ITU-R mobile generations, IEEE and other standardization bodies also develop wireless communication technologies, often for higher data rates, higher frequencies, shorter transmission ranges, no support for roaming between access points and a relatively limited multiple access scheme. The first gigabit IEEE standard was IEEE 802.11ac, commercially available since 2013, soon to be followed by the multigigabit standard WiGig or IEEE 802.11ad.

Debate

Based on the above observations, some sources suggest that a new generation of 5G standards may be introduced in the early 2020s. However, significant debate continued, on what 5G is about exactly. Prior to 2012, some industry representatives expressed skepticism toward 5G. 3GPP held a conference in September 2015 to plan development of the new standard.

New mobile generations are typically assigned new frequency bands and wider spectral bandwidth per frequency channel (1G up to 30 kHz, 2G up to 200 kHz, 3G up to 5 MHz, and 4G up to 20 MHz), but skeptics argue that there is little room for larger channel bandwidths and new frequency bands suitable for land-mobile radio. The higher frequencies would overlap with K-band transmissions of communication satellites. From users' point of view, previous mobile generations have implied substantial increase in peak bitrate (i.e. physical layer net bitrates for short-distance communication), up to 1 gigabit per second to be offered by 4G.

If 5G appears and reflects these prognoses, then the major difference, from a user point of view, between 4G and 5G must be something other than faster speed (increased peak bit rate). For example, higher number of simultaneously connected devices, higher system spectral efficiency (data volume per area unit), lower battery consumption, lower outage probability (better coverage), high bit rates in larger portions of the coverage area, lower latencies, higher number of supported devices, lower infrastructure deployment costs, higher versatility and scalability, or higher reliability of communication. Those are the objectives in several of the research papers and projects below.

GSMHistory.com has recorded three very distinct 5G network visions that had emerged by 2014:

- A super-efficient mobile network that delivers a better performing network for lower investment cost. It addresses the mobile network operators' pressing need to see the unit cost of data transport falling at roughly the same rate as the volume of data demand is rising. It would be a leap forward in efficiency based on the IET Demand Attentive Network (DAN) philosophy.

- A super-fast mobile network comprising the next generation of small cells densely clustered to give a contiguous coverage over at least urban areas and getting the world to the final frontier of true "wide-area mobility." It would require access to spectrum under 4 GHz perhaps via the world's first global implementation of Dynamic Spectrum Access.

- A converged fiber-wireless network that uses, for the first time for wireless Internet access, the millimeter wave bands (20 – 60 GHz) so as to allow very-wide-bandwidth radio channels able to support data-access speeds of up to 10 Gbit/s. The connection essentially comprises "short" wireless links on the end of local fiber optic cable. It would be more a "nomadic" service (like Wi-Fi) rather than a wide-area "mobile" service.

In its white paper, *5G Empowering Vertical Industries*, 5G PPP, the collaborative research programme organized as part of the European Commission's Horizon 2020 programme, suggests that to support the main vertical sectors in Europe - namely automotive, transportation, healthcare, energy, manufacturing, and media and entertainment - the most important 5G infrastructure performance requirements are a latency below 5 ms, support for device densities of up to 100 devices/m² and reliable coverage area, and that a successful 5G deployment will integrate telecommunication technologies including mobile, fixed, optical and satellite (both GEO and MEO).

Research & Development Projects

In 2008, the South Korean IT R&D program of "5G mobile communication systems based on beam-division multiple access and relays with group cooperation" was formed.

In 2012, the UK Government announced the establishment of a 5G Innovation Centre at the University of Surrey – the world's first research center set up specifically for 5G mobile research.

In 2012, NYU WIRELESS was established as a multidisciplinary research center, with a focus on 5G wireless research, as well as its use in the medical and computer-science fields. The center is funded by the National Science Foundation and a board of 10 major wireless companies (as of July 2014) that serve on the Industrial Affiliates board of the center. NYU WIRELESS has conducted and published channel measurements that show that millimeter wave frequencies will be viable for multigigabit-per-second data rates for future 5G networks.

In 2012, the European Commission, under the lead of Neelie Kroes, committed 50 million euros for research to deliver 5G mobile technology by 2020. In particular, The METIS 2020 Project was the flagship project that allowed reaching a world-wide consensus on the requirements and key technology components of the 5G. Driven by several telecommunication companies, the METIS overall technical goal was to provide a system concept that supports 1,000 times higher mobile system spectral efficiency, compared to current LTE deployments. In addition, in 2013, another project has started, called 5GrEEn, linked to project METIS and focusing on the design of green 5G mobile networks. Here the goal is to develop guidelines for the definition of a new-generation network with particular emphasis on energy efficiency, sustainability and affordability.

In November 2012, a research project funded by the European Union under the ICT Programme FP7 was launched under the coordination of IMDEA Networks Institute (Madrid, Spain): i-JOIN (Interworking and JOINt Design of an Open Access and Backhaul Network Architecture for Small Cells based on Cloud Networks). iJOIN introduces the novel concept of the radio access network (RAN) as a service (RANaaS), where RAN functionality is flexibly centralized through an open IT platform based on a cloud infrastructure. iJOIN aims for a joint design and optimization of access and backhaul, operation and management algorithms, and architectural elements, integrating small cells, heterogeneous backhaul and centralized processing. Additionally to the development of technology candidates across PHY, MAC, and the network layer, iJOIN will study the requirements, constraints and implications for existing mobile networks, specifically 3GPP LTE-A.

In January 2013, a new EU project named CROWD (Connectivity management for eneRgy Optimised Wireless Dense networks) was launched under the technical supervision of IMDEA Networks Institute, to design sustainable networking and software solutions for the deployment of very dense, heterogeneous wireless networks. The project targets sustainability targeted in terms of cost effectiveness and energy efficiency. Very high density means 1000x higher than current density (users per square meter). Heterogeneity involves multiple dimensions, from coverage radius to technologies (4G/LTE vs. Wi-Fi), to deployments (planned vs. unplanned distribution of radio base stations and hot spots).

In September 2013, the Cyber-Physical System (CPS) Lab at Rutgers University, NJ, started to work on dynamic provisioning and allocation under the emerging cloud radio-access network (C-RAN). They have shown that the dynamic demand-aware provisioning in the cloud will decrease the energy consumption while increasing the resource utilization. They also have implemented a test bed for feasibility of C-RAN and developed new cloud-based techniques for interference cancellation. Their project is funded by the National Science Foundation.

In November 2013, Chinese telecom equipment vendor Huawei said it will invest $600 million in research for 5G technologies in the next five years. The company's 5G research initiative does not include investment to productize 5G technologies for global telecom operators. Huawei will be testing 5G technology in Malta.

In 2015, Huawei and Ericsson are testing 5G-related technologies in rural areas in northern Netherlands.

In July 2015, the METIS-II and 5GNORMA European projects were launched. The METIS-II project builds on the successful METIS project and will develop the overall 5G radio access network design and to provide the technical enablers needed for an efficient integration and use of the various 5G technologies and components currently developed. METIS-II will also provide the 5G collaboration framework within 5G-PPP for a common evaluation of 5G radio access network concepts and prepare concerted action towards regulatory and standardisation bodies. On the other hand, the key objective of 5G NORMA is to develop a conceptually novel, adaptive and future-proof 5G mobile network architecture. The architecture is enabling unprecedented levels of network customisability, ensuring stringent performance, security, cost and energy requirements to be met; as well as providing an API-driven architectural openness, fuelling economic growth through over-the-top innovation. With 5G NORMA, leading players in the mobile ecosystem aim to underpin Europe's leadership position in 5G.

Additionally, in July 2015, the European research project mmMAGIC was launched. The mmMAGIC project will develop new concepts for mobile radio access technology (RAT) for mmwave band deployment. This is a key component in the 5G multi-RAT ecosystem and will be used as a foundation for global standardization. The project will enable ultrafast mobile broadband services for mobile users, supporting UHD/3D streaming, immersive applications and ultra-responsive cloud services. A new radio interface, including novel network management functions and architecture components will be designed taking as guidance 5G PPP's KPI and exploiting the use of novel adaptive and cooperative beam-forming and tracking techniques to address the specific challenges of mm-wave mobile propagation. The ambition of the project is to pave the way for a European head start in 5G standards and to strengthen European competitiveness. The consortium brings together major infrastructure vendors, major European operators, leading research institutes and universities, measurement equipment vendors and one SME. mmMAGIC is led and coordinated by Samsung. Ericsson acts as technical manager while Intel, Fraunhofer HHI, Nokia, Huawei and Samsung will each lead one of the five technical work packages of the project.

In July 2015, IMDEA Networks launched the Xhaul project, as part of the European H2020 5G Public-Private Partnership (5G PPP). Xhaul will develop an adaptive, sharable, cost-efficient 5G transport network solution integrating the fronthaul and backhaul segments of the network. This transport network will flexibly interconnect distributed 5G radio access and core network functions, hosted on in-network cloud nodes. Xhaul will greatly simplify network operations despite growing technological diversity. It will hence enable system-wide optimisation of Quality of Service (QoS) and energy usage as well as network-aware application development. The Xhaul consortium comprises 21 partners including leading telecom industry vendors, operators, IT companies, small and medium-sized enterprises and academic institutions.

In July 2015, the European 5G research project Flex5Gware was launched. The objective of Flex-

5Gware is to deliver highly reconfigurable hardware (HW) platforms together with HW-agnostic software (SW) platforms targeting both network elements and devices and taking into account increased capacity, reduced energy footprint, as well as scalability and modularity, to enable a smooth transition from 4G mobile wireless systems to 5G. This will enable that 5G HW/SW platforms can meet the requirements imposed by the anticipated exponential growth in mobile data traffic (1000 fold increase) together with the large diversity of applications (from low bit-rate/power for M2M to interactive and high resolution applications).

In July 2015, the SUPERFLUIDITY project, part of the European H2020 Public-Private Partnership (5G PPP) and led by CNIT, an Italian inter-university consortium, was started. The SUPERFLUIDITY consortium comprises telcos and IT players for a total of 18 partners. In physics, superfluidity is a state in which matter behaves like a fluid with zero viscosity. The SUPERFLUIDITY project aims at achieving superfluidity in the Internet: the ability to instantiate services on-the-fly, run them anywhere in the network (core, aggregation, edge) and shift them transparently to different locations. The project tackles crucial shortcomings in today's networks: long provisioning times, with wasteful over-provisioning used to meet variable demand; reliance on rigid and cost-ineffective hardware devices; daunting complexity emerging from three forms of heterogeneity: heterogeneous traffic and sources; heterogeneous services and needs; and heterogeneous access technologies, with multi-vendor network components. SUPERFLUIDITY will provide a converged cloud-based 5G concept that will enable innovative use cases in the mobile edge, empower new business models, and reduce investment and operational costs.

Research

The first widely cited proposal for the use of millimeter wave spectrum for cellular/mobile communications appeared in the IEEE Communications Magazine in June 2011. The first reports of radio channel measurements that validated the ability to use millimeter wave frequencies for urban mobile communication were published in April and May 2013 in the *IEEE Access Journal* and *IEEE Transactions on Antennas and Propagation*, respectively.

The *IEEE Journal on Selected Areas in Communications* published a special issue on 5G in June 2014, including, a comprehensive survey of 5G enabling technologies and solutions. *IEEE Spectrum* has a story about millimeter-wave wireless communications as a viable means to support 5G in its September 2014 issue.

- Radio propagation measurements and channel models for millimeter-wave wireless communication in both outdoor and indoor scenarioes in the 28, 38, 60 and 72–73 GHz bands were published in 2014.

- Massive MIMO: This is a transmission point equipped with a very large number of antennas that simultaneously serve multiple users. With massive MIMO multiple messages for several terminals can be transmitted on the same time-frequency resource, maximizing beamforming gain while minimizing interference.

- Proactive content caching at the edge: While network densification (i.e., adding more cells) is one way to achieve higher capacity and coverage, it becomes evident that the cost of this operation might not be sustainable as the dense deployment of base stations also requires high-speed expensive backhauls. In this regard, assuming that the backhaul is capacity-limited, caching users' contents at the edge of the network (namely at the base stations and user terminals) holds as a solution to offload the backhaul and reduce the access delays to the contents. In any case, caching contents at the edge aim to solve the problem of reducing the end-to-end delay, which is one of the requirements of 5G. Caching can be particularly enabled by leveraging user context information from sources such as mobility and social metrics. The upcoming special issue of IEEE Communications Magazine aims to argue massive content delivery techniques in cache-enabled 5G wireless networks.

- Advanced interference and mobility management, achieved with the cooperation of different transmission points with overlapped coverage, and encompassing the option of a flexible use of resources for uplink and downlink transmission in each cell, the option of direct device-to-device transmission and advanced interference cancellation techniques.

- Efficient support of machine-type devices to enable the Internet of Things with potentially higher numbers of connected devices, as well as novel applications, such as mission-critical control or traffic safety, requiring reduced latency and enhanced reliability.

- Use of millimeter-wave frequencies (e.g. up to 90 GHz) for wireless backhaul and/or access (IEEE rather than ITU generations).

- Pervasive networks providing Internet of things, wireless sensor networks and *ubiquitous computing*: The user can be connected simultaneously to several wireless access technologies and can move seamlessly between them (Media independent handover or vertical handover, IEEE 802.21, also expected to be provided by future 4G releases. These access technologies can be 2.5G, 3G, 4G, or 5G mobile networks, Wi-Fi, WPAN, or any other future access technology. In 5G, the concept may be further developed into multiple concurrent data-transfer paths.

- Multiple-hop networks: A major issue in systems beyond 4G is to make the high bit rates available in a larger portion of the cell, especially to users in an exposed position in between several base stations. In current research, this issue is addressed by cellular repeaters and macro-diversity techniques, also known as group cooperative relay, where users also could be potential cooperative nodes, thanks to the use of direct device-to-device (D2D) communication.

- Wireless network virtualization: Virtualization will be extended to 5G mobile wireless networks. With wireless network virtualization, network infrastructure can be decoupled from the services that it provides, where differentiated services can coexist on the same infrastructure, maximizing its utilization. Consequently, multiple wireless virtual networks operated by different service providers (SPs) can dynamically share the physical substrate wireless networks operated by mobile network operators (MNOs). Since wireless network virtualization enables the sharing of infrastructure and radio spectrum resources, the capital expenses (CapEx) and operation expenses (OpEx) of wireless (radio) access networks (RANs), as well as core networks (CNs), can be reduced significantly. Moreover, mobile virtual network operators (MVNOs) who may provide some specific telecom services (e.g., VoIP, video call, over-the-top services) can help MNOs attract more users, while MNOs can produce more revenue by leasing the isolated virtualized networks to them and evaluating some new services.

- Cognitive radio technology, also known as smart radio. This allows different radio technologies to share the same spectrum efficiently by adaptively finding unused spectrum and adapting the transmission scheme to the requirements of the technologies currently sharing the spectrum. This dynamic radio resource management is achieved in a distributed fashion and relies on software-defined radio.

- Dynamic Adhoc Wireless Networks (DAWN), essentially identical to Mobile ad hoc network (MANET), Wireless mesh network (WMN) or wireless grids, combined with smart antennas, cooperative diversity and flexible modulation.

- Vandermonde-subspace frequency division multiplexing (VFDM): a modulation scheme to allow the co-existence of macro cells and cognitive radio small cells in a two-tiered LTE/4G network.

- IPv6, where a visiting mobile IP care-of address is assigned according to location and connected network.

- Wearable devices with AI capabilities. such as smartwatches and optical head-mounted displays for augmented reality

- One unified global standard.

- *Real wireless world* with no more limitation with access and zone issues.

- *User centric* (or *cell phone developer initiated*) network concept instead of operator-initiated (as in 1G) or system developer initiated (as in 2G, 3G and 4G) standards

- Li-Fi (a portmanteau of *light* and *Wi-Fi*) is a massive MIMO visible light communication network to advance 5G. Li-Fi uses light-emitting diodes to transmit data, rather than radio waves like Wi-Fi.

- *Worldwide wireless web* (WWWW), i.e. comprehensive wireless-based web applications that include full multimedia capability beyond 4G speeds.

History

- In April 2008, NASA partnered with Geoff Brown and Machine-to-Machine Intelligence (M2Mi) Corp to develop 5G communication technology

- In 2008, the South Korean IbjngT R&D program of "5G mobile communication systems based on beam-division multiple access and relays with group cooperation" was formed.

- In August 2012, New York University founded NYU WIRELESS, a multi-disciplinary academic research center that has conducted pioneering work in 5G wireless communications.

- On October 8, 2012, the UK's University of Surrey secured £35M for a new 5G research center, jointly funded by the British government's UK Research Partnership Investment Fund (UKRPIF) and a consortium of key international mobile operators and infrastructure providers, including Huawei, Samsung, Telefonica Europe, Fujitsu Laboratories Europe, Rohde & Schwarz, and Aircom International. It will offer testing facilities to mobile operators keen to develop a mobile standard that uses less energy and less radio spectrum while delivering speeds faster than current 4G with aspirations for the new technology to be ready within a decade.

- On November 1, 2012, the EU project "Mobile and wireless communications Enablers for the Twenty-twenty Information Society" (METIS) starts its activity towards the definition of 5G. METIS achieved an early global consensus on these systems. In this sense, METIS played an important role of building consensus among other external major stakeholders prior to global standardization activities. This was done by initiating and addressing work in relevant global fora (e.g. ITU-R), as well as in national and regional regulatory bodies.

- Also in November 2012, the iJOIN EU project was launched, focusing on "small cell" technology, which is of key importance for taking advantage of limited and strategic resources, such as the radio wave spectrum. According to Günther Oettinger, the European Commissioner for Digital Economy and Society (2014–19), "an innovative utilization of spectrum" is one of the key factors at the heart of 5G success. Oettinger further described it as "the essential resource for the wireless connectivity of which 5G will be the main driver". iJOIN was selected by the European Commission as one of the pioneering 5G research projects to showcase early results on this technology at the Mobile World Congress 2015 (Barcelona, Spain).

- In February 2013, ITU-R Working Party 5D (WP 5D) started two study items: (1) Study on IMT Vision for 2020 and beyond, and; (2) Study on future technology trends for terrestrial IMT systems. Both aiming at having a better understanding of future technical aspects of mobile communications towards the definition of the next generation mobile.

- On May 12, 2013, Samsung Electronics stated that they have developed a "5G" system. The core technology has a maximum speed of tens of Gbit/s (gigabits per second). In testing, the transfer speeds for the "5G" network sent data at 1.056 Gbit/s to a distance of up to 2 kilometres.with the use of an 8*8 MIMO.

- In July 2013, India and Israel have agreed to work jointly on development of fifth generation (5G) telecom technologies.

- On October 1, 2013, NTT (Nippon Telegraph and Telephone), the same company to launch world's first 5G network in Japan, wins Minister of Internal Affairs and Communications Award at CEATEC for 5G R&D efforts

- On November 6, 2013, Huawei announced plans to invest a minimum of $600 million into R&D for next generation 5G networks capable of speeds 100 times faster than modern LTE networks.

- On May 8, 2014, NTT DoCoMo start testing 5G mobile networks with Alcatel Lucent, Ericsson, Fujitsu, NEC, Nokia and Samsung.

- In June 2014, the EU research project CROWD was selected by the European Commission to join the group of "early 5G precursor projects". These projects contribute to the early showcasing of potential technologies for the future ubiquitous, ultra-high bandwidth "5G" infrastructure. CROWD was included in the list of demonstrations at the European Conference on Networks and Communications (EuCNC) organized by the EC in June 2014 (Italy).

- In October 2014, the research project TIGRE5-CM (Integrated technologies for management and operation of 5G networks) is launched with the aim to design an architecture for future generation mobile networks, based on the SDN (Software Defined Networking) paradigm. IMDEA Networks Institute is the project coordinator.

- In November 2014, it was announced that Megafon and Huawei will be developing a 5G network in Russia. A trial network will be available by the end of 2017, just in time for the 2018 World Cup.

- On November 19, 2014, Huawei and SingTel announced the signing of a MoU to launch a joint 5G innovation program.

- On June 22, 2015, Greek government announced to Euro-group council talks that potential licensing 5G and 4G technology would offer 350 million euros earnings, as a result they were criticized for misleading European leaders in producing potential earnings from a technology that is supposed to roll-out after 2020.

- On July 1, 2015, METIS-II project was launched. This project aims at designing the 5G radio access network, building the basis for the multi-service allocation on an holistic cross-layer and cross-air interface framework.

- On September 8, 2015, Verizon announced a roadmap to begin testing 5G in field trials in the United States in 2016.

- On October 1, 2015, the French Operator Orange announced to be about to deploy 5G technologies to begin the first trial in January 2016 in Belfort, a City of Eastern France.

- On January 22, 2016, the Swedish mobile network equipment maker Ericsson said it had partnered with TeliaSonera to develop 5G services based on TeliaSonera's network and Ericsson's 5G technology. The partnership aims to provide 5G services to TeliaSonera customers in Stockholm, Sweden and Tallinn, Estonia in 2018. Sweden has long been a pioneer ICT nation and notably Ericsson and TeliaSonera launched the world's first commercial 4G network in Sweden in 2009.

- On February 22, 2016, NTT DoCoMo and Ericsson succeed in World's first trial to achieve a cumulative 20Gbit/s with two simultaneously connected mobile devices in 5G outdoor trial.

- Also on February 22, 2016, Samsung and Verizon joined to begin trial for 5G.

- On January 29, 2016, Google revealed that they are developing a 5G network called Sky-Bender. They planned to distribute this connection through sun-powered drones.

- In mid-March 2016, the UK government confirmed plans to make the UK a world leader in 5G. Plans for 5G are little more than a footnote in the country's 2016 budget, but it seems the UK government wants it to be a big focus going forward.

- On June 2, 2016, the first comprehensive book on 5G was launched. The book "5G Mobile and Wireless Communications Technology" by Cambridge University Press is edited by Afif Osseiran (Ericsson), Jose F. Monserrat (UPV) and Patrick Marsch (Nokia Bell Labs) and covers everything from the most likely use cases, spectrum aspects, and a wide range of technology options to potential 5G system architectures.

- On October 17, 2016, Qualcomm announced the first 5G modem, the Snapdragon X50, as the first commercial 5G mobile chipset.

General Packet Radio Service

General Packet Radio Service (GPRS) is a packet oriented mobile data service on the 2G and 3G cellular communication system's global system for mobile communications (GSM). GPRS was originally standardized by European Telecommunications Standards Institute (ETSI) in response to the earlier CDPD and i-mode packet-switched cellular technologies. It is now maintained by the 3rd Generation Partnership Project (3GPP).

GPRS usage is typically charged based on volume of data transferred, contrasting with circuit switched data, which is usually billed per minute of connection time. Sometimes billing time is broken down to every third of a minute. Usage above the bundle cap is charged per megabyte, speed limited, or disallowed.

GPRS is a best-effort service, implying variable throughput and latency that depend on the number of other users sharing the service concurrently, as opposed to circuit switching, where a certain quality of service (QoS) is guaranteed during the connection. In 2G systems, GPRS provides data rates of 56–114 kbit/second. 2G cellular technology combined with GPRS is sometimes described as *2.5G*, that is, a technology between the second (2G) and third (3G) generations of mobile telephony. It provides moderate-speed data transfer, by using unused time division multiple access (TDMA) channels in, for example, the GSM system. GPRS is integrated into GSM Release 97 and newer releases.

Technical Overview

The GPRS core network allows 2G, 3G and WCDMA mobile networks to transmit IP packets to external networks such as the Internet. The GPRS system is an integrated part of the GSM network switching subsystem.

Services Offered

GPRS extends the GSM Packet circuit switched data capabilities and makes the following services possible:

- SMS messaging and broadcasting

- "Always on" internet access

- Multimedia messaging service (MMS)

- Push-to-talk over cellular (PoC)

- Instant messaging and presence—wireless village

- Internet applications for smart devices through wireless application protocol (WAP)

- Point-to-point (P2P) service: inter-networking with the Internet (IP)

- Point-to-multipoint (P2M) service: point-to-multipoint multicast and point-to-multipoint group calls

If SMS over GPRS is used, an SMS transmission speed of about 30 SMS messages per minute may be achieved. This is much faster than using the ordinary SMS over GSM, whose SMS transmission speed is about 6 to 10 SMS messages per minute.

Protocols Supported

GPRS supports the following protocols:

- Internet Protocol (IP). In practice, built-in mobile browsers use IPv4 since IPv6 was not yet popular.

- Point-to-Point Protocol (PPP). In this mode PPP is often not supported by the mobile phone operator but if the mobile is used as a modem to the connected computer, PPP is used to tunnel IP to the phone. This allows an IP address to be assigned dynamically (IPCP not DHCP) to the mobile equipment.

- X.25 connections. This is typically used for applications like wireless payment terminals, although it has been removed from the standard. X.25 can still be supported over PPP, or even over IP, but doing this requires either a network-based router to perform encapsulation or intelligence built into the end-device/terminal; e.g., user equipment (UE).

When TCP/IP is used, each phone can have one or more IP addresses allocated. GPRS will store and forward the IP packets to the phone even during handover. The TCP handles any packet loss (e.g. due to a radio noise induced pause).

Hardware

Devices supporting GPRS are divided into three classes:

Class A

Can be connected to GPRS service and GSM service (voice, SMS), using both at the same time. Such devices are known to be available today.

Class B

Can be connected to GPRS service and GSM service (voice, SMS), but using only one or the other at a given time. During GSM service (voice call or SMS), GPRS service is suspended, and then resumed automatically after the GSM service (voice call or SMS) has concluded. Most GPRS mobile devices are Class B.

Class C

Are connected to either GPRS service or GSM service (voice, SMS). Must be switched manually between one or the other service.

A true Class A device may be required to transmit on two different frequencies at the same time, and thus will need two radios. To get around this expensive requirement, a GPRS mobile may implement the dual transfer mode (DTM) feature. A DTM-capable mobile may use simultaneous voice and packet data, with the network coordinating to ensure that it is not required to transmit on two different frequencies at the same time. Such mobiles are considered pseudo-Class A, sometimes referred to as "simple class A". Some networks support DTM since 2007.

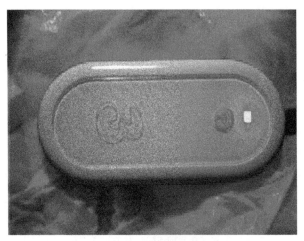

Huawei E220 3G/GPRS Modem

USB 3G/GPRS modems use a terminal-like interface over USB 1.1, 2.0 and later, data formats V.42bis, and RFC 1144 and some models have connector for external antenna. Modems can be added as cards (for laptops) or external USB devices which are similar in shape and size to a computer mouse, or nowadays more like a pendrive.

Addressing

A GPRS connection is established by reference to its access point name (APN). The APN defines the services such as wireless application protocol (WAP) access, short message service (SMS), multimedia messaging service (MMS), and for Internet communication services such as email and World Wide Web access.

In order to set up a GPRS connection for a wireless modem, a user must specify an APN, optionally a user name and password, and very rarely an IP address, provided by the network operator.

GPRS Modems and Modules

GSM module or GPRS modules are similar to modems, but there's one difference: the modem is an external piece of equipment, whereas the GSM module or GPRS module can be integrated within an electrical or electronic equipment. It is an embedded piece of hardware. A GSM mobile, on the other hand, is a complete embedded system in itself. It comes with embedded processors dedicated to provide a functional interface between the user and the mobile network.

Coding Schemes and Speeds

The upload and download speeds that can be achieved in GPRS depend on a number of factors such as:

- the number of BTS TDMA time slots assigned by the operator

- the channel encoding is used.

- the maximum capability of the mobile device expressed as a GPRS multislot class

Multiple Access Schemes

The multiple access methods used in GSM with GPRS are based on frequency division duplex (FDD) and TDMA. During a session, a user is assigned to one pair of up-link and down-link frequency channels. This is combined with time domain statistical multiplexing which makes it possible for several users to share the same frequency channel. The packets have constant length, corresponding to a GSM time slot. The down-link uses first-come first-served packet scheduling, while the up-link uses a scheme very similar to reservation ALOHA (R-ALOHA). This means that slotted ALOHA (S-ALOHA) is used for reservation inquiries during a contention phase, and then the actual data is transferred using dynamic TDMA with first-come first-served.

Channel Encoding

The channel encoding process in GPRS consists of two steps: first, a cyclic code is used to add parity bits, which are also referred to as the Block Check Sequence, followed by coding with a possibly punctured convolutional code. The Coding Schemes CS-1 to CS-4 specify the number of parity bits generated by the cyclic code and the puncturing rate of the convolutional code. In Coding Schemes CS-1 through CS-3, the convolutional code is of rate 1/2, i.e. each input bit is converted into two coded bits. In Coding Schemes CS-2 and CS-3, the output of the convolutional code is punctured to achieve the desired code rate. In Coding Scheme CS-4, no convolutional coding is applied. The following table summarises the options.

1. This is rate at which the RLC/MAC layer protocol data unit (PDU) (called a radio block) is transmitted. As shown in TS 44.060 section 10.0a.1, a radio block consists of MAC header, RLC header, RLC data unit and spare bits. The RLC data unit represents the payload, the rest is overhead. The radio block is coded by the convolutional code specified for a particular Coding Scheme, which yields the same PHY layer data rate for all Coding Schemes.

2. Cited in various sources, e.g. in TS 45.001 table 1. is the bitrate including the RLC/MAC headers, but excluding the uplink state flag (USF), which is part of the MAC header, yielding a bitrate that is 0.15 kbit/s lower.

3. The net bitrate here is the rate at which the RLC/MAC layer payload (the RLC data unit) is transmitted. As such, this bit rate excludes the header overhead from the RLC/MAC layers.

The least robust, but fastest, coding scheme (CS-4) is available near a base transceiver station (BTS), while the most robust coding scheme (CS-1) is used when the mobile station (MS) is further away from a BTS.

Using the CS-4 it is possible to achieve a user speed of 20.0 kbit/s per time slot. However, using this scheme the cell coverage is 25% of normal. CS-1 can achieve a user speed of only 8.0 kbit/s per time slot, but has 98% of normal coverage. Newer network equipment can adapt the transfer speed automatically depending on the mobile location.

In addition to GPRS, there are two other GSM technologies which deliver data services: circuit-switched data (CSD) and high-speed circuit-switched data (HSCSD). In contrast to the shared nature of GPRS, these instead establish a dedicated circuit (usually billed per minute). Some applications such as video calling may prefer HSCSD, especially when there is a continuous flow of data between the endpoints.

The following table summarises some possible configurations of GPRS and circuit switched data services.

Multislot Class

The multislot class determines the speed of data transfer available in the Uplink and Downlink directions. It is a value between 1 and 45 which the network uses to allocate radio channels in the uplink and downlink direction. Multislot class with values greater than 31 are referred to as high multislot classes.

A multislot allocation is represented as, for example, 5+2. The first number is the number of downlink timeslots and the second is the number of uplink timeslots allocated for use by the mobile station. A commonly used value is class 10 for many GPRS/EGPRS mobiles which uses a maximum of 4 timeslots in downlink direction and 2 timeslots in uplink direction. However simultaneously a maximum number of 5 simultaneous timeslots can be used in both uplink and downlink. The network will automatically configure for either 3+2 or 4+1 operation depending on the nature of data transfer.

Some high end mobiles, usually also supporting UMTS, also support GPRS/EDGE multislot class 32. According to 3GPP TS 45.002 (Release 12), Table B.1, mobile stations of this class support 5 timeslots in downlink and 3 timeslots in uplink with a maximum number of 6 simultaneously used timeslots. If data traffic is concentrated in downlink direction the network will configure the connection for 5+1 operation. When more data is transferred in the uplink the network can at any time change the constellation to 4+2 or 3+3. Under the best reception conditions, i.e. when the best EDGE modulation and coding scheme can be used, 5 timeslots can carry a bandwidth of 5*59.2 kbit/s = 296 kbit/s. In uplink direction, 3 timeslots can carry a bandwidth of 3*59.2 kbit/s = 177.6 kbit/s.

Multislot Classes for GPRS/EGPRS

Attributes of a Multislot Class

Each multislot class identifies the following:

- the maximum number of Timeslots that can be allocated on uplink
- the maximum number of Timeslots that can be allocated on downlink
- the total number of timeslots which can be allocated by the network to the mobile
- the time needed for the MS to perform adjacent cell signal level measurement and get ready to transmit
- the time needed for the MS to get ready to transmit
- the time needed for the MS to perform adjacent cell signal level measurement and get ready to receive
- the time needed for the MS to get ready to receive.

The different multislot class specification is detailed in the Annex B of the 3GPP Technical Specification 45.002 (Multiplexing and multiple access on the radio path)

Usability

The maximum speed of a GPRS connection offered in 2003 was similar to a modem connection in an analog wire telephone network, about 32–40 kbit/s, depending on the phone used. Latency is very high; round-trip time (RTT) is typically about 600–700 ms and often reaches 1s. GPRS is typically prioritized lower than speech, and thus the quality of connection varies greatly.

Devices with latency/RTT improvements (via, for example, the extended UL TBF mode feature) are generally available. Also, network upgrades of features are available with certain operators. With these enhancements the active round-trip time can be reduced, resulting in significant increase in application-level throughput speeds.

History of GPRS

GPRS opened in 2000 as a packet-switched data service embedded to the channel-switched cellular radio network GSM. GPRS extends the reach of the fixed Internet by connecting mobile terminals worldwide.

The CELLPAC protocol developed 1991-1993 was the trigger point for starting in 1993 specification of standard GPRS by ETSI SMG. Especially, the CELLPAC Voice & Data functions introduced in a 1993 ETSI Workshop contribution anticipate what was later known to be the roots of GPRS. This workshop contribution is referenced in 22 GPRS related US-Patents. Successor systems to GSM/GPRS like W-CDMA (UMTS) and LTE rely on key GPRS functions for mobile Internet access as introduced by CELLPAC.

According to a study on history of GPRS development Bernhard Walke and his student Peter Decker are the inventors of GPRS – the first system providing worldwide mobile Internet access.

Enhanced Data Rates for GSM Evolution

Enhanced Data rates for GSM Evolution (EDGE) (also known as Enhanced GPRS (EGPRS), or IMT Single Carrier (IMT-SC), or Enhanced Data rates for Global Evolution) is a digital mobile phone technology that allows improved data transmission rates as a backward-compatible extension of GSM. EDGE is considered a pre-3G radio technology and is part of ITU's 3G definition. EDGE was deployed on GSM networks beginning in 2003 – initially by Cingular (now AT&T) in the United States.

EDGE sign shown in notification bar on an Android-based smartphone.

EDGE is standardized also by 3GPP as part of the GSM family. A variant, so called Compact-EDGE, was developed for use in a portion of Digital AMPS network spectrum.

Through the introduction of sophisticated methods of coding and transmitting data, EDGE delivers higher bit-rates per radio channel, resulting in a threefold increase in capacity and performance compared with an ordinary GSM/GPRS connection.

EDGE can be used for any packet switched application, such as an Internet connection.

Evolved EDGE continues in Release 7 of the 3GPP standard providing reduced latency and more than doubled performance e.g. to complement High-Speed Packet Access (HSPA). Peak bit-rates of up to 1 Mbit/s and typical bit-rates of 400 kbit/s can be expected.

Technology

EDGE/EGPRS is implemented as a bolt-on enhancement for 2.5G GSM/GPRS networks, making it easier for existing GSM carriers to upgrade to it. EDGE is a superset to GPRS and can function on any network with GPRS deployed on it, provided the carrier implements the necessary upgrade. EDGE requires no hardware or software changes to be made in GSM core networks. EDGE-compatible transceiver units must be installed and the base station subsystem needs to be upgraded to support EDGE. If the operator already has this in place, which is often the case today, the network can be upgraded to EDGE by activating an optional software feature. Today EDGE is supported by all major chip vendors for both GSM and WCDMA/HSPA.

Transmission Techniques

In addition to Gaussian minimum-shift keying (GMSK), EDGE uses higher-order PSK/8 phase shift keying (8PSK) for the upper five of its nine modulation and coding schemes. EDGE produces a 3-bit word for every change in carrier phase. This effectively triples the gross data rate offered by

GSM. EDGE, like GPRS, uses a rate adaptation algorithm that adapts the modulation and coding scheme (MCS) according to the quality of the radio channel, and thus the bit rate and robustness of data transmission. It introduces a new technology not found in GPRS, Incremental Redundancy, which, instead of retransmitting disturbed packets, sends more redundancy information to be combined in the receiver. This increases the probability of correct decoding.

EDGE can carry a bandwidth up to 500 kbit/s (with end-to-end latency of less than 150 ms) for 4 timeslots (theoretical maximum is 473.6 kbit/s for 8 timeslots) in packet mode. This means it can handle four times as much traffic as standard GPRS. EDGE meets the International Telecommunications Union's requirement for a 3G network, and has been accepted by the ITU as part of the IMT-2000 family of 3G standards. It also enhances the circuit data mode called HSCSD, increasing the data rate of this service.

EDGE Modulation and Coding Scheme (MCS)

The channel encoding process in GPRS as well as EGPRS/EDGE consists of two steps: first, a cyclic code is used to add parity bits, which are also referred to as the Block Check Sequence, followed by coding with a possibly punctured convolutional code. In GPRS, the Coding Schemes CS-1 to CS-4 specify the number of parity bits generated by the cyclic code and the puncturing rate of the convolutional code. In GPRS Coding Schemes CS-1 through CS-3, the convolutional code is of rate 1/2, i.e. each input bit is converted into two coded bits. In Coding Schemes CS-2 and CS-3, the output of the convolutional code is punctured to achieve the desired code rate. In GPRS Coding Scheme CS-4, no convolutional coding is applied.

In EGPRS/EDGE, the Modulation and Coding Schemes MCS-1 to MCS-9 take the place of the Coding Schemes of GPRS, and additionally specify which modulation scheme is used, GMSK or 8PSK. MCS-1 through MCS-4 use GMSK and have performance similar (but not equal) to GPRS, while MCS-5 through MCS-9 use 8PSK. In all EGPRS Modulation and Coding Schemes, a convolutional code of rate 1/3 is used, and puncturing is used to achieve the desired code rate. In contrast to GRPS, the Radio Link Control (RLC) and Media Access Control (MAC) headers and the payload data are coded separately in EGPRS. The headers are coded more robustly than the data.

1. This is rate at which the RLC/MAC layer protocol data unit (PDU) (called a radio block) is transmitted. As shown in TS 44.060 section 10.0a.1, a radio block consists of MAC header, RLC header, RLC data unit and spare bits. The RLC data unit represents the payload, the rest is overhead. The radio block is coded by the convolutional code specified for a particular Coding Scheme, which yields the same PHY layer data rate for all Coding Schemes.

2. Cited in various sources, e.g. in TS 45.001 table 1. is the bitrate including the RLC/MAC headers, but excluding the uplink state flag (USF), which is part of the MAC header, yielding a bitrate that is 0.15 kbit/s lower.

3. The net bitrate here is the rate at which the RLC/MAC layer payload (the RLC data unit) is transmitted. As such, this bit rate excludes the header overhead from the RLC/MAC layers.

Evolved EDGE

Evolved EDGE, also called EDGE Evolution, is a bolt-on extension to the GSM mobile telephony standard, which improves on EDGE in a number of ways. Latencies are reduced by lowering the Transmission Time Interval by half (from 20 ms to 10 ms). Bit rates are increased up to 1 Mbit/s peak bandwidth and latencies down to 80 ms using dual carrier, higher symbol rate and higher-order modulation (32QAM and 16QAM instead of 8PSK), and turbo codes to improve error correction. This results in real world downlink speeds of up to 600kbit/s. Further the signal quality is improved using dual antennas improving average bit-rates and spectrum efficiency.

The main intention of increasing the existing EDGE throughput is that many operators would like to upgrade their existing infrastructure rather than invest on new network infrastructure. Mobile operators have invested billions in GSM networks, many of which are already capable of supporting EDGE data speeds up to 236.8 kbit/s. With a software upgrade and a new device compliant with Evolved EDGE (like an Evolved EDGE smart phone) for the user, these data rates can be boosted to speeds approaching 1 Mbit/s (i.e. 98.6 kbit/s per timeslot for 32QAM). Many service providers may not invest on a completely new technology like 3G networks.

Considerable research and development happened throughout the world for this new technology. A successful trial by Nokia Siemens and "one of China's leading operators" has been archieved in a live environment. With the introduction for more advanced wireless technologies like UMTS and LTE, which also focus on a network coverage layer on low frequencies and the upcoming phase-out and shutdown of 2G mobile networks, it is very unlikely that Evolved EDGE will ever see any deployment on live networks. Up to now (as of 2016) there are no commercial networks which support the Evolved EDGE standard (3GPP Rel-7).

Technology

Reduced Latency

With Evolved EDGE come three major features designed to reduce latency over the air interface.

In EDGE, a single RLC data block (ranging from 23 to 148 bytes of data) is transmitted over four frames, using a single time slot. On average, this requires 20 ms for one way transmission. Under the RTTI scheme, one data block is transmitted over two frames in two timeslots, reducing the latency of the air interface to 10 ms.

In addition, Reduced Latency also implies support of Piggy-backed ACK/NACK (PAN), in which a bitmap of blocks not received is included in normal data blocks. Using the PAN field, the receiver may report missing data blocks immediately, rather than waiting to send a dedicated PAN message.

A final enhancement is RLC-non persistent mode. With EDGE, the RLC interface could operate in either acknowledged mode, or unacknowledged mode. In unacknowledged mode, there is no retransmission of missing data blocks, so a single corrupt block would cause an entire upper-layer IP packet to be lost. With non-persistent mode, an RLC data block may be retransmitted if it is less than a certain age. Once this time expires, it is considered lost, and subsequent data blocks may then be forwarded to upper layers.

Downlink Dual Carrier

With Downlink Dual Carrier, the handheld is able to receive on two different frequency channels at the same time, doubling the downlink throughput. In addition, if second receiver is present then the handheld is able to receive on an additional timeslot in single-carrier mode, because it may overlap the tuning of one receiver with other tasks.

Higher Modulation Schemes

Both uplink and downlink throughput is improved by using 16 or 32 QAM (Quadrature Amplitude Modulation), along with turbo codes and higher symbol rates.

Networks

The Global mobile Suppliers Association (GSA) states that, as of May 2013, there were 604 GSM/EDGE networks in 213 countries, from a total of 606 mobile network operator commitments in 213 countries.

References

- Afif Osseiran; Jose F. Monserrat; Patrick Marsch (June 2016). 5G Mobile and Wireless Communications Technology. Cambridge University Press. ISBN 9781107130098. Retrieved 20 July 2016.

- Liyanage, Madhusanka (2015). Software Defined Mobile Networks (SDMN): Beyond LTE Network Architecture. UK: Wiley Publishers. pp. 1–438. ISBN 978-1-118-90028-4.

- Federal Reserve Bank of Minneapolis Community Development Project. "Consumer Price Index (estimate) 1800–". Federal Reserve Bank of Minneapolis. Retrieved October 21, 2016.

- Tom, Wheeler. "Leading Towards Next Generation "5G" Mobile Services". Federal Communications Commission. Federal Communications Commission. Retrieved 25 July 2016.

- "The METIS-II Project – Mobile and Wireless Communication Enablers for the 2020 Information Society". METIS. 1 July 2015. Retrieved 20 July 2016.

- "The world's first academic research center combining Wireless, Computing, and Medical Applications". Nyu Wireless. 2014-06-20. Retrieved 2016-01-14.

- "NYU Wireless' Rappaport envisions a 5G, millimeter-wave future - FierceWirelessTech". Fiercewireless.com. 2014-01-13. Retrieved 2016-01-14.

- Alleven, Monica (2015-01-14). "NYU Wireless says U.S. falling behind in 5G, presses FCC to act now on mmWave spectrum". Fiercewireless.com. Retrieved 2016-01-14.

- Harris, Mark (29 January 2016). "Project Skybender: Google's secretive 5G internet drone tests revealed". The Guardian. Retrieved 31 January 2016.

- Allied Newspapers Ltd. "Update 2: Agreement for 5G technology testing signed; 'You finally found me' - Sai Mizzi Liang". timesofmalta.com. Retrieved 2016-01-14.

- "Communications, Caching, and Computing for Content-Centric Mobile Networks | IEEE Communications Society". Comsoc.org. 2016-01-01. Retrieved 2016-01-14.

- "Wireless News Briefs — February 15, 2008". WirelessWeek. February 15, 2008. Archived from the original on August 19, 2015. Retrieved September 14, 2008.

- E. Bastug; M. Bennis; M. Kountouris; M. Debbah (August 2014). "Cache-enabled small cell networks: modeling and tradeoffs". EURASIP Journal on Wireless Communications and Networking. Springer. 2015 (1): 41. arXiv:1405.3477 [cs.IT]. Bibcode:2014arXiv1405.3477B. Retrieved 8 November 2015.

- T. S. Rappaport, et. al., "Wideband Millimeter-Wave Propagation Measurements and Channel Models for Future Wireless Communication System Design," IEEE Trans. Comm., Vol. 63, No. 9, Sept. 2015, pp. 3029-3056.

- G. MacCartney, et. al., "Indoor Office Wideband Millimeter-Wave Propagation Measurements and Channel Models at 28 and 73 GHz for Ultra-Dense 5G Wireless Networks," IEEE Access, Vol. 3, 2388-2424, October 2015.

- C. Liang; F. Richard Yu (2014). "Wireless Network Virtualization: A Survey, Some Research Issues and Challenges". IEEE Communications Surveys & Tutorials. Retrieved 3 November 2014.

- "SingTel and Huawei Ink MOU to Launch 5G Joint Innoviation Program". Huawei. 19 November 2014. Retrieved 21 November 2014.

Mobile Computing Management

Mobile computing management helps in the administration of mobile devices. It is a way of ensuring the productivity of employees and making sure they do not violate corporate policies. The section serves as a source to understand the major aspects of mobile computing management.

Mobile Device Management

Mobile device management (MDM) is an industry term for the administration of mobile devices, such as smartphones, tablet computers, laptops and desktop computers. MDM is usually implemented with the use of a third party product that has management features for particular vendors of mobile devices.

Overview

MDM is a way to ensure employees stay productive and do not breach corporate policies. Many organizations control activities of their employees using MDM products/services. MDM primarily deals with corporate data segregation, securing emails, securing corporate documents on device, enforcing corporate policies, integrating and managing mobile devices including laptops and handhelds of various categories. MDM implementations may be either on-premises or cloud-based.

MDM functionality can include over-the-air distribution of applications, data and configuration settings for all types of mobile devices, including mobile phones, smartphones, tablet computers, ruggedized mobile computers, mobile printers, mobile POS devices, etc. Most recently laptops and desktops have been added to the list of systems supported as Mobile Device Management becomes more about basic device management and less about the mobile platform itself. MDM tools are leveraged for both company-owned and employee-owned (BYOD) devices across the enterprise or mobile devices owned by consumers. Consumer Demand for BYOD is now requiring a greater effort for MDM and increased security for both the devices and the enterprise they connect to, especially since employers and employees have different expectations concerning the types of restrictions that should be applied to mobile devices.

By controlling and protecting the data and configuration settings of all mobile devices in a network, MDM can reduce support costs and business risks. The intent of MDM is to optimize the functionality and security of a mobile communications network while minimizing cost and downtime.

With mobile devices becoming ubiquitous and applications flooding the market, mobile monitoring is growing in importance. Numerous vendors help mobile device manufacturers, content portals and developers test and monitor the delivery of their mobile content, applications and services. This testing of content is done in real time by simulating the actions of thousands of customers and detecting and correcting bugs in the applications.

Implementation

Typically solutions include a server component, which sends out the management commands to the mobile devices, and a client component, which runs on the managed device and receives and implements the management commands. In some cases, a single vendor provides both the client and the server, while in other cases the client and server come from different sources.

The management of mobile devices has evolved over time. At first it was necessary to either connect to the handset or install a SIM in order to make changes and updates; scalability was a problem.

One of the next steps was to allow a client-initiated update, similar to when a user requests a Windows Update.

Central remote management, using commands sent over the air, is the next step. An administrator at the mobile operator, an enterprise IT data center or a handset OEM can use an administrative console to update or configure any one handset, group or groups of handsets. This provides scalability benefits particularly useful when the fleet of managed devices is large in size.

Device management software platforms ensure that end-users benefit from plug and play data services for whatever device they are using. Such a platform can automatically detect devices in the network, sending them settings for immediate and continued usability. The process is fully automated, keeps a history of used devices and sends settings only to subscriber devices which were not previously set, sometimes at speeds reaching 50 over-the-air settings update files per second. Device management systems can deliver this function by filtering IMEI/IMSI pairs.

Device Management Specifications

- The Open Mobile Alliance (OMA) specified a platform-independent device management protocol called OMA Device Management. The specification meets the common definitions of an open standard, meaning the specification is freely available and implementable. It is supported by several mobile devices, such as PDAs and mobile phones.

- Smart message is text SMS-based provisioning protocol (ringtones, calendar entries but service settings also supported like: ftp, telnet, SMSC number, email settings, etc...)

- OMA Client Provisioning is a binary SMS-based service settings provisioning protocol.

- Nokia-Ericsson OTA is binary SMS-based service settings provisioning protocol, designed mainly for older Nokia and Ericsson mobile phones.

Over-the-air programming (OTA) capabilities are considered a main component of mobile network operator and enterprise-grade mobile device management software. These include the ability to remotely configure a single mobile device, an entire fleet of mobile devices or any IT-defined set of mobile devices; send software and OS updates; remotely lock and wipe a device, which protects the data stored on the device when it is lost or stolen; and remote troubleshooting. OTA commands are sent as a binary SMS message. Binary SMS is a message including binary data.

Mobile device management software enables corporate IT departments to manage the many mobile devices used across the enterprise; consequently, over-the-air capabilities are in high demand. Enterprises using OTA SMS as part of their MDM infrastructure demand high quality in the sending of OTA messages, which imposes on SMS gateway providers a requirement to offer a high level of quality and reliability.

Use in the Enterprise

As the bring your own device (BYOD) approach becomes increasingly popular across mobile service providers, MDM lets corporations provide employees with access to the internal networks using a device of their choice, whilst these devices are managed remotely with minimal disruption to employees' schedules.

MDM for Mobile Security

All MDM products are built with an idea of Containerization. The MDM Container is secured using latest crypto techniques (AES-256 or more preferred). All the corporate data like email, documents, enterprise application are encrypted and processed inside the container. This ensures that corporate data is separated from user's personal data on the device. Additionally, encryption for entire device and/or SD Card can also be enforced depending on MDM product capability.

Secure email: MDM products allow organization to integrate their existing email setup to be easily integrated with MDM environment. Almost all MDM products support easy integration with Exchange Server (2003/2007/2010), Office365, Lotus Notes, BlackBerry Enterprise Server (BES) and others. This provided flexibility of configuring Email-over-air. Secure Docs: It is frequently seen that, employees copy attachments downloaded from corporate email to their personal devices and then misuse it. MDM can easily restrict/disable clipboard usage in/out of Secure Container; forwarding attachments to external domains can be restricted, downloading/saving attachments on SD Card. This ensures corporate data is not left insecure.

Secure browser: Using secure browser can avoid many potential security risks. Every MDM solution comes with built-in custom browser. Administrator can disable native browsers to force user to use Secure Browser, which is also inside the MDM container. URL filtering can be enforced to add additional productivity measure.

Secure app catalog: Organization can distribute, manage, and upgrade applications on employee's device using App Catalogue. It allows applications to be pushed on user device directly from the App Store or push an enterprise developed private application through the App Catalogue. This provides an option for the organization to deploy devices in Kiosk Mode or Lock-Down Mode.

Additional MDM Features

There are plenty of other features depending on which MDM product is chosen:

- Policy Enforcing: There are multiple types of policies which can be enforced on MDM users.
    1. Personal Policy: According to corporate environment, highly customizable

2. Device Platform specific: policies for advanced management of Android, iOS, Windows and Blackberry devices.

3. Compliance Policies/Rules

- VPN configuration

- Application Catalogue

- Pre-defined Wi-Fi and Hotspot settings

- Jailbreak/Root detection

- Remote Wipe of corporate data

- Remote Wipe of entire device

- Device remote locking

- Remote messaging/buzz

- Disabling native apps on device

SaaS Versus on-premises Solutions

Present day MDM solutions offer both software as a service (SaaS) and on-premises models. In the rapidly evolving industry such as mobile, SaaS (cloud-based) systems are quicker to set up, offer easier updates with lower capital costs compared to on-premises solutions which require costly hardware, need regular software maintenance, and incur higher capital costs.

For security in cloud computing, the US Government has compliance audits such as Federal Information Security Management Act of 2002 (FISMA) which cloud providers can go through to meet security standards.

The primary policy approach taken by Federal agencies to build relationships with cloud service providers is Federal Risk and Authorization Management Program (FedRAMP) accreditation and certification, designed in part to protect FISMA Low and Moderate systems.

More on MDM, MAM and MEM

Mobile device management (MDM) is like adding an extra layer of security and ensuring a way to monitor device related activities. MDM provides device platform specific features like device encryption, platform specific policies, SD Card encryption. Geo-location tracking, connectivity profiles (VPN, Wi-Fi, Bluetooth) and plenty other features are part of MDM Suite.

Mobile application management (MAM) is done by application wrapping i.e. injection arbitrary encryption code in the mobile application source. This is necessary for commercial applications or applications being developed in-house for Enterprise use. Additionally, white-listing/black-listing of application can be done. Features like Application Catalogue allow admin to push applications remotely to the devices for instant install, push remote updates and also remote removal of apps.

Mobile email management (MEM) ensures your corporate emails are containerized using advanced proprietary/free encryption algorithms. MEM ensures all emails remain inside the secure container, so that attackers get encrypted data even if they try to compromise the device data using USB cable on a system. Heavy restrictions on clipboard, attachments and trusted domains can be enforced. Nothing can move in-out of the secure container as clipboard is disabled. Even the attachments are downloaded and saved inside the secure container. To view the attachments there is secure document reader as well as secure document editor available in MDM solutions. Adding trusted domains will ensure that data from corporate email is not leaked to malicious/suspicious domains.

Mobile Application Management

Mobile application management (MAM) describes software and services responsible for provisioning and controlling access to internally developed and commercially available mobile apps used in business settings on both company-provided and "bring your own" smartphones and tablet computers.

Mobile application management provides granular controls at the application level that enable administrators to manage and secure app data. MAM differs from mobile device management (MDM), which focuses on controlling the entire device and requires that users enroll their device and install a service agent.

While some enterprise mobility management (EMM) suites include a MAM function, their capabilities may be limited in comparison to stand-alone MAM solutions because EMM suites require a device management profile in order to enable app management capabilities.

History

Enterprise mobile application management has been driven by the widespread adoption and use of mobile applications in business settings. In 2010 International Data Corporation reported that smartphone use in the workplace will double between 2009 and 2014.

The "bring your own device" (BYOD) phenomenon is a factor behind mobile application management, with personal PC, smartphone and tablet use in business settings (vs. business-owned devices) rising from 31 percent in 2010 to 41 percent in 2011. When an employee brings a personal device into an enterprise setting, mobile application management enables the corporate IT staff to download required applications, control access to business data, and remove locally cached business data from the device if it is lost, or when its owner no longer works with the company.

Use of mobile devices in the workplace is also being driven from above. According to Forrester Research, businesses now see mobile as an opportunity to drive innovation across a wide range of business processes. Forrester issued a forecast in August 2011 predicting that the "mobile management services market" would reach $6.6 billion by 2015 – a 69 percent increase over a previous forecast issued six months earlier.

Citing the plethora of mobile devices in the enterprise – and a growing demand for mobile apps from employees, line-of-business decision-makers, and customers – the report states that organizations are broadening their "mobility strategy" beyond mobile device management to "managing a growing number of mobile applications."

App Wrapping

App wrapping was initially a favored method of applying policy to applications as part of mobile application management solutions.

App wrapping sets up a dynamic library and adds to an existing binary that controls certain aspects of an application. For instance, at startup, you can change an app so that it requires authentication using a local passkey. Or you could intercept a communication so that it would be forced to use your company's virtual private network (VPN) or prevent that communication from reaching a particular application that holds sensitive data.

Increasingly, the likes of Apple and Samsung are overcoming the issue of app wrapping. Aside from the fact that app wrapping is a legal grey zone, and may not meet its actual aims, it is not possible to adapt the entire operating system to deal with numerous wrapped apps. In general, wrapped apps available in the app stores have also not proven to be successful due to their inability to perform without MDM.

System Features

An end-to-end MAM solution provides the ability to: control the provisioning, updating and removal of mobile applications via an enterprise app store, monitor application performance and usage, and remotely wipe data from managed applications. Core features of mobile application management systems include:

• App delivery (Enterprise App Store)	• App version management
• App updating	• App configuration management
• App performance monitoring	• Push services
• User authentication	• Reporting and tracking
• Crash log reporting	• Usage analytics
• User & group access control	• Event management
	• App wrapping

References

- Steele, Colin. "Mobile device management vs. mobile application management". SearchMobile Computing. TechTarget. Retrieved 18 November 2015.

- Faas, Ryan. "BYOD Failure - Five Big Reasons Why Employees Don't Want to Use Their iPhones, iPads At Work". Cult of Mac. Retrieved 18 November 2015.

- Silva, Chris; Wong, Jason (30 June 2014). Use the Mobile App Mix to Choose an Enterprise App Store Strategy. Gartner. p. 6. Retrieved 18 November 2015.

- Ellis, Lisa, Jeffrey Saret, and Peter Weed (2012). "BYOD: From company-issued to employee-owned devices" (PDF). Telecom, Media & High Tech Extranet: No. 20 Recall. Retrieved 15 May 2014.

Mobile Phone: An Overview

Mobile phones are portable phones that a person can carry wherever they want to, and can receive and make calls while moving around. The applications of mobile phones mentioned in this text are text messaging, camera phone, Internet access, Bluetooth, mobile cloud computing etc. This section is an overview of the subject matter incorporating all the major aspects of mobile phones.

Mobile Phone

A mobile phone is a portable telephone that can make and receive calls over a radio frequency carrier while the user is moving within a telephone service area. The radio frequency link establishes a connection to the switching systems of a mobile phone operator, which provides access to the public switched telephone network (PSTN). Most modern mobile telephone services use a cellular network architecture, and therefore mobile telephones are often also called *cellular telephones* or *cell phones*. In addition to telephony, 2000s-era mobile phones support a variety of other services, such as text messaging, MMS, email, Internet access, short-range wireless communications (infrared, Bluetooth), business applications, gaming, and digital photography. Mobile phones which offer these and more general computing capabilities are referred to as smartphones.

Evolution of mobile phones, to an early smartphone

The first handheld mobile phone was demonstrated by John F. Mitchell and Martin Cooper of Motorola in 1973, using a handset weighing c. 4.4 lbs (2 kg). In 1983, the DynaTAC 8000x was the first commercially available handheld mobile phone. From 1983 to 2014, worldwide mobile phone subscriptions grew to over seven billion, penetrating 100% of the global population and reaching even the bottom of the economic pyramid. In first quarter of 2016, the top smartphone manufacturers were Samsung, Apple and Huawei (and "[s]martphone sales represented 78 percent of total mobile phone sales").

History

Martin Cooper of Motorola made the first publicized handheld mobile phone call on a prototype DynaTAC model on April 4, 1973. This is a reenactment in 2007.

A handheld mobile radio telephone service was envisioned in the early stages of radio engineering. In 1917, Finnish inventor Eric Tigerstedt filed a patent for a "pocket-size folding telephone with a very thin carbon microphone". Early predecessors of cellular phones included analog radio communications from ships and trains. The race to create truly portable telephone devices began after World War II, with developments taking place in many countries. The advances in mobile telephony have been traced in successive "generations", starting with the early "0G" (zeroth generation) services, such as Bell System's Mobile Telephone Service and its successor, the Improved Mobile Telephone Service. These "0G" systems were not cellular, supported few simultaneous calls, and were very expensive.

The first handheld mobile cell phone was demonstrated by Motorola in 1973. The first commercial automated cellular network was launched in Japan by Nippon Telegraph and Telephone in 1979. This was followed in 1981 by the simultaneous launch of the Nordic Mobile Telephone (NMT) system in Denmark, Finland, Norway and Sweden. Several other countries then followed in the early to mid-1980s. These first-generation (1G) systems could support far more simultaneous calls, but still used analog technology.

In 1991, the second-generation (2G) digital cellular technology was launched in Finland by Radiolinja on the GSM standard. This sparked competition in the sector as the new operators challenged the incumbent 1G network operators.

Ten years later, in 2001, the third generation (3G) was launched in Japan by NTT DoCoMo on the WCDMA standard. This was followed by 3.5G, 3G+ or turbo 3G enhancements based on the high-speed packet access (HSPA) family, allowing UMTS networks to have higher data transfer speeds and capacity.

The Motorola DynaTAC 8000X. First commercially available handheld cellular mobile phone, 1984.

By 2009, it had become clear that, at some point, 3G networks would be overwhelmed by the growth of bandwidth-intensive applications, such as streaming media. Consequently, the industry began looking to data-optimized fourth-generation technologies, with the promise of speed improvements up to ten-fold over existing 3G technologies. The first two commercially available technologies billed as 4G were the WiMAX standard, offered in North America by Sprint, and the LTE standard, first offered in Scandinavia by TeliaSonera.

Features

All mobile phones have a variety of features in common, but manufacturers seek product differentiation by adding functions to attract consumers. This competition has led to great innovation in mobile phone development over the past 20 years.

The common components found on all phones are:

- A battery, providing the power source for the phone functions.

- An input mechanism to allow the user to interact with the phone. The most common input mechanism is a keypad, but touch screens are also found in most smartphones.

- A screen which echoes the user's typing, displays text messages, contacts and more.

- Basic mobile phone services to allow users to make calls and send text messages.

- All GSM phones use a SIM card to allow an account to be swapped among devices. Some CDMA devices also have a similar card called a R-UIM.

- Individual GSM, WCDMA, iDEN and some satellite phone devices are uniquely identified by an International Mobile Equipment Identity (IMEI) number.

Low-end mobile phones are often referred to as feature phones, and offer basic telephony. Handsets with more advanced computing ability through the use of native software applications became known as smartphones.

Several phone series have been introduced to address specific market segments, such as the RIM BlackBerry focusing on enterprise/corporate customer email needs, the Sony-Ericsson 'Walkman' series of music/phones and 'Cyber-shot' series of camera/phones, the Nokia Nseries of multimedia phones, the Palm Pre, the HTC Dream and the Apple iPhone.

Sound Quality

In sound quality, smartphones and feature phones vary little. Some audio-quality enhancing features, such as Voice over LTE and HD Voice, have appeared and are often available on newer smartphones. Sound quality can remain a problem with both, as this depends not so much on the phone itself, as on the quality of the network and, in long distance calls, the bottlenecks/choke points met along the way. As such, for long-distance calls even features such as Voice over LTE and HD Voice may not improve things. In some cases smartphones can improve audio quality even on long-distance calls, by using a VoIP phone service, with someone else's WiFi/internet connection. Some cellphones have small speakers so that the user can use a speakerphone feature and talk to a person on the phone without holding it to their ear. The small speakers can also be used to listen to digital audio files of music or speech or watch videos with an audio component, without holding the phone close to the ear.

Text Messaging

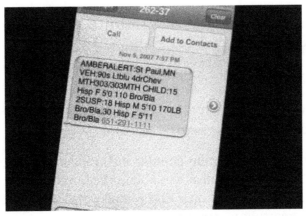

A text message (SMS).

The most commonly used data application on mobile phones is Short Message Service (SMS) text messaging. The first SMS message was sent from a computer to a mobile phone in 1992 in the UK, while the first person-to-person SMS from phone to phone was sent in Finland in 1993. The first mobile news service, delivered via SMS, was launched in Finland in 2000, and subsequently many organizations provided "on-demand" and "instant" news services by SMS. Multimedia Messaging Service (MMS) was introduced in 2001.

SIM Card

GSM feature phones require a small microchip called a Subscriber Identity Module or SIM card, in order to function. The SIM card is approximately the size of a small postage stamp and is usually placed underneath the battery in the rear of the unit. The SIM securely stores the service-subscriber key (IMSI) and the K_i used to identify and authenticate the user of the mobile phone. The SIM card allows users to change phones by simply removing the SIM card from one mobile phone and inserting it into another mobile phone or broadband telephony device, provided that this is not prevented by a SIM lock.

Typical mobile phone SIM card.

The first SIM card was made in 1991 by Munich smart card maker Giesecke & Devrient for the Finnish wireless network operator Radiolinja.

Multi-card Hybrid Phones

A hybrid mobile phone can hold up to four SIM cards. SIM and R-UIM cards may be mixed together to allow both GSM and CDMA networks to be accessed.

From 2010 onwards, such phones became popular in India and Indonesia and other emerging markets, and this was attributed to the desire to obtain the lowest on-net calling rate. In Q3 2011, Nokia shipped 18 million of its low-cost dual SIM phone range in an attempt to make up for lost ground in the higher-end smartphone market.

Kosher Phones

There are Jewish orthodox religious restrictions which, by some interpretations, standard mobile telephones overstep. To deal with this problem, some rabbinical organizations have recommended that phones with text-messaging capability not be used by children. Phones with restricted features are known as kosher phones and have rabbinical approval for use in Israel and elsewhere by observant Orthodox Jews. Although these phones are intended to prevent immodesty, some vendors report good sales to adults who prefer the simplicity of the devices. Some phones are approved for use by essential workers (such as health, security and public service workers) on the sabbath, even though the use of any electrical device is generally prohibited during this time.

Mobile Phone Operators

The world's largest individual mobile operator by number of subscribers is China Mobile, which has over 500 million mobile phone subscribers. Over 50 mobile operators have over ten million subscribers each, and over 150 mobile operators had at least one million subscribers by the end of 2009. In 2014, there were more than seven billion mobile phone subscribers worldwide, a number that is expected to keep growing.

Growth in mobile phone subscribers per country from 1980 to 2009.

Manufacturers

Prior to 2010, Nokia was the market leader. However, since then competition has emerged in the Asia Pacific region, from brands such as Micromax, Nexian and i-Mobile, which have chipped away at Nokia's market share. Android-powered smartphones have also gained momentum across the region at the expense of Nokia. In India, Nokia's market share dropped significantly to around 31% from 56% in the same period. Its share was displaced by Chinese and Indian vendors of low-end mobile phones.

In Q1 2012, according to Strategy Analytics, Samsung surpassed Nokia, selling 93.5 million units as against Nokia's 82.7 million units. In 2012 Standard & Poor's downgraded Nokia to "junk" status, at BB+/B, with negative outlook due to high loss and an expected further decline owing to insufficient growth in Lumia smartphone sales to offset a rapid decline in revenue from Symbi-an-based smartphones that was forecast for subsequent quarters.

In Q3 2014, the top ten manufacturers were Samsung (20.6%), Nokia (9.5%), Apple Inc. (8.4%), LG (4.2%), Huawei (3.6%), TCL Communication (3.5), Xiaomi (3.5%), Lenovo (3.3%), ZTE (3.0%) and Micromax (2.2%).

Other manufacturers outside the top five include TCL Communication, Lenovo, Sony Mobile Communications, Motorola and LG Electronics. Smaller current and past players include Audiovox (now UTStarcom), BenQ-Siemens, BlackBerry, Casio, CECT, Coolpad, Fujitsu, HTC, Just5, Intex, Karbonn Mobiles, Kyocera, Lumigon, LYF, Micromax Mobile, Mitsubishi Electric, Modu, NEC, Neonode, OnePlus, Openmoko, Panasonic, Palm, Pantech Wireless Inc., Philips, Sagem, Sanyo, Sharp, Sierra Wireless, SK Teletech, Trium and Toshiba.

Use

General

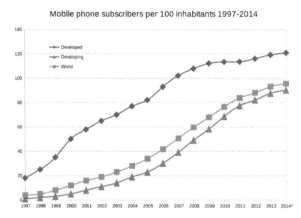

Mobile phone subscribers per 100 inhabitants. 2014 figure is estimated.

Mobile phones are used for a variety of purposes, such as keeping in touch with family members, for conducting business, and in order to have access to a telephone in the event of an emergency. Some people carry more than one mobile phone for different purposes, such as for business and personal use. Multiple SIM cards may be used to take advantage of the benefits of different calling plans. For example, a particular plan might provide for cheaper local calls, long-distance calls, international calls, or roaming.

The mobile phone has been used in a variety of diverse contexts in society. For example:

- A study by Motorola found that one in ten mobile phone subscribers have a second phone that is often kept secret from other family members. These phones may be used to engage in such activities as extramarital affairs or clandestine business dealings.

- Some organizations assist victims of domestic violence by providing mobile phones for use in emergencies. These are often refurbished phones.

- The advent of widespread text-messaging has resulted in the cell phone novel, the first literary genre to emerge from the cellular age, via text messaging to a website that collects the novels as a whole.

- Mobile telephony also facilitates activism and public journalism being explored by Reuters and Yahoo! and small independent news companies such as Jasmine News in Sri Lanka.

- The United Nations reported that mobile phones have spread faster than any other form of technology and can improve the livelihood of the poorest people in developing countries, by providing access to information in places where landlines or the Internet are not available, especially in the least developed countries. Use of mobile phones also spawns a wealth of micro-enterprises, by providing such work as selling airtime on the streets and repairing or refurbishing handsets.

- In Mali and other African countries, people used to travel from village to village to let friends and relatives know about weddings, births and other events. This can now be avoided in areas with mobile phone coverage, which are usually more extensive than areas with just land line penetration.

- The TV industry has recently started using mobile phones to drive live TV viewing through mobile apps, advertising, social TV, and mobile TV. It is estimated that 86% of Americans use their mobile phone while watching TV.

- In some parts of the world, mobile phone sharing is common. It is prevalent in urban India, as families and groups of friends often share one or more mobile phones among their members. There are obvious economic benefits, but often familial customs and traditional gender roles play a part. It is common for a village to have access to only one mobile phone, perhaps owned by a teacher or missionary, which is available to all members of the village for necessary calls.

Smartphones

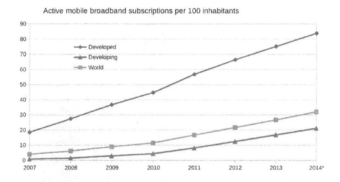

Active mobile broadband subscriptions per 100 inhabitants.

Smartphone as camera

Smartphones have a number of distinguishing features. The International Telecommunication Union measures those with Internet connection, which it calls *Active Mobile-Broadband subscriptions* (which includes tablets, etc.). In the developed world, smartphones have now overtaken the usage of earlier mobile systems. However, in the developing world, they account for only 20% of mobile telephony.

For Distributing Content

In 1998, one of the first examples of distributing and selling media content through the mobile phone was the sale of ringtones by Radiolinja in Finland. Soon afterwards, other media content appeared, such as news, video games, jokes, horoscopes, TV content and advertising. Most early content for mobile phones tended to be copies of legacy media, such as banner advertisements or TV news highlight video clips. Recently, unique content for mobile phones has been emerging, from ringtones and ringback tones to mobisodes, video content that has been produced exclusively for mobile phones.

In 2006, the total value of mobile-phone-paid media content exceeded Internet-paid media content and was worth US$31 billion. The value of music on mobile phones was worth US$9.3 billion in 2007, and gaming was worth over US$5 billion in 2007.

While Driving

A New York driver using two hand held mobile phones at once.

Mobile phone use while driving, including talking on the phone, texting, or operating other phone features, is common but controversial. It is widely considered dangerous due to distracted driving. Being distracted while operating a motor vehicle has been shown to increase the risk of accidents. In September 2010, the US National Highway Traffic Safety Administration (NHTSA) reported that 995 people were killed by drivers distracted by cell phones. In March 2011 a US insurance company, State Farm Insurance, announced the results of a study which showed 19% of drivers surveyed accessed the Internet on a smart phone while driving. Many jurisdictions prohibit the use of mobile phones while driving. In Egypt, Israel, Japan, Portugal and Singapore, both handheld and hands-free use of a mobile phone (which uses a speakerphone) is banned. In other countries including the UK and France and in many U.S. states, only handheld phone use is banned, while hands-free use is permitted.

A 2011 study reported that over 90% of college students surveyed text (initiate, reply or read) while driving. The scientific literature on the dangers of driving while sending a text message from a mobile phone, or *texting while driving*, is limited. A simulation study at the University of Utah found a sixfold increase in distraction-related accidents when texting.

Due to the increasing complexity of mobile phones, they are often more like mobile computers in their available uses. This has introduced additional difficulties for law enforcement officials when attempting to distinguish one usage from another in drivers using their devices. This is more apparent in countries which ban both handheld and hands-free usage, rather than those which ban handheld use only, as officials cannot easily tell which function of the mobile phone is being used simply by looking at the driver. This can lead to drivers being stopped for using their device illegally for a phone call when, in fact, they were using the device legally, for example, when using the phone's incorporated controls for car stereo, GPS or satnav.

A sign along Bellaire Boulevard in Southside Place, Texas (Greater Houston) states that using mobile phones while driving is prohibited from 7:30 am to 9:30 am and from 2:00 pm to 4:15 pm

A 2010 study reviewed the incidence of mobile phone use while cycling and its effects on behaviour and safety. In 2013 a national survey in the US reported the number of drivers who reported using their cellphones to access the Internet while driving had risen to nearly one of four. A study conducted by the University of Illinois examined approaches for reducing inappropriate and problematic use of mobile phones, such as using mobile phones while driving.

Accidents involving a driver being distracted by talking on a mobile phone have begun to be prosecuted as negligence similar to speeding. In the United Kingdom, from 27 February 2007, motorists who are caught using a hand-held mobile phone while driving will have three penalty points added to their license in addition to the fine of £60. This increase was introduced to try to stem the increase in drivers ignoring the law. Japan prohibits all mobile phone use while driving, including use of hands-free devices. New Zealand has banned hand held cellphone use since 1 November 2009. Many states in the United States have banned texting on cell phones while driving. Illinois became the 17th American state to enforce this law. As of July 2010, 30 states had banned texting while driving, with Kentucky becoming the most recent addition on July 15.

Public Health Law Research maintains a list of distracted driving laws in the United States. This database of laws provides a comprehensive view of the provisions of laws that restrict the use of mobile communication devices while driving for all 50 states and the District of Columbia between 1992, when first law was passed, through December 1, 2010. The dataset contains information on 22 dichotomous, continuous or categorical variables including, for example, activities regulated (e.g., texting versus talking, hands-free versus handheld), targeted populations, and exemptions.

Mobile Banking and Payments

In many countries, mobile phones are used to provide mobile banking services, which may include the ability to transfer cash payments by secure SMS text message. Kenya's M-PESA mobile banking service, for example, allows customers of the mobile phone operator Safaricom to hold cash balances which are recorded on their SIM cards. Cash can be deposited or withdrawn from M-PESA accounts at Safaricom retail outlets located throughout the country, and can be transferred electronically from person to person and used to pay bills to companies.

Mobile payment system.

Branchless banking has also been successful in South Africa and the Philippines. A pilot project in Bali was launched in 2011 by the International Finance Corporation and an Indonesian bank, Bank Mandiri.

Another application of mobile banking technology is Zidisha, a US-based nonprofit micro-lending platform that allows residents of developing countries to raise small business loans from Web users worldwide. Zidisha uses mobile banking for loan disbursements and repayments, transferring funds from lenders in the United States to borrowers in rural Africa who have mobile phones and can use the Internet.

Mobile payments were first trialled in Finland in 1998 when two Coca-Cola vending machines in Espoo were enabled to work with SMS payments. Eventually, the idea spread and in 1999, the Philippines launched the country's first commercial mobile payments systems with mobile operators Globe and Smart.

Some mobile phones can make mobile payments via direct mobile billing schemes, or through contactless payments, if the phone and the point of sale support near field communication (NFC). Enabling contactless payments through NFC-equipped mobile phones requires the co-operation of manufacturers, network operators and retail merchants.

Tracking and Privacy

Mobile phones are commonly used to collect location data. While the phone is turned on, the geographical location of a mobile phone can be determined easily (whether it is being used or not) using a technique known as multilateration to calculate the differences in time for a signal to travel from the mobile phone to each of several cell towers near the owner of the phone.

The movements of a mobile phone user can be tracked by their service provider and, if desired, by law enforcement agencies and their governments. Both the SIM card and the handset can be tracked.

China has proposed using this technology to track the commuting patterns of Beijing city residents. In the UK and US, law enforcement and intelligence services use mobile phoness to perform surveillance operations. They possess technology that enables them to activate the microphones in mobile phones remotely in order to listen to conversations which take place near the phone.

Hackers are able to track a phone's location, read messages, and record calls, just by knowing the phone number.

Thefts

According to the Federal Communications Commission, one out of three robberies involve the theft of a cellular phone. Police data in San Francisco show that half of all robberies in 2012 were thefts of cellular phones. An online petition on Change.org, called *Secure our Smartphones*, urged smartphone manufacturers to install kill switches in their devices to make them unusable if stolen. The petition is part of a joint effort by New York Attorney General Eric Schneiderman and San Francisco District Attorney George Gascón, and was directed to the CEOs of the major smartphone manufacturers and telecommunication carriers.

On Monday, 10 June 2013, Apple announced that it would install a "kill switch" on its next iPhone operating system, due to debut in October 2013.

All mobile phones have a unique identifier called IMEI. Anyone can report their phone as lost or stolen with their Telecom Carrier, and the IMEI would be blacklisted with a central registry. Telecom carriers, depending upon local regulation can or must implement blocking of blacklisted phones in their network. There are however a number of ways to circumvent a blacklist. One method is to send the phone to a country where the telecom carriers are not required to implement the blacklisting and sell it there, another involves altering the phones IMEI number. Even so, blacklisted phones typically has less value on the second hand market if the phones original IMEI is blacklisted.

Educational and Social Impact

A study by the London School of Economics found that banning mobile phones in schools could increase pupils' academic performance, providing benefits equal to one extra week of schooling per year.

Health Effects

The effect of mobile phone radiation on human health is the subject of recent interest and study, as a result of the enormous increase in mobile phone usage throughout the world. Mobile phones use electromagnetic radiation in the microwave range, which some believe may be harmful to human health. A large body of research exists, both epidemiological and experimental, in non-human animals and in humans. The majority of this research shows no definite causative relationship between exposure to mobile phones and harmful biological effects in humans. This is often paraphrased simply as the balance of evidence showing no harm to humans from mobile phones, although a significant number of individual studies do suggest such a relationship, or are inconclusive. Other digital wireless systems, such as data communication networks, produce similar radiation.

On 31 May 2011, the World Health Organization stated that mobile phone use may possibly represent a long-term health risk, classifying mobile phone radiation as "possibly carcinogenic to humans" after a team of scientists reviewed studies on mobile phone safety. The mobile phone is in category 2B, which ranks it alongside coffee and other possibly carcinogenic substances.

Some recent studies have found an association between mobile phone use and certain kinds of brain and salivary gland tumors. Lennart Hardell and other authors of a 2009 meta-analysis of 11 studies from peer-reviewed journals concluded that cell phone usage for at least ten years "approximately doubles the risk of being diagnosed with a brain tumor on the same ('ipsilateral') side of the head as that preferred for cell phone use".

One study of past mobile phone use cited in the report showed a "40% increased risk for gliomas (brain cancer) in the highest category of heavy users (reported average: 30 minutes per day over a 10-year period)". This is a reversal of the study's prior position that cancer was unlikely to be caused by cellular phones or their base stations and that reviews had found no convincing evidence for other health effects. However, a study published 24 March 2012 in the *British Medical Journal* questioned these estimates, because the increase in brain cancers has not paralleled the increase in mobile phone use. Certain countries, including France, have warned against the use of mobile phones by minors in particular, due to health risk uncertainties. Mobile pollution by transmitting electromagnetic waves can be decreased up to 90% by adopting the circuit as designed in mobile phone (MS) and mobile exchange (BTS, MSC etc.).

In May 2016 preliminary findings of a long-term study by the U.S. government suggested that radio-frequency (RF) radiation, the type emitted by cellphones, can cause cancer.

Future Evolution

5G is a technology and term used in research papers and projects to denote the next major phase in mobile telecommunication standards beyond the 4G/IMT-Advanced standards. The term 5G is not officially used in any specification or official document yet made public by telecommunication companies or standardization bodies such as 3GPP, WiMAX Forum or ITU-R. New standards beyond 4G are currently being developed by standardization bodies, but they are at this time seen as under the 4G umbrella, not for a new mobile generation. Deloitte is predicting a collapse in wireless performance to come as soon as 2016, as more devices using more and more services compete for limited bandwidth for their operation.

Environmental Impact

A mobile phone repair kiosk in Hong Kong.

Studies have shown that around 40-50% of the environmental impact of mobile phones occurs during the manufacture of their printed wiring boards and integrated circuits.

The average user replaces their mobile phone every 11 to 18 months, and the discarded phones then contribute to electronic waste. Mobile phone manufacturers within Europe are subject to the WEEE directive, and Australia has introduced a mobile phone recycling scheme.

Apple Inc. has realized how their products when not recycled impact the environment and waste valuable resources. Apple's Liam was introduced to the world, an advanced robotic disassembler and sorter designed by Apple Engineers in California specifically for recycling outdated or broken iPhones. Reuses and recycles parts from traded in products.

Conflict Minerals

Demand for metals used in mobile phones and other electronics fuelled the Second Congo War, which claimed almost 5.5 million lives. In a 2012 news story, *The Guardian* reported: "In unsafe mines deep underground in eastern Congo, children are working to extract minerals essential for the electronics industry. The profits from the minerals finance the bloodiest conflict since the second world war; the war has lasted nearly 20 years and has recently flared up again. ... For the last 15 years, the Democratic Republic of the Congo has been a major source of natural resources for the mobile phone industry."

The company Fairphone has attempted to develop a mobile phone that does not contain conflict minerals.

Applications of Mobile Phone

Text Messaging

Text messaging, or texting, is the act of composing and sending electronic messages, typically consisting of alphabetic and numeric characters, between two or more users of mobile phones, fixed devices (e.g., desktop computers) or portable devices (e.g., tablet computers or smartphones). While text messages are usually sent over a phone network, due to the convergence between the telecommunication and broadcasting industries in the 2000s, text messages may also be sent via a cable network or Local Area Network. The term originally referred to messages sent using the Short Message Service (SMS). It has grown beyond alphanumeric text to include multimedia messages (known as MMS) containing digital images, videos, and sound content, as well as ideograms known as emoji (happy faces and other icons).

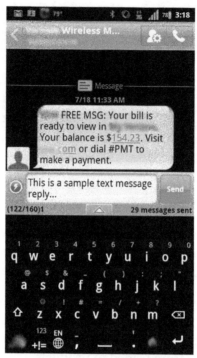

A text message as it appears on the display screen of a smartphone.

As of 2016, text messages are used by youth and adults for personal, family and social purposes and in business, government and non-governmental organizations for communication between colleagues. As with e-mailing, in the 2010s, the sending of short informal messages has become an accepted part of many cultures. This makes texting a quick and easy way to communicate with friends and colleagues, including in contexts where a phone call would be impolite or inappropriate (e.g., calling very late at night or when one knows the other person is busy with family or work activities). Like e-mail and voice mail, and unlike landline or mobile phone calls (in which the caller hopes to speak directly with the recipient), texting does not require the caller and recipient to both be free at the same moment; this permits communication even between busy individuals. Text messages can also be used to interact with automated systems, for example, to order products or services from e-commerce websites, or to participate in online contests. Advertisers and service providers use direct text marketing to send messages to mobile phone users about promotions, payment due dates, and other notifications instead of using postal mail, e-mail, or voicemail.

Terminology

The service is referred to by different colloquialisms depending on the region. It may simply be referred to as a "text" in North America, the United Kingdom, Australia, New Zealand and the Philippines, an "SMS" in most of mainland Europe, or an "MMS" or "SMS" in the Middle East, Africa, and Asia. The sender of a text message is commonly referred to as a "texter".

History

In 1933 RCA Communications, New York introduced the first "telex" service. The first messages over RCA transatlantic circuits went between New York and London. The first year of operation saw seven million words or 300,000 radiograms transmitted. Radio has long sent alphanumeric messages via radiotelegraphy. The University of Hawaii began using radio to send digital information as early as 1971, using ALOHAnet. Friedhelm Hillebrand conceptualised SMS in 1984 while working for Deutsche Telekom. Sitting at a typewriter at home, Hillebrand typed out random sentences and counted every letter, number, punctuation, and space. Almost every time, the messages contained fewer than 160 characters, thus giving the basis for the limit one could type via text messaging. With Bernard Ghillebaert of France Télécom, he developed a proposal for the GSM group meeting in February 1985 in Oslo. The first technical solution evolved in a GSM subgroup under the leadership of Finn Trosby. It was further developed under the leadership of Kevin Holley and Ian Harris. SMS forms an integral part of SS7 (Signalling System No. 7). Here it is a "state" with a 160 character data, coded in the ITU-T "T.56" text format, that has "sequence lead in" to determine different language codes, and can have special character codes that allows (for example) sending simple graphs as text. This was part of ISDN (Integrated Services Digital Network) and since GSM (Groupe Spécial Mobile) is based on this, made its way to the mobile phone. Messages could be sent and received on ISDN phones, and these can send SMS to any GSM phone. The possibility of doing something is one thing, implementing it another, but systems existed from 1988 that sent SMS messages to mobile phones (compare ND-NOTIS).

SMS messaging was used for the first time on 3 December 1992, when Neil Papworth, a 22-year-old test engineer for Sema Group in the UK (now Airwide Solutions), used a personal computer to send the text message "Merry Christmas" via the Vodafone network to the phone of Richard Jarvis who was at a party in Newbury, Berkshire which had been organised to celebrate the event. Modern SMS text messaging is usually messaging from one mobile phone to another mobile phone. Radiolinja became the first network to offer a commercial person-to-person SMS text messaging service in 1994. When Radiolinja's domestic competitor, Telecom Finland (now part of TeliaSonera) also launched SMS text messaging in 1995 and the two networks offered cross-network SMS functionality, Finland became the first nation where SMS text messaging was offered on a competitive as well as on a commercial basis. GSM was not allowed in the United States and the radio frequencies were blocked and awarded to US "Carriers" to use US technology. Hence there is no "development" in the US in mobile messaging service. The GSM in the US had to use a frequency allocated for private communication services (PCS) – what the ITU frequency régime had blocked for DECT – Digital Enhanced Cordless Telecommunications – 1000-feet range picocell, but survived. American Personal Communications (APC), the first GSM carrier in America, provided the first text-messaging service in the United States. Sprint Telecommunications Venture, a partnership of Sprint Corp. and three large cable-TV companies, owned 49 percent of APC. The Sprint venture was the largest single buyer at a government-run spectrum auction that raised $7.7 billion in 2005 for PCS licenses. APC operated under the brand name of Sprint Spectrum and launched its service on November 15, 1995 in Washington, D.C. and in Baltimore, Maryland. Vice President Al Gore in Washington, D.C. made the initial phone-call to launch the network, calling Mayor Kurt Schmoke in Baltimore.

Initial growth of text messaging was slow, with customers in 1995 sending on average only 0.4

message per GSM customer per month. One factor in the slow take-up of SMS was that operators were slow to set up charging systems, especially for prepaid subscribers, and to eliminate billing fraud, which was possible by changing SMSC settings on individual handsets to use the SMSCs of other operators. Over time, this issue was eliminated by switch-billing instead of billing at the SMSC and by new features within SMSCs to allow blocking of foreign mobile users sending messages through it. SMS is available on a wide range of networks, including 3G networks. However, not all text-messaging systems use SMS; some notable alternate implementations of the concept include J-Phone's SkyMail and NTT Docomo's Short Mail, both in Japan. E-mail messaging from phones, as popularized by NTT Docomo's i-mode and the RIM BlackBerry, also typically use standard mail protocols such as SMTP over TCP/IP. As of 2007 text messaging was the most widely used mobile data service, with 74% of all mobile phone users worldwide, or 2.4 billion out of 3.3 billion phone subscribers, at the end of 2007 being active users of the Short Message Service. In countries such as Finland, Sweden and Norway, over 85% of the population use SMS. The European average is about 80%, and North America is rapidly catching up with over 60% active users of SMS by end of 2008. The largest average usage of the service by mobile phone subscribers occurs in the Philippines, with an average of 27 texts sent per day per subscriber.

Uses

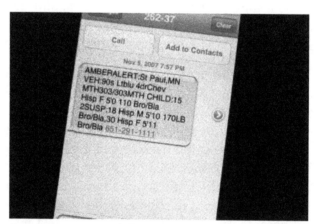

A text message on an iPhone announcing an AMBER Alert

Text messaging is most often used between private mobile phone users, as a substitute for voice calls in situations where voice communication is impossible or undesirable (e.g., during a school class or a work meeting). Texting is also used to communicate very brief massages, such as informing someone that you will be late or reminding a friend or colleague about a meeting. As with e-mail, informality and brevity have become an accepted part of text messaging. Some text messages such as SMS can also be used for the remote controlling of home appliances. It is widely used in domotics systems. Some amateurs have also built own systems to control (some of) their appliances via SMS. Other methods such as group messaging, which was patented in 2012 by the GM of Andrew Ferry, Devin Peterson, Justin Cowart, Ian Ainsworth, Patrick Messinger, Jacob Delk, Jack Grande, Austin Hughes, Brendan Blake, and Brooks Brasher are used to involve more than two people into a text messaging conversation. A Flash SMS is a type of text message that appears directly on the main screen without user interaction and is not automatically stored in the inbox. It can be useful in cases such as an emergency (e.g., fire alarm) or confidentiality (e.g., one-time password).

Short message services are developing very rapidly throughout the world. SMS is particularly popular in Europe, Asia, United States, Australia and New Zealand and is also gaining influence in Africa. Popularity has grown to a sufficient extent that the term *texting* (used as a verb meaning the act of mobile phone users sending short messages back and forth) has entered the common lexicon. Young Asians consider SMS as the most popular mobile phone application. Fifty percent of American teens send fifty text messages or more per day, making it their most frequent form of communication. In China, SMS is very popular and has brought service providers significant profit (18 billion short messages were sent in 2001). It is a very influential and powerful tool in the Philippines, where the average user sends 10–12 text messages a day. The Philippines alone sends on average over 1 billion text messages a day, more than the annual average SMS volume of the countries in Europe, and even China and India. SMS is hugely popular in India, where youngsters often exchange lots of text messages, and companies provide alerts, infotainment, news, cricket scores updates, railway/airline booking, mobile billing, and banking services on SMS.

Texting became popular in the Philippines in 1998. In 2001, text messaging played an important role in deposing former Philippine president Joseph Estrada. Similarly, in 2008, text messaging played a primary role in the implication of former Detroit Mayor Kwame Kilpatrick in an SMS sex scandal. Short messages are particularly popular among young urbanites. In many markets, the service is comparatively cheap. For example, in Australia, a message typically costs between A$0.20 and $0.25 to send (some prepaid services charge $0.01 between their own phones), compared with a voice call, which costs somewhere between $0.40 and $2.00 per minute (commonly charged in half-minute blocks). The service is enormously profitable to the service providers. At a typical length of only 190 bytes (including protocol overhead), more than 350 of these messages per minute can be transmitted at the same data rate as a usual voice call (9 kbit/s). There are also free SMS services available, which are often sponsored and allow sending SMS from a PC connected to the internet. Mobile service providers in New Zealand, such as Vodafone and Telecom NZ, provide up to 2000 SMS messages for NZ$10 per month. Users on these plans send on average 1500 SMS messages every month. Text messaging has become so popular that advertising agencies and advertisers are now jumping into the text messaging business. Services that provide bulk text message sending are also becoming a popular way for clubs, associations, and advertisers to reach a group of optin subscribers quickly.

Research suggests that Internet-based mobile messaging will have grown to equal the popularity of SMS in 2013, with nearly 10 trillion messages being sent through each technology. Services such as Facebook Messenger, WhatsApp and Viber have led to a decline in the use of SMS in parts of the world.

Applications

Microblogging

Of many texting trends, a system known as microblogging has surfaced, which consists of a miniaturised blog, inspired mainly by people's tendency to jot down informal thoughts and post them online. They consist of websites like Twitter and its Chinese equivalent Weibo (微博). As of 2016, both of these websites were popular.

Emergency Services

In some countries, text messages can be used to contact emergency services. In the UK, text messages can be used to call emergency services only after registering with the emergency SMS service. This service is primarily aimed at people who, by reason of disability, are unable to make a voice call. It has recently been promoted as a means for walkers and climbers to call emergency services from areas where a voice call is not possible due to low signal strength. In the US, there is a move to require both traditional operators and Over-the-top messaging providers to support texting to 911. In Asia, SMS is used for tsunami warnings and in Europe, SMS is used to inform individuals of imminent disaster. Since the location of a handset is known, systems can alert everyone in an area that the events has made impossible to pass through e.g. an avalanche.

Reminders of Hospital Appointments

SMS messages are used in some countries as reminders of hospital appointments. Missed outpatient clinic appointments cost the National Health Service (England) more than £600 million ($980 million) a year SMS messages are thought to be more cost effective, swifter to deliver, and more likely to receive a faster response than letters. A recent study by Sims and colleagues (2012) examined the outcomes of 24,709 outpatient appointments scheduled in mental health services in South-East London. The study found that SMS message reminders could reduce the number of missed psychiatric appointments by 25–28%, representing a potential national yearly saving of over £150 million.

Commercial Uses

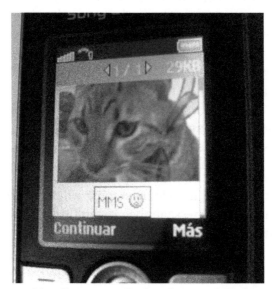

A multimedia message displayed on a mobile phone

Short Codes

Short codes are special telephone numbers, shorter than full telephone numbers, that can be used to address SMS and MMS messages from mobile phones or fixed phones. There are two types of short codes: dialing and messaging.

Text Messaging Gateway Providers

SMS gateway providers facilitate the SMS traffic between businesses and mobile subscribers, being mainly responsible for carrying mission-critical messages, SMS for enterprises, content delivery and entertainment services involving SMS, e.g., TV voting. Considering SMS messaging performance and cost, as well as the level of text messaging services, SMS gateway providers can be classified as resellers of the text messaging capability of another provider's SMSC or offering the text messaging capability as an operator of their own SMSC with SS7. SMS messaging gateway providers can provide gateway-to-mobile (Mobile Terminated–MT) services. Some suppliers can also supply mobile-to-gateway (text-in or Mobile Originated/MO services). Many operate text-in services on shortcodes or mobile number ranges, whereas others use lower-cost geographic text-in numbers.

Premium Content

SMS is widely used for delivering digital content, such as news alerts, financial information, pictures, GIFs, logos and ringtones. Such messages are also known as premium-rated short messages (PSMS). The subscribers are charged extra for receiving this premium content, and the amount is typically divided between the mobile network operator and the value added service provider (VASP), either through revenue share or a fixed transport fee. Services like 82ASK and Any Question Answered have used the PSMS model to enable rapid response to mobile consumers' questions, using on-call teams of experts and researchers.

Premium short messages are increasingly being used for "real-world" services. For example, some vending machines now allow payment by sending a premium-rated short message, so that the cost of the item bought is added to the user's phone bill or subtracted from the user's prepaid credits. Recently, premium messaging companies have come under fire from consumer groups due to a large number of consumers racking up huge phone bills. A new type of free-premium or hybrid-premium content has emerged with the launch of text-service websites. These sites allow registered users to receive free text messages when items they are interested in go on sale, or when new items are introduced. An alternative to inbound SMS is based on long numbers (international mobile number format, e.g., +44 7624 805000, or geographic numbers that can handle voice and SMS, e.g., 01133203040), which can be used in place of short codes or premium-rated short messages for SMS reception in several applications, such as TV voting, product promotions, and campaigns. Long numbers are internationally available, as well as enabling businesses to have their own number, rather than short codes, which are usually shared across a lot of brands. Additionally, Long numbers are non-premium inbound numbers.

In Workplaces

The use of text messaging for workplace purposes has grown significantly during the mid-2000s (decade). As companies seek competitive advantages, many employees are using new technology, collaborative applications, and real-time messaging such as SMS, instant messaging, and mobile communications to connect with teammates and customers. Some practical uses of text messaging include the use of SMS for confirming delivery or other tasks, for instant communication between a service provider and a client (e.g., stock broker and investor), and for sending alerts. Several universities have implemented a system of texting students and faculties campus alerts. One such example is Penn State. As text messaging has proliferated in business, so too have regulations governing its use. One regulation specifically governing the use of text messaging in financial-services firms engaged in stocks, equities, and securities trading is *Regulatory Notice 07-59, Supervision of Electronic Communications, December 2007*, issued to member firms by the Financial Industry Regulatory Authority. In 07-59, FINRA noted that "electronic communications", "e-mail", and "electronic correspondence" may be used interchangeably and can include such forms of electronic messaging as instant messaging and text messaging. Industry has had to develop new technology to allow companies to archive their employees' text messages.

Security, confidentiality, reliability and speed of SMS are among the most important guarantees industries such as financial services, energy and commodities trading, health care and enterprises demand in their mission-critical procedures. One way to guarantee such a quality of text messaging lies in introducing SLAs (Service Level Agreement), which are common in IT contracts. By providing measurable SLAs, corporations can define reliability parameters and set up a high quality of their services. Just one of many SMS applications that has proven highly popular and successful in the financial-services industry is mobile receipts. In January 2009, Mobile Marketing Association (MMA) published the *Mobile Banking Overview* for financial institutions in which it discussed the advantages and disadvantages of mobile channel platforms such as Short Message Services (SMS), Mobile Web, Mobile Client Applications, SMS with Mobile Web and Secure SMS.

Mobile interaction services are an alternative way of using SMS in business communications with greater certainty. Typical business-to-business applications are telematics and Machine-to-Machine, in which two applications automatically communicate with each other. Incident alerts are also common, and staff communications are also another use for B2B scenarios. Businesses can use SMS for time-critical alerts, updates and reminders, mobile campaigns, content and entertainment applications. Mobile interaction can also be used for consumer-to-business interactions, such as media voting and competitions, and for consumer-to-consumer interaction, for example, with mobile social networking, chatting and dating.

Text messaging is widely used on business settings; as well, it is used in a number of civil service and non-governmental organization workplaces. The U.S. And Canadian civil service both adopted Blackberry smartphones in the 2000s.

Online SMS Services

There are a growing number of websites that allow users to send free SMS messages online. Some websites provide free SMS for promoting premium business packages.

Worldwide Use

Europe

Europe follows next behind Asia in terms of the popularity of the use of SMS. In 2003, an average of 16 billion messages were sent each month. Users in Spain sent a little more than fifty messages per month on average in 2003. In Italy, Germany and the United Kingdom, the figure was around 35–40 SMS messages per month. In each of these countries, the cost of sending an SMS message varies from €0.04–0.23, depending on the payment plan (with many contractual plans including all or a number of texts for free). In the United Kingdom, text messages are charged between £0.05–0.12. Curiously, France has not taken to SMS in the same way, sending just under 20 messages on average per user per month. France has the same GSM technology as other European countries, so the uptake is not hampered by technical restrictions.

SMS is used to send "welcome" messages to mobile phones roaming between countries. Here, T-Mobile welcomes a Proximus subscriber to the UK, and BASE welcomes an Orange UK customer to Belgium.

In the Republic of Ireland, 1.5 billion messages are sent every quarter, on average 114 messages per person per month. In the United Kingdom over 1 billion text messages are sent every week. The Eurovision Song Contest organized the first pan-European SMS voting in 2002, as a part of the voting system (there was also a voting over traditional landline phone lines). In 2005, the Eurovision Song Contest organized the biggest televoting ever (with SMS and phone voting). During roaming, that is, when a user connects to another network in different country from his own, the prices may be higher, but in July 2009, EU legislation went into effect limiting this price to €0.11.

Finland

Mobile-service providers in Finland offer contracts in which users can send 1000 text messages a month for €10. In Finland, which has very high mobile phone ownership rates, some TV channels began "SMS chat", which involved sending short messages to a phone number, and the messages would be shown on TV. Chats are always moderated, which prevents users from sending offensive material to the channel. The craze evolved into quizzes and strategy games and then faster-paced games designed for television and SMS control. Games require users to register their nickname and send short messages to control a character onscreen. Messages usually cost 0.05 to 0.86 Euro apiece, and games can require the player to send dozens of messages. In December 2003, a Finnish TV channel, MTV3, put a Santa Claus character on air reading aloud text messages sent in by viewers. On 12 March 2004, the first entirely "interactive" TV channel, VIISI, began operation in Finland. However, SBS Finland Oy took over the channel and turned it into a music channel named *The Voice* in November 2004. In 2006, the Prime Minister of Finland, Matti Vanhanen, made the news when he allegedly broke up with his girlfriend with a text message. In 2007, the first book written solely in text messages, *Viimeiset viestit* (*Last Messages*), was released by Finnish author Hannu Luntiala. It is about an executive who travels through Europe and India.

United States

In the United States, text messaging is very popular; as reported by CTIA in December 2009, the 286 million US subscribers sent 152.7 billion text messages per month, for an average of 534 messages per subscriber per month. The Pew Research Center found in May 2010 that 72% of U.S. adult cellphone users send and receive text messages. In the U.S., SMS is often charged both at the sender and at the destination, but, unlike phone calls, it cannot be rejected or dismissed. The reasons for lower uptake than other countries are varied. Many users have unlimited "mobile-to-mobile" minutes, high monthly minute allotments, or unlimited service. Moreover, "push to talk" services offer the instant connectivity of SMS and are typically unlimited. The integration between competing providers and technologies necessary for cross-network text messaging was not initially available. Some providers originally charged extra for texting, reducing its appeal. In the third quarter of 2006, at least 12 billion text messages were sent on AT&T's network, up almost 15% from the preceding quarter. In the U.S., while texting is mainly popular among people from 13–22 years old, it is also increasing among adults and business users. The age that a child receives his/her first cell phone has also decreased, making text messaging a popular way of communicating. The number of texts sent in the US has gone up over the years as the price has gone down to an average of $0.10 per text sent and received. To convince more customers to buy unlimited text messaging plans, some major cellphone providers have increased the price to send and receive text messages from $.15 to $.20 per message. This is over $1,300 per megabyte. Many providers offer unlimited plans, which can result in a lower rate per text, given sufficient volume.

Japan

Japan was among the first countries to adopt short messages widely, with pioneering non-GSM services including J-Phone's SkyMail and NTT Docomo's Short Mail. Japanese adolescents first began text messaging, because it was a cheaper form of communication than the other available forms. Thus, Japanese theorists created the selective interpersonal relationship theory, claiming that mobile phones can change social networks among young people (classified as 13- to 30-year-olds). They theorized this age group had extensive but low-quality relationships with friends, and mobile-phone usage may facilitate improvement in the quality of their relationships. They concluded this age group prefers "selective interpersonal relationships in which they maintain particular, partial, but rich relations, depending on the situation." The same studies showed participants rated friendships in which they communicated face-to-face and through text messaging as being more intimate than those in which they communicated solely face-to-face. This indicates participants make new relationships with face-to-face communication at an early stage, but use text messaging to increase their contact later on. It is also interesting to note that as the relationships between participants grew more intimate, the frequency of text messaging also increased. However, short messaging has been largely rendered obsolete by the prevalence of mobile Internet e-mail, which can be sent to and received from any e-mail address, mobile or otherwise. That said, while usually presented to the user simply as a uniform "mail" service (and most users are unaware of the distinction), the operators may still internally transmit the content as short messages, especially if the destination is on the same network.

China

Text messaging is popular and cheap in China. About 700 billion messages were sent in 2007. Text message spam is also a problem in China. In 2007, 353.8 billion spam messages were sent, up 93% from the previous year. It is about 12.44 messages per week per person. It is routine that the People's Republic of China government monitor text messages across the country for illegal content.

Among Chinese migrant workers with little formal education, it is common to refer to SMS manuals when text messaging. These manuals are published as cheap, handy, smaller-than-pocket-size booklets that offer diverse linguistic phrases to utilize as messages.

Philippines

SMS was introduced to selected markets in the Philippines in 1995. In 1998, Philippine mobile-service providers launched SMS more widely across the country, with initial television marketing campaigns targeting hearing-impaired users. The service was initially free with subscriptions, but Filipinos quickly exploited the feature to communicate for free instead of using voice calls, which they would be charged for. After telephone companies realized this trend, they began charging for SMS. The rate across networks is 1 peso per SMS (about US$0.023). Even after users were charged for SMS, it remained cheap, about one-tenth of the price of a voice call. This low price led to about five million Filipinos owning a cell phone by 2001. Because of the highly social nature of Philippine culture and the affordability of SMS compared to voice calls, SMS usage shot up. Filipinos used texting not only for social messages but also for political purposes, as it allowed the Filipinos to express their opinions on current events and political issues. It became a powerful tool for Filipinos in promoting or denouncing issues and was a key factor during the 2001 EDSA II revolution, which overthrew then-President Joseph Estrada, who was eventually found guilty of corruption. According to 2009 statistics, there are about 72 million mobile-service subscriptions (roughly 80% of the Filipino population), with around 1.39 billion SMS messages being sent daily. Because of the large amount of text messages being sent, the Philippines became known as the "text capital of the world" during the late 1990s until the early 2000s.

New Zealand

There are three mobile network companies in New Zealand. Spark NZ, (formally Telecom NZ), was the first telecommunication company in New Zealand. In 2011, Spark was broken into two companies, with Chorus Ltd taking the landline infrastructure and Spark NZ providing services including over their mobile network. Vodafone NZ acquired mobile network provider Bellsouth New Zealand in 1998 and has 2.32 million customers as at July 2013 Vodafone launched the first Text messaging service in 1999 and has introduced innovative TXT services like Safe TXT and Call-Me 2degrees Mobile Ltd launched in August 2009. In 2005, around 85% of the adult population had a mobile phone. In general, texting is more popular than making phone calls, as it is viewed as less intrusive and therefore more polite.

Africa

Text messaging will become a key revenue driver for mobile network operators in Africa over the next couple of years. Today, text messaging is already slowly gaining influence in the African market. One such person used text messaging to spread the word about HIV and AIDS. Also, in September 2009, a multi-country campaign in Africa used text messaging to expose stock-outs of essential medicines at public health facilities and put pressure on governments to address the issue.

Social Effects

The advent of text messaging made possible new forms of interaction that were not possible before. A person may now carry out a conversation with another user without the constraint of being expected to reply within a short amount of time and without needing to set time aside to engage in conversation. With voice calling, both participants need to be free at the same time.Mobile phone users can maintain communication during situations in which a voice call is impractical, impossible, or unacceptable, such as during a school class or work meeting. Texting has provided a venue for participatory culture, allowing viewers to vote in online and TV polls, as well as receive information while they are on the move. Texting can also bring people together and create a sense of community through "Smart Mobs" or "Net War", which create "people power".

Effect on Language

This sticker seen in Paris satirizes the popularity of communication in SMS shorthand. In French: "Is that you? / It's me! / Do you love me? / Shut up!"

The small phone keypad and the rapidity of typical text message exchanges has caused a number spelling abbreviations: as in the phrase "txt msg", "u" (an abbreviation for "you"), "HMU", or use of CamelCase, such as in "ThisIsVeryLame". To avoid the even more limited message lengths allowed when using Cyrillic or Greek letters, speakers of languages written in those alphabets often use the Latin alphabet for their own language. In certain languages utilizing diacritic marks, such as Polish, SMS technology created an entire new variant of written language: characters normally written with diacritic marks (e.g., q, $ę$, $ś$, $ż$ in Polish) are now being written without them (as a, e, s, z) to enable using cell phones without Polish script or to save space in Unicode messages. Historically, this language developed out of shorthand used in bulletin board systems and later in Internet chat rooms, where users would abbreviate some words to allow a response to be typed more quickly, though the amount of time saved was often inconsequential. However, this became much more pronounced in SMS, where mobile phone users either have a numeric keyboard (with older cellphones) or a small QWERTY keyboard (for 2010s-era smartphones), so more effort is required to type each character, and there is sometimes a limit on the number of characters that may be sent. In Mandarin Chinese, numbers that sound similar to words are used in place of those words. For example, the numbers 520 in Chinese (*wǔ èr líng*) sound like the words for "I love you" (*wǒ ài ni*). The sequence 748 (*qī sì bā*) sounds like the curse "go to hell" (*qù sǐ ba*).

Predictive text software, which attempts to guess words (Tegic's T9 as well as iTap) or letters (Eatoni's LetterWise) reduces the labour of time-consuming input. This makes abbreviations not only less necessary, but slower to type than regular words that are in the software's dictionary. However, it makes the messages longer, often requiring the text message to be sent in multiple parts and, therefore, costing more to send. The use of text messaging has changed the way that people talk and write essays, some believing it to be harmful. Children today are receiving cell phones at an age as young as eight years old; more than 35 percent of children in second and third grade have their own mobile phone. Because of this, the texting language is integrated into the way that students think from an earlier age than ever before. In November 2006, New Zealand Qualifications Authority approved the move that allowed students of secondary schools to use mobile phone text language in the end-of-the-year-exam papers. Highly publicized reports, beginning in 2002, of the use of text language in school assignments caused some to become concerned that the quality of written communication is on the decline, and other reports claim that teachers and professors are beginning to have a hard time controlling the problem. However, the notion that text language is widespread or harmful is refuted by research from linguistic experts.

An article in *The New Yorker* explores how text messaging has anglicized some of the world's languages. The use of diacritic marks is dropped in languages such as French, as well as symbols in Ethiopian languages. In his book, *Txtng: the Gr8 Db8* (which translates as "Texting: the Great Debate"), David Crystal states that texters in all eleven languages use "lol" ("laughing out loud"), "u", "brb" ("be right back"), and "gr8" ("great"), all English-based shorthands. The use of pictograms and logograms in texts are present in every language. They shorten words by using symbols to represent the word or symbols whose name sounds like a syllable of the word such as in 2day or b4. This is commonly used in other languages as well. Crystal gives some examples in several languages such as Italian *sei*, "six", is used for *sei*, "you are". Example: dv 6 = dove sei ("where are you") and French *sept* "seven" = *cassette* ("casette"). There is also the use of numeral sequences, substituting for several syllables of a word and creating whole phrases using numerals. For example, in French, a12c4 can be said as *à un de ces quatres*, "see you around" (literally: "to one of these four *[days]*"). An example of using symbols in texting and borrowing from English is the use of @. Whenever it is used in texting, its intended use is with the English pronunciation. Crystal gives the example of the Welsh use of @ in @F, pronounced ataf, meaning "to me". In character-based languages such as Chinese and Japanese, numbers are assigned syllables based on the shortened form of the pronunciation of the number, sometimes the English pronunciation of the number. In this way, numbers alone can be used to communicate whole passages, such as in Chinese, "8807701314520" can be literally translated as "Hug hug you, kiss kiss you, whole life, whole life I love you." English influences worldwide texting in variation but still in combination with the individual properties of languages.

American popular culture is also recognized in shorthand. For example, Homer Simpson translates into: ~(_8^(|). Crystal also suggests that texting has led to more creativity in the English language, giving people opportunities to create their own slang, emoticons, abbreviations, acronyms, etc. The feeling of individualism and freedom makes texting more popular and a more efficient way to communicate. Crystal has also been quoted in saying that "In a logical world, text messaging should not have survived." But text messaging didn't just come out of nowhere. It originally began as a messaging system that would send out emergency information. But it gained immediate popularity with the public. What followed is the SMS we see today, which is a very quick and efficient way of sharing information from person to person. Work by Richard Ling has shown that texting has a gendered dimension and it plays into the development of teen identity. In addition we text to a very small number of other persons. For most people, half of their texts go to 3 – 5 other people.

Research by Rosen et al. (2009) found that those young adults who used more language-based textisms (shortcuts such as LOL, 2nite, etc.) in daily writing produced worse formal writing than those young adults who used fewer linguistic textisms in daily writing. However, the exact opposite was true for informal writing. This suggests that perhaps the act of using textisms to shorten communication words leads young adults to produce more informal writing, which may then help them to be better "informal" writers. Due to text messaging, teens are writing more, and some teachers see that this comfort with language can be harnessed to make better writers. This new form of communication may be encouraging students to put their thoughts and feelings into words and this may be able to be used as a bridge, to get them more interested in formal writing.

Joan H. Lee in her thesis *What does txting do 2 language: The influences of exposure to messaging and print media on acceptability constraints* (2011) associates exposure to text messaging with more rigid acceptability constraints. The thesis suggests that more exposure to the colloquial, Generation Text language of text messaging contributes to being less accepting of words. In contrast, Lee found that students with more exposure to traditional print media (such as books and magazines) were more accepting of both real and fictitious words. The thesis, which garnered international media attention, also presents a literature review of academic literature on the effects of text messaging on language. Texting has also been shown to have had no effect or some positive effects on literacy. According to Plester, Wood and Joshi and their research done on the study of 88 British 10–12-year-old children and their knowledge of text messages, "textisms are essentially forms of phonetic abbreviation" that show that "to produce and read such abbreviations arguably requires a level of phonological awareness (and orthographic awareness) in the child concerned."

Texting While Driving

A driver with attention divided between a mobile phone and the road ahead

Texting while driving leads to increased distraction behind the wheel and can lead to an increased risk of an accident. In 2006, Liberty Mutual Insurance Group conducted a survey with more than 900 teens from over 26 high schools nationwide. The results showed that 87% of students found texting to be "very" or "extremely" distracting. A study by AAA found that 46% of teens admitted to being distracted behind the wheel due to texting. One example of distraction behind the wheel is the 2008 Chatsworth train collision, which killed 25 passengers. The engineer had sent 45 text messages while operating the train. A 2009 experiment with *Car and Driver* editor Eddie Alterman (that took place at a deserted air field, for safety reasons) compared texting with drunk driving. The experiment found that texting while driving was more dangerous than being drunk. While being legally drunk added four feet to Alterman's stopping distance while going 70 mph, reading an e-mail on a phone added 36 feet, and sending a text message added 70 feet. In 2009, the Virginia Tech Transportation Institute released the results of an 18-month study that involved placing cameras inside the cabs of more than 100 long-haul trucks, which recorded the drivers over a combined driving distance of three million miles. The study concluded that when the drivers were texting, their risk of crashing was 23 times greater than when not texting.

Texting While Walking

Due to the proliferation of smart phone applications performed while walking, "texting while walking" or "wexting" is the increasing practice of people being transfixed to their mobile device without looking in any direction but their personal screen while walking. First coined reference in 2015 in New York from Rentrak's chief client officer when discussing time spent with media and various media usage metrics. Text messaging among pedestrians leads to increased cognitive distraction and reduced situation awareness, and may lead to increases in unsafe behavior leading to injury and death. Recent studies conducted on cell phone use while walking showed that cell phone users recall fewer objects when conversing, walk slower, have altered gait and are more unsafe when crossing a street. Additionally, some gait analyses showed that stance phase during overstepping motion, longitudinal and lateral deviation increased during cell phone operation but step length and clearance did not; a different analysis did find increased step clearance and reduced step length.

It is unclear which processes may be affected by distraction, which types of distraction may affect which cognitive processes, and how individual differences may affect the influence of distraction. Lamberg and Muratori believe that engaging in a dual-task, such as texting while walking, may interfere with working memory and result in walking errors. Their study demonstrated that participants engaged in text messaging were unable to maintain walking speed or retain accurate spatial information, suggesting an inability to adequately divide their attention between two tasks. According to them, the addition of texting while walking with vision occluded increases the demands placed on the working memory system resulting in gait disruptions.

Texting on a phone distracts participants, even when the texting task used is a relatively simple one. Stavrinos et al. investigated the effect of other cognitive tasks, such as engaging in conversations or cognitive tasks on a phone, and found that participants actually have reduced visual awareness. This finding was supported by Licence et al., who conducted a similar study. For example, texting pedestrians may fail to notice unusual events in their environment, such as a unicycling clown. These findings suggest that tasks that require the allocation of cognitive resources can affect visual attention even when the task itself does not require the participants to avert their eyes from their environment. The act of texting itself seems to impair pedestrians' visual awareness. It appears that the distraction produced by texting is a combination of both a cognitive and visual perceptual distraction. A study conducted by Licence et al. supported some of these findings, particularly that those who text while walking significantly alter their gait. However, they also found that the gait pattern texters adopted was slower and more "protective", and consequently did not increase obstacle contact or tripping in a typical pedestrian context.

Sexting

Sexting is slang for the act of sending sexually explicit or suggestive content between mobile devices using SMS. A genre of texting, it contains either text, images, or video that is intended to be sexually arousing. A portmanteau of *sex* and *texting*, sexting was reported as early as 2005 in *The Sunday Telegraph Magazine*, constituting a trend in the creative use of SMS to excite another with alluring messages throughout the day.

Although sexting often takes place consensually between two people, it can also occur against the wishes of a person who is the subject of the content. A number of instances have been reported in which the recipients of sexting have shared the content of the messages with others, with less intimate intentions, such as to impress their friends or embarrass their sender. Celebrities such as Miley Cyrus, Vanessa Hudgens, and Adrienne Bailon have been victims of such abuses of sexting. A 2008 survey by The National Campaign to Prevent Teen and Unplanned Pregnancy and CosmoGirl.com suggested a trend of sexting and other seductive online content being readily shared between teens. One in five teen girls surveyed (22 percent)—and 11 percent of teen girls aged 13–16 years old—say they have electronically sent, or posted online, nude or semi-nude images of themselves. One-third (33 percent) of teen boys and one-quarter (25 percent) of teen girls say they were shown private nude or semi-nude images. According to the survey, sexually suggestive messages (text, e-mail, and instant messaging) were even more common than images, with 39 percent of teens having sent or posted such messages, and half of teens (50 percent) having received them. A 2012 study that has received wide international media attention was conducted at the University of Utah Department of Psychology by Donald S. Strassberg, Ryan Kelly McKinnon, Michael Sustaíta and Jordan Rullo. They surveyed 606 teenagers ages 14–18 and found that nearly 20 percent of the students said they had sent a sexually explicit image of themselves via cell phone, and nearly twice as many said that they had received a sexually explicit picture. Of those receiving such a picture, over 25 percent indicated that they had forwarded it to others.

In addition, of those who had sent a sexually explicit picture, over a third had done so despite believing that there could be serious legal and other consequences if they got caught. Students who had sent a picture by cell phone were more likely than others to find the activity acceptable. The authors conclude: "These results argue for educational efforts such as cell phone safety assemblies, awareness days, integration into class curriculum and teacher training, designed to raise awareness about the potential consequences of sexting among young people." Sexting becomes a legal issue when teens (under 18) are involved, because any nude photos they may send of themselves would put the recipients in possession of child pornography.

In Schools

Text messaging has affected students academically by creating an easier way to cheat on exams. In December 2002, a dozen students were caught cheating on an accounting exam through the use of text messages on their mobile phones. In December 2002, Hitotsubashi University in Japan failed 26 students for receiving e-mailed exam answers on their mobile phones. The number of students caught using mobile phones to cheat on exams has increased significantly in recent years. According to Okada (2005), most Japanese mobile phones can send and receive long text messages of between 250 and 3000 characters with graphics, video, audio, and Web links. In England, 287 school and college students were excluded from exams in 2004 for using mobile phones during exams.

Some teachers and professors claim that advanced texting features can lead to students cheating on exams. Students in high school and college classrooms are using their mobile phones to send and receive texts during lectures at high rates. Further, published research has established that students who text during college lectures have impaired memories of the lecture material compared to students who do not. For example, in one study, the number of irrelevant text messages sent and received during a lecture covering the topic of developmental psychology was related to students' memory of the lecture.

Two girls text during class at school

Bullying

Spreading rumors and gossip by text message, using text messages to bully individuals, or forwarding texts that contain defamatory content is an issue of great concern for parents and schools. Text "bullying" of this sort can cause distress and damage reputations. In some cases, individuals who are bullied online have committed suicide. Harding and Rosenberg (2005) argue that the urge to forward text messages can be difficult to resist, describing text messages as "loaded weapons".

Influence on Perceptions of the Student

When a student sends an email that contains phonetic abbreviations and acronyms that are common in text messaging (e.g., "gr8" instead of "great"), it can influence how that student is subsequently evaluated. In a study by Lewandowski and Harrington (2006), participants read a student's email sent to a professor that either contained text-messaging abbreviations (gr8, How R U?) or parallel text in standard English (great, How are you?), and then provided impressions of the sender. Students who used abbreviations in their email were perceived as having a less favorable personality and as putting forth less effort on an essay they submitted along with the email. Specifically, abbreviation users were seen as less intelligent, responsible, motivated, studious, dependable, and hard-working. These findings suggest that the nature of a student's email communication can influence how others perceive the student and their work.

Law and Crime

Text messaging has been a subject of interest for police forces around the world. One of the issues of concern to law enforcement agencies is the use of encrypted text messages. In 2003, a British company developed a program called Fortress SMS which used 128 bit AES encryption to protect SMS messages. Police have also retrieved deleted text messages to aid them in solving crimes. For example, Swedish police retrieved deleted texts from a cult member who claimed she committed a double murder based on forwarded texts she received. Police in Tilburg, Netherlands, started an SMS alert program, in which they would send a message to ask citizens to be vigilant when a burglar was on the loose or a child was missing in their neighborhood. Several thieves have been caught and children have been found using the SMS Alerts. The service has been expanding to other cities. A Malaysian–Australian company has released a multi-layer SMS security program. Boston police are now turning to text messaging to help stop crime. The Boston Police Department asks citizens to send texts to make anonymous crime tips. A Malaysian court had ruled that it is legal to divorce through the use of text messaging, as long as the sender is clear and unequivocal.

Social Unrest

Texting has been used on a number of occasions with the result of the gathering of large aggressive crowds. SMS messaging drew a crowd to Cronulla Beach in Sydney resulting in the 2005 Cronulla riots. Not only were text messages circulating in the Sydney area, but in other states as well (*Daily Telegraph*). The volume of such text messages and e-mails also increased in the wake of the riot. The crowd of 5000 at stages became violent, attacking certain ethnic groups. Sutherland Shire Mayor directly blamed heavily circulated SMS messages for the unrest. NSW police considered whether people could be charged over the texting. Retaliatory attacks also used SMS.

The Narre Warren Incident, when a group of 500 party goers attended a party at Narre Warren in Melbourne, Australia, and rioted in January 2008, also was a response of communication being spread by SMS and Myspace. Following the incident, the Police Commissioner wrote an open letter asking young people to be aware of the power of SMS and the Internet. In Hong Kong, government officials find that text messaging helps socially because they can send multiple texts to the community. Officials say it is an easy way of contacting community or individuals for meetings or events. Texting was used to coordinate gatherings during the 2009 Iranian election protests.

Between 2009 and 2012 the U.S. secretly created and funded a Twitter-like service for Cubans called ZunZuneo, initially based on mobile phone text message service and later with an internet interface. The service was funded by the U.S. Agency for International Development through its Office of Transition Initiatives, who utiliized contractors and front companies in the Cayman Islands, Spain and Ireland. A longer term objective was to organize "smart mobs" that might "re-negotiate the balance of power between the state and society." A database about the subscribers was created, including gender, age, and "political tendencies". At its peak ZunZuneo had 40,000 Cuban users, but the service closed as financially unsustainable when U.S. funding was stopped.

In Politics

Text messaging has affected the political world. American campaigns find that text messaging is a much easier, cheaper way of getting to the voters than the door-to-door approach. Mexico's president-elect Felipe Calderón launched millions of text messages in the days immediately preceding his narrow win over Andres Manuel Lopez Obradór. In January 2001, Joseph Estrada was forced to resign from the post of president of the Philippines. The popular campaign against him was widely reported to have been co-ordinated with SMS chain letters. A massive texting campaign was credited with boosting youth turnout in Spain's 2004 parliamentary elections. In 2008, Detroit Mayor Kwame Kilpatrick and his Chief of Staff at the time became entangled in a sex scandal stemming from the exchange of over 14,000 text messages that eventually led to his forced resignation, conviction of perjury, and other charges. Text messaging has been used to turn down other political leaders. During the 2004 U.S. Democratic and Republican National Conventions, protesters used an SMS-based organizing tool called TXTmob to get to opponents. In the last day before the 2004 presidential elections in Romania, a message against Adrian Năstase was largely circulated, thus breaking the laws that prohibited campaigning that day. Text messaging has helped politics by promoting campaigns.

A text message that (he says) promises 500 Libyan dinars ($400) to anyone who "makes noise" in support of Gaddafi in the coming days

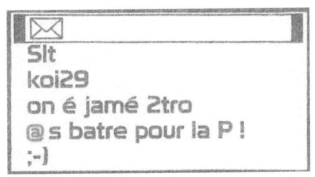

A recruitment ban in French SMS language: «Slt koi29 on é jamé 2tro @ s batre pour la P. ;-)» = «*Salut! Quoi de neuf? On n'est jamais de trop à se battre pour la Paix!*»

On 20 January 2001, President Joseph Estrada of the Philippines became the first head of state in history to lose power to a smart mob. More than one million Manila residents assembled at the site of the 1986 People Power peaceful demonstrations that has toppled the Marcos regime. These people have organized themselves and coordinated their actions through text messaging. They were able to bring down a government without having to use any weapons or violence. Through text messaging, their plans and ideas were communicated to others and successfully implemented. Also, this move encouraged the military to withdraw their support from the regime, and as a result, the Estrada government fell. People were able to converge and unite with the use of their cell phones. "The rapid assembly of the anti-Estrada crowd was a hallmark of early smart mob technology, and the millions of text messages exchanged by the demonstrators in 2001 was, by all accounts, a key to the crowds esprit de corps."

Use in Healthcare

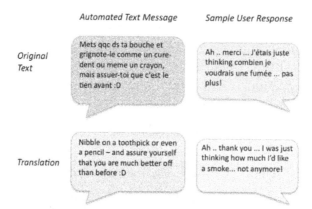

	Automated Text Message	Sample User Response
Original Text	Mets qqc ds ta bouche et grignote-le comme un cure-dent ou meme un crayon, mais assuer-toi que c'est le tien avant :D	Ah .. merci ... J'étais juste thinking combien je voudrais une fumée ... pas plus!
Translation	Nibble on a toothpick or even a pencil – and assure yourself that you are much better off than before :D	Ah .. thank you ... I was just thinking how much I'd like a smoke... not anymore!

Some health organizations manage text messaging services to help people avoid smoking

Text messaging is a rapidly growing trend in Healthcare.[when?] "One survey found that 73% of physicians text other physicians about work- similar to the overall percentage of the population that texts." A 2006 study of reminder messages sent to children and adolescents with type 1 diabetes mellitus showed favorable changes in adherence to treatment. A risk is that these physicians could be violating the Health Insurance Portability and Accountability Act. Where messages could be saved to a phone indefinitely, patient information could be subject to theft or loss, and could be seen by other unauthorized persons. The HIPAA privacy rule requires that any text message involving a medical decision must be available for the patient to access, meaning that any texts that are not documented in an EMR system could be a HIPAA violation.

Medical Concerns

The excessive use of the thumb for pressing keys on mobile devices has led to a high rate of a form of repetitive strain injury termed "BlackBerry thumb". (Although this refers to strain developed on older Blackberry devices, which had a scroll wheel on the side of the phone.) An inflammation of the tendons in the thumb caused by constant text-messaging is also called text-messager's thumb, or texting tenosynovitis. Texting has also been linked as a secondary source in numerous traffic collisions, in which police investigations of mobile phone records have found that many drivers have lost control of their cars while attempting to send or retrieve a text message. Increasing cases of Internet addiction are now also being linked to text messaging, as mobile phones are now more likely to have e-mail and Web capabilities to complement the ability to text.

Etiquette

Texting etiquette refers to what is considered appropriate texting behavior. These expectations may concern different areas, such as the context in which a text was sent and received/read, who each participant was with when the participant sent or received/read a text message or what constitutes impolite text messages. At the website of The Emily Post Institute, the topic of *texting* has spurred several articles with the "do's and dont's" regarding the new form of communication. One example from the site is: "Keep your message brief. No one wants to have an entire conversation with you by texting when you could just call him or her instead." Another example is: "Don't use all Caps. Typing a text message in all capital letters will appear as though you are shouting at the recipient, and should be avoided."

Expectations for etiquette may differ depending on various factors. For example, expectations for appropriate behavior have been found to differ markedly between the U.S. and India. Another example is generational differences. In *The M-Factor: How the Millennial Generation Is Rocking the Workplace*, Lynne Lancaster and David Stillman note that younger Americans often do not consider it rude to answer their cell or begin texting in the middle of a face-to-face conversation with someone else, while older people, less used to the behavior and the accompanying lack of eye contact or attention, find this to be disruptive and ill-mannered. With regard to texting in the workplace, Plantronics studied how we communicate at work and found that 58% of US knowledge workers have increased the use of text messaging for work in the past five years. The same study found that 33% of knowledge workers felt text messaging was critical or very important to success and productivity at work.

Challenges

Spam

In 2002, an increasing trend towards spamming mobile phone users through SMS prompted cellular-service carriers to take steps against the practice, before it became a widespread problem. No major spamming incidents involving SMS had been reported as of March 2007, but the existence of mobile phone spam has been noted by industry watchdogs including *Consumer Reports* magazine and the Utility Consumers' Action Network (UCAN). In 2005, UCAN brought a case against Sprint for spamming its customers and charging $0.10 per text message. The case was settled in 2006 with Sprint agreeing not to send customers Sprint advertisements via SMS. SMS expert Acision (formerly LogicaCMG Telecoms) reported a new type of SMS malice at the end of 2006, noting the first instances of SMiShing (a cousin to e-mail phishing scams). In SMiShing, users receive SMS messages posing to be from a company, enticing users to phone premium-rate numbers or reply with personal information. Similar concerns were reported by PhonepayPlus, a consumer watchdog in the United Kingdom, in 2012.

Pricing Concerns

Concerns have been voiced over the excessive cost of off-plan text messaging in the United States. AT&T Mobility, along with most other service providers, charges texters 20 cents per message if they do not have a messaging plan or if they have exceeded their allotted number of texts. Given that an SMS message is at most 160 bytes in size, this cost scales to a cost of $1,310 per megabyte sent via text message. This is in sharp contrast with the price of unlimited data plans offered by the same carriers, which allow the transmission of hundreds of megabytes of data for monthly prices of about $15 to $45 in addition to a voice plan. As a comparison, a one-minute phone call uses up the same amount of network capacity as 600 text messages, meaning that if the same cost-per-traffic formula were applied to phone calls, cell phone calls would cost $120 per minute. With service providers gaining more customers and expanding their capacity, their overhead costs should be decreasing, not increasing. In 2005, text messaging generated nearly 70 billion dollars in revenue, as reported by Gartner,industry analysts, three times as much as Hollywood box office sales in 2005. World figures showed that over a trillion text messages were sent in 2005.

Although major cellphone providers deny any collusion, fees for out-of-package text messages have increased, doubling from 10 to 20 cents in the United States between 2007 and 2008 alone. On 16 July 2009, Senate hearings were held to look into any breach of the Sherman Antitrust Act. The same trend is visible in other countries, though increasingly widespread flatrate plans, for example in Germany, do make text messaging easier, text messages sent abroad still result in higher costs.

Increasing Competition

While text messaging is still a growing market, traditional SMS are becoming increasingly challenged by alternative messaging services which are available on smartphones with data connections. These services are much cheaper and offer more functionality like exchanging of multimedia content (e.g. photos, videos or audio notes) and group messaging. Especially in western countries some of these services attract more and more users.

Security Concerns

Consumer SMS should not be used for confidential communication. The contents of common SMS messages are known to the network operator's systems and personnel. Therefore, consumer SMS is not an appropriate technology for secure communications. To address this issue, many companies use an SMS gateway provider based on SS7 connectivity to route the messages. The advantage of this international termination model is the ability to route data directly through SS7, which gives the provider visibility of the complete path of the SMS. This means SMS messages can be sent directly to and from recipients without having to go through the SMS-C of other mobile operators. This approach reduces the number of mobile operators that handle the message; however, it should not be considered as an end-to-end secure communication, as the content of the message is exposed to the SMS gateway provider.

An alternative approach is to use end-to-end security software that runs on both the sending and receiving device, where the original text message is transmitted in encrypted form as a consumer SMS. By using key rotation, the encrypted text messages stored under data retention laws at the network operator cannot be decrypted even if one of the devices is compromised. A problem with this approach is that communicating devices needs to run compatible software. Failure rates without backward notification can be high between carriers.. International texting can be unreliable depending on the country of origin, destination and respective operators (US: "carriers"). Differences in the character sets used for coding can cause a text message sent from one country to another to become unreadable.

In Popular Culture

Records and Competition

The *Guinness Book of World Records* has a world record for text messaging, currently held by Sonja Kristiansen of Norway. Kristiansen keyed in the official text message, as established by Guinness, in 37.28 seconds. The message is, "The razor-toothed piranhas of the genera Serrasalmus and Pygocentrus are the most ferocious freshwater fish in the world. In reality, they seldom attack a human." In 2005, the record was held by a 24-year-old Scottish man, Craig Crosbie, who completed the same message in 48 seconds, beating the previous time by 19 seconds. *The Book of Alternative Records* lists Chris Young of Salem, Oregon, as the world-record holder for the fastest 160-character text message where the contents of the message are not provided ahead of time. His record of 62.3 seconds was set on 23 May 2007.

Elliot Nicholls of Dunedin, New Zealand, currently holds the world record for the fastest blind-folded text messaging. A record of a 160-letter text in 45 seconds while blindfolded was set on 17 November 2007, beating the old record of 1-minute 26 seconds set by an Italian in September 2006. Ohio native Andrew Acklin is credited with the world record for most text messages sent or received in a single month, with 200,052. His accomplishments were first in the World Records Academy and later followed up by *Ripley's Believe It Or Not 2010: Seeing Is Believing*. He has been acknowledged by The Universal Records Database for the most text messages in a single month; however, this has since been broken twice and now is listed as 566607 messages by Mr. Fred Lindgren.

In January 2010, LG Electronics sponsored an international competition, the LG Mobile World Cup, to determine the fastest pair of texters. The winners were a team from South Korea, Ha Mok-min and Bae Yeong-ho. On 6 April 2011, SKH Apps released an iPhone app, iTextFast, to allow consumers to test their texting speed and practice the paragraph used by *Guinness Book of World Records*. The current best time listed on Game Center for that paragraph is 34.65 seconds.

Morse Code

A few competitions have been held between expert Morse code operators and expert SMS users. Several mobile phones have Morse code ring tones and alert messages. For example, many Nokia mobile phones have an option to beep "S M S" in Morse code when it receives a short message. Some of these phones could also play the Nokia slogan "Connecting people" in Morse code as a message tone. There are third-party applications available for some mobile phones that allow Morse input for short messages.

Tattle Texting

"Tattle texting" can mean either of two different texting trends:

Arena Security

Many sports arenas now offer a number where patrons can text report security concerns, like drunk or unruly fans, or safety issues like spills. These programs have been praised by patrons and security personnel as more effective than traditional methods. For instance, the patron doesn't need to leave his seat and miss the event in order to report something important. Also, disruptive fans can be reported with relative anonymity. "Text Tattling" also gives security personnel a useful tool to prioritize messages. For instance, a single complaint in one section about an unruly fan can be addressed when convenient, while multiple complaints by several different patrons can be acted upon immediately.

Smart Cars

In this context, "tattle texting" refers to an automatic text sent by the computer in an automobile, because a preset condition was met. The most common use for this is for parents to receive texts from the car their child is driving, alerting them to speeding or other issues.Employers can also use the service to monitor their corporate vehicles. The technology is still new and (currently) only available on a few car models.

Common conditions that can be chosen to send a text are:

- Speeding With the use of GPS, stored maps, and speed limit information, the onboard computer can determine if the driver is exceeding the current speed limit. The device can store this information and/or send it to another recipient.

- Range Parents/employers can set a maximum range from a fixed location after which a "tattle text" is sent. Not only can this keep children close to home and keep employees from using corporate vehicles inappropriately, but it can also be a crucial tool for quickly identifying stolen vehicles, car jackings, and kidnappings.

Camera Phone

A camera phone is a mobile phone which is able to capture photographs. Most camera phones also record video. The first camera phone was sold in 2000 in Japan, a J-Phone model, although some argue that the SCH-V200 and Kyocera VP-210 Visual Phone, both introduced months earlier in South Korea and Japan respectively, are the first camera phones.

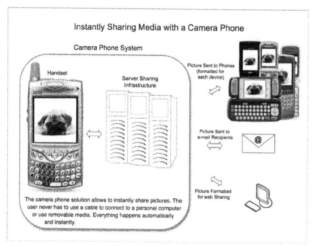

Camera phone allows instant, automatic photo sharing. There is no need for a cable or removable media to connect to a personal computer.

Most camera phones are simpler than separate digital cameras. Their usual fixed-focus lenses and smaller sensors limit their performance in poor lighting. Lacking a physical shutter, some have a long shutter lag. Photoflash is typically provided by an LED source which illuminates less intensely over a much longer exposure time than a bright and near-instantaneous flash strobe. Optical zoom and tripod screws are rare and none has a hot shoe for attaching an external flash. Some also lack a USB connection or a removable memory card. Most have Bluetooth and WiFi, and can make geotagged photographs.

Some of the more expensive camera phones have only a few of these technical disadvantages, but with bigger image sensors (a few are up to 1"), their capabilities approach those of low-end point-

and-shoot cameras. In the smartphone era, the steady sales increase of camera phones caused point-and-shoot camera sales to peak about 2010 and decline thereafter. Most model lines improve their cameras every year or two.

Most smartphones only have a menu choice to start a camera application program and an on-screen button to activate the shutter. Some also have a separate camera button, for quickness and convenience. A few camera phones are designed to resemble separate low-end digital compact cameras in appearance and to some degree in features and picture quality, and are branded as both mobile phones and cameras.

The principal advantages of camera phones are cost and compactness; indeed for a user who carries a mobile phone anyway, the addition is negligible. Smartphones that are camera phones may run mobile applications to add capabilities such as geotagging and image stitching. A few high end phones can use their touch screen to direct their camera to focus on a particular object in the field of view, giving even an inexperienced user a degree of focus control exceeded only by seasoned photographers using manual focus. However, the touch screen, being a general purpose control, lacks the agility of a separate camera's dedicated buttons and dial(s).

Technology

Nearly all camera phones use CMOS image sensors, due to largely reduced power consumption compared to CCD type cameras, which are also used, but not in today's camera phones. Some of today's camera phones even use more expensive Back Side Illuminated CMOS which use energy lesser than CMOS, although more expensive than CMOS and CCD.

Images are usually saved in the JPEG file format, except for some high-end camera phones which have also RAW feature and the Android (operating system) 5.0 Lollipop has facility of it. The wireless infrastructure manages the sharing. The lower power consumption prevents the camera from quickly depleting the phone's battery. In any case, an external battery or flash can be employed, to improve performance.

Samsung Galaxy S5 camera, including the floating element group, suspended by ceramic bearings and a small rare earth magnet.

As camera phone technology has progressed over the years, the lens design has evolved from a simple double Gauss or Cooke triplet to many molded plastic aspheric lens elements made with varying dispersion and refractive indexes. The latest generation of phone cameras also apply distortion (optics), vignetting, and various optical aberration corrections to the image before it is compressed into a .jpeg format.'

Image showing the six molded elements in the Samsung Galaxy S5

Most camera phones have a digital zoom feature. A few have optical zoom. An #external camera can be added, coupled wirelessly to the phone by Wi-Fi. They are compatible with most smartphones.

Android Lollipop 5.0 allows apps to take RAW format pictures. Windows Phones can be configured to operate as a camera even if the phone is asleep.

Directory

Phones usually store pictures and video in a directory called /DCIM in the internal memory. Some can store this media in external memory (Secure digital card or USB on the go pen drive).

History

The Nokia N8 smartphone is the first Nokia smartphone with a 12-megapixel autofocus lens, and is one of the few camera phones (the first was Nokia N82) to feature Carl Zeiss optics with xenon flash.

The camera phone, like many complex systems, is the result of converging and enabling technologies. There are dozens of relevant patents dating back as far as 1956. Compared to digital cameras of the 1990s, a consumer-viable camera in a mobile phone would require far less power and a higher level of camera electronics integration to permit the miniaturization.

The CMOS active pixel sensor "camera-on-a-chip" developed by Dr. Eric Fossum and his team in the early 1990s achieved the first step of realizing the modern camera phone as described in a

March 1995 Business Week article. While the first camera phones (e.g. J-SH04), as successfully marketed by J-Phone in Japan, used CCD sensors and not CMOS sensors, more than 90% of camera phones sold today use CMOS image sensor technology.

Over the years there have been many videophones and cameras that have included communication capability. Some devices experimented with integration of the device to communicate wirelessly with Internet, which would allow instant media sharing with anyone anywhere. For example, in 1995 Apple experimented with the Apple Videophone/PDA. There were several digital cameras with cellular phone transmission capability shown by companies such as Kodak, Olympus in the early 1990s. There was also a digital camera with cellular phone designed by Shosaku Kawashima of Canon in Japan in May 1997.

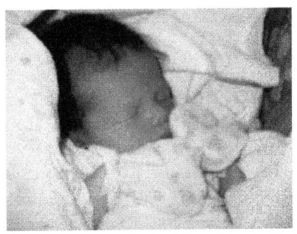

June 11, 1997, Santa Cruz, CA: Image taken by Philippe Kahn after his daughter's birth
and instantly shared with more than 2000 people around the world.

On June 11, 1997, Philippe Kahn shared instantly the first pictures from the maternity ward where his daughter Sophie was born. He wirelessly transmitted his cell phone pictures to more than 2,000 family, friends and associates around the world. Kahn's wireless sharing software and camera integrated into his cell phone augured the birth of instant visual communications. Kahn's cell phone transmission is the first known publicly shared picture via a cell phone.

Typical camera phone photograph

In Japan, two competing projects were run by Sharp and Kyocera in 1997. Both had cell phones with integrated cameras. However, the Kyocera system was designed as a peer-to-peer video-phone as opposed to the Sharp project which was initially focused on sharing instant pictures. That was made possible when the Sharp devices was coupled to the Sha-mail infrastructure designed in collaboration with American technologist, Kahn. The Kyocera team was led by Mr. Kazumi Saburi.

In 1995, work by James Greenwold of Bureau Of Technical Services, in Chippewa Falls, WI, was developing a pocket video camera for surveillance purposes. By 1999, the Tardis recorder was in prototype and being used by the government. Bureau Of Technical Services, advanced further by the patent # 6,845,215,B1 on Body-Carryable, digital Storage medium, Audio/Video recording Assembly.

Cameras on cell phones proved popular right from the start, as indicated by the fact that the J-Phone in Japan had more than half of its subscribers using cell phone cameras in two years. The world soon followed. By 2003, more camera phones were sold worldwide than stand-alone digital cameras. In 2005, Nokia became the world's most sold digital camera brand. In 2006, half of the world's mobile phones had a built-in camera.

In 2006, Thuraya released the first satellite phone with an integrated camera. The Thuraya SG-2520 was manufactured by a Korean company called APSI and ran Windows CE.

By 2007, the first cell phones and other consumer products appeared using the Tardis technology to make the move from still cameras to full motion video.

In 2008, Nokia sold more camera phones than Kodak sold film based simple cameras, thus becoming the biggest manufacturer of any kind of camera.

In 2010, the worldwide number of camera phones totaled more than a billion. Most mobile phones, even inexpensive ones, were being sold with a camera. High end camera phones usually had a relatively good lens and high resolution, but a small sensor.

Twelve-megapixel camera phones have been produced by at least two companies. To highlight the capabilities of the Nokia N8 (Big CMOS Sensor) camera, Nokia created a short film, *The Commuter*, in October 2010. The seven-minute film was shot entirely on the phone's 720p camera. A 14-megapixel smartphone with 3x optical zoom was announced in 2010.

Apple's iPhone started including a 12-megapixel camera with the iPhone 6S and 6S Plus in 2015 (it had been 8-megapixel on models released in previous years), and continues to do so with iPhone 7 and iPhone 7 Plus, that where introduced in 2016, while the has latter includes a dual-lens camera and includes a 2× optical zoom.

In 2012, Nokia announced Nokia 808 PureView. It features a 41-megapixel 1/1.2-inch sensor and a high-resolution f/2.4 Zeiss all-aspherical one-group lens. It also features Nokia's PureView Pro technology, a pixel oversampling technique that reduces an image taken at full resolution into a lower resolution picture, thus achieving higher definition and light sensitivity, and enables lossless zoom.

In mid-2013, Nokia announced the Nokia Lumia 1020. It had an improved version of the 41-megapixel sensor and ran Windows Phone 8 unlike the 808 PureView which was the last phone to run Nokia's Symbian OS.

Multimedia Messaging Service

Camera phones can share pictures almost instantly and automatically via a sharing infrastructure integrated with the carrier network. Early developers including Philippe Kahn envisioned a technology that would enable service providers to "collect a fee every time anyone snaps a photo." The resulting technologies, Multimedia Messaging Service and Sha-Mail were developed parallel to and in competition to open Internet based mobile communication provided by GPRS and later 3G networks.

The closed sharing infrastructure was critical and explains the early successes of J-Phone, DoCoMo in Japan, Sprint, and other carriers worldwide.

The first commercial camera phone complete with infrastructure was the J-SH04, made by Sharp Corporation, had an integrated CCD sensor, with the Sha-Mail (Picture-Mail in Japanese) infrastructure developed in collaboration with Kahn's LightSurf venture, and marketed from 2001 by J-Phone in Japan today owned by Softbank.

The first commercial deployment in North America of camera phones was in 2004. The Sprint wireless carriers deployed over one million camera phone manufactured by Sanyo and launched by the PictureMail infrastructure (Sha-Mail in English) developed and managed by LightSurf.

Users of early camera phones were held captive by the MMS business model. While phones had internet connectivity, working web browsers and email-programs, the phone menu offered no way of including a photo in an email or uploading it to a web site. Connecting cables or removable media that would enable the local transfer of pictures were also usually missing.

Modern smartphones have more connectivity and transfer options with photograph attachment features.

Manufacturers

Major manufacturers include Toshiba, Sharp, Nokia, Sanyo, Samsung, Motorola, Siemens, Sony Mobile, Apple, and LG Electronics. Resolution is typically in the range of one tenth to one half as many megapixels as contemporary low end compact digital cameras.

Major manufacturers of cameras for phones include Toshiba, ST Micro, Sharp, Omnivision, and Aptina (Now part of ON Semiconductor).

External Camera

During 2003 as camera phones were gaining popularity in Europe some phones without cameras had support for MMS and external cameras that could be connected with a small cable or directly to the data port at the base of the phone. The external cameras were comparable in quality to those fitted on regular camera phones at the time, typically offering VGA resolution.

In 2013-2014 Sony and other manufacturers announced add-on camera modules for smartphones called lens-style cameras. They have larger sensors and lenses than those in a camera phone but lack a viewfinder, display and most controls. They can be mounted to an Android or iOS phone or tablet and use its display and controls. Lens-style cameras include:

- Sony SmartShot QX series, announced and released in mid 2013. They include the DSC-QX-100/B, the large Sony ILCE-QX1, and the small Sony DSC-QX30.

- Kodak PixPro smart lens camera series, announced in 2014.

- Vivicam smart lens camera series from Vivitar/Sakar, announced in 2014.

- HTC RE HTC also announced an external camera module for smartphones, the module can capture 16 MP still shots and cap capture Full HD videos. The RE Module is waterproof and dustproof so you can take it anywhere as u like.

External cameras for thermal imaging also became available in late 2014.

Social Impact

Personal photography allows people to capture and construct personal and group memory, maintain social relationships as well as expressing their identity. The hundreds of millions of camera phones sold every year provide the same opportunities, yet these functions are altered and allow for a different user experience. As mobile phones are constantly carried, camera phones allow for capturing moments at any time. Mobile communication also allows for immediate transmission of content (for example via Multimedia Messaging Services), which cannot be reversed or regulated. Brooke Knight observes that "the carrying of an external, non-integrated camera (like a DSLR) always changes the role of the wearer at an event, from participant to photographer". The cameraphone user, on the other hand, can remain a participant in whatever moment they photograph. Photos taken on a cameraphone serve to prove the physical presence of the photographer. The immediacy of sharing and the liveness that comes with it allows the photographs shared through cameraphones to emphasize their indexing of the photographer.

Taking a photograph with cell phone.

While phones have been found useful by tourists and for other common civilian purposes, as they are cheap, convenient, and portable; they have also posed controversy, as they enable secret photography. A user may pretend to be simply talking on the phone or browsing the internet, drawing no suspicion while photographing a person or place in non-public areas where photography is restricted, or perform photography against that person's wishes. At the same time, camera phones have enabled every citizen to exercise her or his freedom of speech by being able to quickly communicate to others what she or he has seen with their own eyes. In most democratic free countries, there are no restrictions against photography in public and thus camera phones enable new forms of citizen journalism, fine art photography, and recording one's life experiences for facebooking or blogging.

Camera phones have also been very useful to street photographers and social documentary photographers as they enable them to take pictures of strangers in the street without them noticing, thus allowing the artist/photographer to get close to her or his subjects and take more liveful photos. While most people are suspect of secret photography, artists who do street photography (like Henri Cartier-Bresson did), photojournalists and photographers documenting people in public (like the photographers who documented the Great Depression in 1930s America) must often work unnoticed as their subjects are often unwilling to be photographed or are not aware of legitimate uses of secret photography like those photos that end up in fine art galleries and journalism.

As a network-connected device, megapixel camera phones are playing significant roles in crime prevention, journalism and business applications as well as individual uses. They can also be used for activities such as voyeurism, invasion of privacy, and copyright infringement. Because they can be used to share media almost immediately, they are a potent personal content creation tool. On January 17, 2007, New York City Mayor Michael Bloomberg announced a plan to encourage people to use their camera-phones to capture crimes happening in progress or dangerous situations and send them to emergency responders. Through the program, people will be able to send their images or video directly to 911.

Enforcing bans on camera phones has proven nearly impossible. They are small and numerous and their use is easy to hide or disguise, making it hard for law enforcement and security personnel to detect or stop use. Total bans on camera phones would also raise questions about freedom of speech and the freedom of the press, since camera phone ban would prevent a citizen or a journalist (or a citizen journalist) from communicating to others a newsworthy event that could be captured with a camera phone.

From time to time, organizations and places have prohibited or restricted the use of camera phones and other cameras because of the privacy, security, and copyright issues they pose. Such places include the Pentagon, federal and state courts, museums, schools, theaters, and local fitness clubs. Saudi Arabia, in April 2004, banned the sale of camera phones nationwide for a time before reallowing their sale in December 2004 (although pilgrims on the Hajj were allowed to bring in camera phones). There is the occasional anecdote of camera phones linked to industrial espionage and the activities of paparazzi (which are legal but often controversial), as well as some hacking into wireless operators' network.

Camera phones have also been used to discreetly take photographs in museums, performance halls, and other places where photography is prohibited. However, as sharing can be instantaneous, even if the action is discovered, it is too late, as the image is already out of reach, unlike a photo taken by a digital camera that only stores images locally for later transfer (however, as the newer digital cameras support Wi-Fi, a photographer can perform photography with a DSLR and instantly post the photo on the internet through the mobile phone's Wi-Fi and 3G capabilities).

In 2010, in Ireland the annual "RTÉ 60 second short award" was won by 15-year-old Laura Gaynor, who made her winning cartoon,"Piece of Cake" on her Sony Ericsson C510 camera phone.

In 2012, Director/writer Eddie Brown Jr, made the reality thriller *Camera Phone* which is one of the first commercial produced movies using camera phones as the story's prospective. The film is a reenactment of an actual case and they changed the names to protect those involved.

Some modern camera phones (in 2013-2014) have big sensors, thus allowing a street photographer or any other kind of photographer to take photos of similar quality to a semi-pro camera.

Apart from street photographers and social documentary photographers or cinematographers, camera phones have also been used successfully by war photographers. The small size of the camera phone allows a war photographer to secretly film the men and women who fight in a war, without them realizing that they have been photographed, thus the camera phone allows the war photographer to document wars while maintaining her or his safety.

Notable Events Involving Camera Phones

- The 2004 Indian Ocean earthquake was the first global news event where the majority of the first day news footage was no longer provided by professional news crews, but rather by citizen journalists, using primarily camera phones.

- On November 17, 2006, during a performance at the Laugh Factory comedy club, comedian Michael Richards was recorded responding to hecklers with racial slurs by a member of the audience using a camera phone. The video was widely circulated in television and internet news broadcasts.

- On December 30, 2006, the execution of former Iraqi dictator Saddam Hussein was recorded by a video camera phone, and made widely available on the Internet. A guard was arrested a few days later.

- Camera phone video and photographs taken in the immediate aftermath of the 7 July 2005 London bombings were featured worldwide. CNN executive Jonathan Klein predicts camera phone footage will be increasingly used by news organizations.

- Camera phone digital images helped to spread the 2009 Iranian election protests.

- Camera phones recorded the BART Police shooting of Oscar Grant.

Camera as an Interaction Device

The cameras of smartphones are used as input devices in numerous research projects and commercial applications. A commercially successful example is the use of QR Codes attached to physical objects. QR Codes can be sensed by the phone using its camera and provide an according link to related digital content, usually a URL. Another approach is using camera images to recognize objects. Content-based image analysis is used to recognize physical objects such as advertisement posters to provide information about the object. Hybrid approaches use a combination of unobstrusive visual markers and image analysis. An example is to estimate the pose of the camera phone to create a real-time overlay for a 3D paper globe.

Some smartphones can provide an augmented reality overlay for 2D objects and to recognize multiple objects on the phone using a stripped down object recognition algorithm as well as using GPS and compass. A few can translate text from a foreign language. Auto-geotagging can show where a picture is taken, promoting interactions and allowing a photo to be mapped with others for comparison.

Besides the usual back camera, almost all smartphones have a front camera of lesser performance facing the user for purposes including videoconferencing and mainly for self-portraiture (selfie).

Laws

Camera phones, or more specifically, widespread use of such phones as cameras by the general public, has increased exposure to laws relating to public and private photography. The laws that relate to other types of cameras also apply to camera phones. There are no special laws for camera phones. Privacy is a major concern these days, more people are worried these days as compared to ten or even twenty years ago.

Comparison Photos

With camera phone

With camera

Multimedia Messaging Service

Multimedia Messaging Service (MMS) is a standard way to send messages that include multimedia content to and from mobile phones over a cellular network. Users and providers may refer to such a message as a PXT, a picture message, or a multimedia message. The MMS standard extends the core SMS (Short Message Service) capability, allowing the exchange of text messages greater than 160 characters in length. Unlike text-only SMS, MMS can deliver a variety of media, including up to forty seconds of video, one image, a slideshow of multiple images, or audio.

The most common use involves sending photographs from camera-equipped handsets. Media companies have utilized MMS on a commercial basis as a method of delivering news and entertainment content, and retailers have deployed it as a tool for delivering scannable coupon codes, product images, videos, and other information.

The 3GPP and WAP groups fostered the development of the MMS standard, which is now continued by the Open Mobile Alliance (OMA).

History

Multimedia messaging services were first developed[when?] as a captive technology which enabled service providers to "collect a fee every time anyone snaps a photo."

Early MMS deployments were plagued by technical issues and frequent consumer disappointments. In recent years, MMS deployment by major technology companies have solved many of the early challenges through handset detection, content optimization, and increased throughput.

China was one of the early markets to make MMS a major commercial success, partly as the penetration rate of personal computers was modest but MMS-capable camera phones spread rapidly. The chairman and CEO of China Mobile said at the GSM Association Mobile Asia Congress in 2009 that MMS in China was now a mature service on par with SMS text messaging.

Europe's most advanced MMS market has been Norway, and in 2008, the Norwegian MMS usage level passed 84% of all mobile phone subscribers. Norwegian mobile subscribers sent on average one MMS per week.

Between 2010 and 2013, MMS traffic in the U.S. increased by 70% from 57 billion to 96 billion messages sent. This is due in part to the wide adoption of smartphones.

Technical Description

MMS messages are delivered in a different way from SMS. The first step is for the sending device to encode the multimedia content in a fashion similar to sending a MIME message (MIME content formats are defined in the MMS Message Encapsulation specification). The message is then forwarded to the carrier's MMS store and forward server, known as the MMSC (Multimedia Messaging Service Centre). If the receiver is on a carrier different from the sender, then the MMSC acts as a relay, and forwards the message to the MMSC of the recipient's carrier using the internet.

Once the recipient's MMSC has received a message, it first determines whether the receiver's handset is "MMS capable", that it supports the standards for receiving MMS. If so, the content is extracted and sent to a temporary storage server with an HTTP front-end. An SMS "control message" containing the URL of the content is then sent to the recipient's handset to trigger the receiver's WAP browser to open and receive the content from the embedded URL. Several other messages are exchanged to indicate the status of the delivery attempt. Before delivering content, some MMSCs also include a conversion service that will attempt to modify the multimedia content into a format suitable for the receiver. This is known as "content adaptation".

If the receiver's handset is not MMS capable, the message is usually delivered to a web-based service from where the content can be viewed from a normal internet browser. The URL for the content is usually sent to the receiver's phone in a normal text message. This behavior is usually known as a "legacy experience" since content can still be received by a phone number, even if the phone itself does not support MMS.

The method for determining whether a handset is MMS capable is not specified by the standards. A database is usually maintained by the operator, and in it each mobile phone number is marked as being associated with a legacy handset or not. This method is unreliable, however, because customers can independently change their handsets, and many of these databases are not updated dynamically.

MMS does not utilize operator-maintained "data" plans to distribute multimedia content, which is only used if the operator clicks links inside the message.

E-mail and web-based gateways to the MMS system are common. On the reception side, the content servers can typically receive service requests both from WAP and normal HTTP browsers, so delivery via the web is simple. For sending from external sources to handsets, most carriers allow a MIME encoded message to be sent to the receiver's phone number using a special e-mail address combining the recipient's public phone number and a special domain name, which is typically carrier-specific.

Challenges

There are some interesting challenges with MMS that do not exist with SMS:

Handset configuration can cause problems sending and receiving MMS messages.

- Content adaptation: Multimedia content created by one brand of MMS phone may not be entirely compatible with the capabilities of the recipient's MMS phone. In the MMS architecture, the recipient MMSC is responsible for providing for *content adaptation* (e.g., image resizing, audio codec transcoding, etc.), if this feature is enabled by the mobile network operator. When content adaptation is supported by a network operator, its MMS subscribers enjoy compatibility with a larger network of MMS users than would otherwise be available.

- Distribution lists: Current MMS specifications do not include distribution lists nor methods by which large numbers of recipients can be conveniently addressed, particularly by content providers, called *Value-added service providers* (VASPs) in 3GPP. Since most SMSC vendors have adopted FTP as an ad-hoc method by which large distribution lists are transferred to the SMSC prior to being used in a bulk-messaging SMS submission, it is expected that MMSC vendors will also adopt FTP.

- Bulk messaging: The flow of *peer-to-peer* MMS messaging involves several over-the-air transactions that become inefficient when MMS is used to send messages to large numbers of subscribers, as is typically the case for VASPs. For example, when one MMS message is submitted to a very large number of recipients, it is possible to receive a *delivery report* and *read-reply report* for each and every recipient. Future MMS specification work is likely to optimize and reduce the transactional overhead for the bulk-messaging case.

- Handset Configuration: Unlike SMS, MMS requires a number of handset parameters to be set. Poor handset configuration is often blamed as the first point of failure for many users. Service settings are sometimes preconfigured on the handset, but mobile operators are now looking at new device management technologies as a means of delivering the necessary settings for data services (MMS, WAP, etc.) via over-the-air programming (OTA).

- WAP Push: Few mobile network operators offer direct connectivity to their MMSCs for content providers. This has resulted in many content providers using WAP push as the only method available to deliver 'rich content' to mobile handsets. WAP push enables 'rich content' to be delivered to a handset by specifying the URL (via binary SMS) of a pre-compiled MMS, hosted on a content provider's Web server. A consequence is that the receiver who pays WAP per kb or minute (as opposed to a flat monthly fee) pays for receiving the MMS, as opposed to only paying for sending one, and also paying a different rate.

Although the standard does not specify a maximum size for a message, 300 kB is the current recommended size used by networks[which?] due to some limitations on the WAP gateway side.

Interfaces

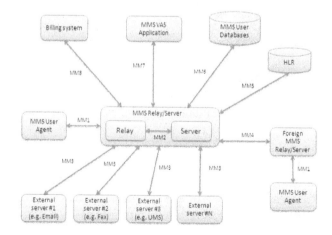

MMSC Reference Architecture

- MM1: the 3GPP interface between MMS User Agent and MMS Center (MMSC, the combination of the MMS Relay & Server)

- MM2: the 3GPP interface between MMS Relay and MMS Server

- MM3: the 3GPP interface between MMSC and external servers

- MM4: the 3GPP interface between different MMSCs

- MM5: the 3GPP interface between MMSC and HLR

- MM6: the 3GPP interface between MMSC and user databases

- MM7: the 3GPP interface between MMS VAS applications and MMSC

- MM8: the 3GPP interface between MMSC and the billing systems

- MM9: the 3GPP interface between MMSC and an online charging system

- MM10: the 3GPP interface between MMSC and a message service control function

- MM11: the 3GPP interface between MMSC and an external transcoder

Email

Electronic mail is a method of exchanging digital messages between computer users; Email first entered substantial use in the 1960s and by the 1970s had taken the form now recognised as email. Email operates across computer networks, which in the 2010s is primarily the Internet. Some early email systems required the author and the recipient to both be online at the same time, in common with instant messaging. Today's email systems are based on a store-and-forward model. Email servers accept, forward, deliver, and store messages. Neither the users nor their computers are required to be online simultaneously; they need to connect only briefly, typically to a mail server, for as long as it takes to send or receive messages.

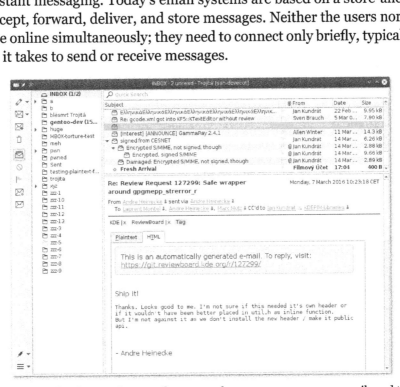

This screenshot shows the "Inbox" page of an email system, where users can see new emails and take actions, such as reading, deleting, saving, or responding to these messages.

The at sign, a part of every SMTP email address.

Originally an ASCII text-only communications medium, Internet email was extended by Multi-purpose Internet Mail Extensions (MIME) to carry text in other character sets and multimedia content attachments. International email, with internationalized email addresses using UTF-8, has been standardized, but as of 2016 not widely adopted. . Indian population started using Internationalized email address in 8 languages on .bharat (in 8 linguistic scripts) as probably the world's first IDN Email service user provided by Data Xgen Technologies. The history of modern, global Internet email services reaches back to the early ARPANET, with standards for encoding email messages proposed as early as 1973 (RFC 561). An email message sent in the early 1970s looks very similar to a basic text email sent today. Email played an important part in creating the Internet, and the conversion from ARPANET to the Internet in the early 1980s produced the core of the current services. The ARPANET initially used extensions to the File Transfer Protocol (FTP) to exchange network email, but this is now done with the Simple Mail Transfer Protocol (SMTP), first published as Internet standard 10 (RFC 821) in 1982.

Terminology

Historically, the term *electronic mail* was used generically for any electronic document transmission. For example, several writers in the early 1970s used the term to describe fax document transmission. As a result, it is difficult to find the first citation for the use of the term with the more specific meaning it has today.

Electronic mail has been most commonly called email or e-mail since around 1993, but various variations of the spelling have been used:

- *email* is the most common form used online, and is required by IETF Requests for Comments and working groups and increasingly by style guides. This spelling also appears in most dictionaries.

- *e-mail* has long been the form that appears most frequently in edited, published American English and British English writing as reflected in the Corpus of Contemporary American English data and style guides.

- *mail* was the form used in the original RFC. The service is referred to as *mail*, and a single piece of electronic mail is called a *message*.

- *EMail* is a traditional form that has been used in RFCs for the "Author's Address" and is expressly required "for historical reasons".

- *E-mail* is sometimes used, capitalizing the initial *E* as in similar abbreviations like *E-piano*, *E-guitar*, *A-bomb*, and *H-bomb*.

Origin

The AUTODIN network, first operational in 1962, provided a message service between 1,350 terminals, handling 30 million messages per month, with an average message length of approximately 3,000 characters. Autodin was supported by 18 large computerized switches, and was connected to the United States General Services Administration Advanced Record System, which provided similar services to roughly 2,500 terminals.

Host-based Mail Systems

With the introduction of MIT's Compatible Time-Sharing System (CTSS) in 1961 multiple users could log in to a central system from remote dial-up terminals, and to store and share files on the central disk. Informal methods of using this to pass messages were developed and expanded:

- 1965 – MIT's CTSS MAIL.

Developers of other early systems developed similar email applications:

- 1962 – 1440/1460 Administrative Terminal System

- 1968 – ATS/360

- 1971 – *SNDMSG*, a local inter-user mail program incorporating the experimental file transfer program, *CPYNET*, allowed the first networked electronic mail

- 1972 – Unix mail program

- 1972 – APL Mailbox by Larry Breed

- 1974 – The PLATO IV Notes on-line message board system was generalized to offer 'personal notes' in August 1974.

- 1978 – *Mail* client written by Kurt Shoens for Unix and distributed with the Second Berkeley Software Distribution included support for aliases and distribution lists, forwarding, formatting messages, and accessing different mailboxes. It used the Unix *mail* client to send messages between system users. The concept was extended to communicate remotely over the Berkley Network.

- 1979 – *EMAIL* written by V.A. Shiva Ayyadurai to emulate the interoffice mail system of the University of Medicine and Dentistry of New Jersey

- 1979 – MH Message Handling System developed at RAND provided several tools for managing electronic mail on Unix.

- 1981 – PROFS by IBM

- 1982 – ALL-IN-1 by Digital Equipment Corporation

These original messaging systems had widely different features and ran on systems that were incompatible with each other. Most of them only allowed communication between users logged into the same host or "mainframe", although there might be hundreds or thousands of users within an organization.

LAN Email Systems

In the early 1980s, networked personal computers on LANs became increasingly important. Server-based systems similar to the earlier mainframe systems were developed. Again, these systems initially allowed communication only between users logged into the same server infrastructure. Examples include:

- cc:Mail

- Lantastic

- WordPerfect Office

- Microsoft Mail

- Banyan VINES

- Lotus Notes

Eventually these systems too could link different organizations as long as they ran the same email system and proprietary protocol.

Email Networks

To facilitate electronic mail exchange between remote sites and with other organizations, telecommunication links, such as dialup modems or leased lines, provided means to transport email globally, creating local and global networks. This was challenging for a number of reasons, including the widely different email address formats in use.

- In 1971 the first ARPANET email was sent, and through RFC 561, RFC 680, RFC 724, and finally 1977's RFC 733, became a standardized working system.

- PLATO IV was networked to individual terminals over leased data lines prior to the implementation of personal notes in 1974.

- Unix mail was networked by 1978's uucp, which was also used for USENET newsgroup postings, with similar headers.

- BerkNet, the Berkeley Network, was written by Eric Schmidt in 1978 and included first in the Second Berkeley Software Distribution. It provided support for sending and receiving messages over serial communication links. The Unix mail tool was extended to send messages using BerkNet.

- The delivermail tool, written by Eric Allman in 1979 and 1980 (and shipped in 4BSD), provided support for routing mail over different networks, including Arpanet, UUCP, and BerkNet. (It also provided support for mail user aliases).

- The mail client included in 4BSD (1980) was extended to provide interoperability between a variety of mail systems.

- BITNET (1981) provided electronic mail services for educational institutions. It was based on the IBM VNET email system.

- 1983 – MCI Mail Operated by MCI Communications Corporation. This was the first commercial public email service to use the internet. MCI Mail also allowed subscribers to send regular postal mail (overnight) to non-subscribers.

- In 1984, IBM PCs running DOS could link with FidoNet for email and shared bulletin board posting.

Email Address Internationalization

Globally countries started adopting IDN registrations for supporting country specific scripts (non-english) for domain names. In 2010 Egypt, the Russian Federation, Saudi Arabia, and the United Arab Emirates started offering IDN registrations. Govt. of India also started .bharat (in 8 languages) IDN registration in 2014 and in 2016 XgenPlus also started providing email address in Hindi, Gujrati, Marathi, Bangali, Tamil, Telgu, Punjabi and Urdu.

Attempts at Interoperability

Early interoperability among independent systems included:

- ARPANET, a forerunner of the Internet, defined protocols for dissimilar computers to exchange email.

- uucp implementations for Unix systems, and later for other operating systems, that only had dial-up communications available.

- CSNET, which initially used the UUCP protocols via dial-up to provide networking and mail-relay services for non-ARPANET hosts.

- Novell developed the Message Handling System (MHS) protocol, but abandoned it after purchasing the non-MHS WordPerfect Office (renamed Groupwise).

- The Coloured Book protocols ran on UK academic networks until 1992.

- X.400 in the 1980s and early 1990s was promoted by major vendors, and mandated for government use under GOSIP, but abandoned by all but a few in favor of Internet SMTP by the mid-1990s.

From SNDMSG to MSG

In the early 1970s, Ray Tomlinson updated an existing utility called SNDMSG so that it could copy messages (as files) over the network. Lawrence Roberts, the project manager for the ARPANET development, took the idea of READMAIL, which dumped all "recent" messages onto the user's terminal, and wrote a programme for TENEX in TECO macros called *RD*, which permitted access to individual messages. Barry Wessler then updated RD and called it *NRD*.

Marty Yonke rewrote NRD to include reading, access to SNDMSG for sending, and a help system, and called the utility *WRD*, which was later known as *BANANARD*. John Vittal then updated this version to include three important commands: *Move* (combined save/delete command), *Answer* (determined to whom a reply should be sent) and *Forward* (sent an email to a person who was not already a recipient). The system was called *MSG*. With inclusion of these features, MSG is considered to be the first integrated modern email programme, from which many other applications have descended.

ARPANET Mail

Experimental email transfers between separate computer systems began shortly after the creation of the ARPANET in 1969. Ray Tomlinson is generally credited as having sent the first email across a network, initiating the use of the "@" sign to separate the names of the user and the user's machine in 1971, when he sent a message from one Digital Equipment Corporation DEC-10 computer to another DEC-10. The two machines were placed next to each other. Tomlinson's work was quickly adopted across the ARPANET, which significantly increased the popularity of email.

Initially addresses were of the form, *username@hostname* but were extended to "username@ host.domain" with the development of the Domain Name System (DNS).

As the influence of the ARPANET spread across academic communities, gateways were developed to pass mail to and from other networks such as CSNET, JANET, BITNET, X.400, and FidoNet. This often involved addresses such as:

> hubhost!middlehost!edgehost!user@uucpgateway.somedomain.example.com

which routes mail to a user with a "bang path" address at a UUCP host.

Operation

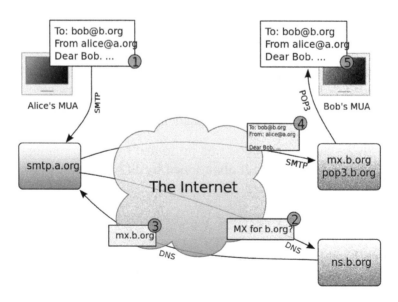

The diagram to the right shows a typical sequence of events that takes place when sender Alice transmits a message using a mail user agent (MUA) addressed to the email address of the recipient.

1. The MUA formats the message in email format and uses the submission protocol, a profile of the Simple Mail Transfer Protocol (SMTP), to send the message to the local mail submission agent (MSA), in this case *smtp.a.org*.

2. The MSA determines the destination address provided in the SMTP protocol (not from the message header), in this case *bob@b.org*. The part before the @ sign is the *local part* of the address, often the username of the recipient, and the part after the @ sign is a domain name. The MSA resolves a domain name to determine the fully qualified domain name of the mail server in the Domain Name System (DNS).

3. The DNS server for the domain *b.org* (*ns.b.org*) responds with any MX records listing the mail exchange servers for that domain, in this case *mx.b.org*, a message transfer agent (MTA) server run by the recipient's ISP.

4. smtp.a.org sends the message to mx.b.org using SMTP. This server may need to forward the message to other MTAs before the message reaches the final message delivery agent (MDA).

5. The MDA delivers it to the mailbox of user *bob*.

6. Bob's MUA picks up the message using either the Post Office Protocol (POP3) or the Internet Message Access Protocol (IMAP).

In addition to this example, alternatives and complications exist in the email system:

- Alice or Bob may use a client connected to a corporate email system, such as IBM Lotus Notes or Microsoft Exchange. These systems often have their own internal email format and their clients typically communicate with the email server using a vendor-specific, proprietary protocol. The server sends or receives email via the Internet through the product's Internet mail gateway which also does any necessary reformatting. If Alice and Bob work for the same company, the entire transaction may happen completely within a single corporate email system.

- Alice may not have a MUA on her computer but instead may connect to a webmail service.

- Alice's computer may run its own MTA, so avoiding the transfer at step 1.

- Bob may pick up his email in many ways, for example logging into mx.b.org and reading it directly, or by using a webmail service.

- Domains usually have several mail exchange servers so that they can continue to accept mail even if the primary is not available.

Many MTAs used to accept messages for any recipient on the Internet and do their best to deliver them. Such MTAs are called *open mail relays*. This was very important in the early days of the Internet when network connections were unreliable. However, this mechanism proved to be exploitable by originators of unsolicited bulk email and as a consequence open mail relays have become rare, and many MTAs do not accept messages from open mail relays.

Message Format

The Internet email message format is now defined by RFC 5322, with multimedia content attachments being defined in RFC 2045 through RFC 2049, collectively called *Multipurpose Internet Mail Extensions* or *MIME*. RFC 5322 replaced the earlier RFC 2822 in 2008, and in turn RFC 2822 in 2001 replaced RFC 822 – which had been the standard for Internet email for nearly 20 years. Published in 1982, RFC 822 was based on the earlier RFC 733 for the ARPANET.

Internet email messages consist of two major sections, the message header and the message body. The header is structured into fields such as From, To, CC, Subject, Date, and other information about the email. In the process of transporting email messages between systems, SMTP communicates delivery parameters and information using message header fields. The body contains the message, as unstructured text, sometimes containing a signature block at the end. The header is separated from the body by a blank line.

Message Header

Each message has exactly one header, which is structured into fields. Each field has a name and a value. RFC 5322 specifies the precise syntax.

Informally, each line of text in the header that begins with a printable character begins a separate field. The field name starts in the first character of the line and ends before the separator character ":". The separator is then followed by the field value (the "body" of the field). The value is continued onto subsequent lines if those lines have a space or tab as their first character. Field names and values are restricted to 7-bit ASCII characters. Non-ASCII values may be represented using MIME encoded words.

Header Fields

Email header fields can be multi-line, and each line should be at most 78 characters long and in no event more than 998 characters long. Header fields defined by RFC 5322 can only contain US-ASCII characters; for encoding characters in other sets, a syntax specified in RFC 2047 can be used. Recently the IETF EAI working group has defined some standards track extensions, replacing previous experimental extensions, to allow UTF-8 encoded Unicode characters to be used within the header. In particular, this allows email addresses to use non-ASCII characters. Such addresses are supported by Google and Microsoft products, and promoted by some governments.

The message header must include at least the following fields:

- *From*: The email address, and optionally the name of the author(s). In many email clients not changeable except through changing account settings.

- *Date*: The local time and date when the message was written. Like the *From:* field, many email clients fill this in automatically when sending. The recipient's client may then display the time in the format and time zone local to him/her.

The message header should include at least the following fields:

- *Message-ID*: Also an automatically generated field; used to prevent multiple delivery and for reference in In-Reply-To.

- *In-Reply-To*: Message-ID of the message that this is a reply to. Used to link related messages together. This field only applies for reply messages.

RFC 3864 describes registration procedures for message header fields at the IANA; it provides for permanent and provisional field names, including also fields defined for MIME, netnews, and HTTP, and referencing relevant RFCs. Common header fields for email include:

- *To*: The email address(es), and optionally name(s) of the message's recipient(s). Indicates primary recipients (multiple allowed), for secondary recipients.

- *Subject*: A brief summary of the topic of the message. Certain abbreviations are commonly used in the subject, including "RE:" and "FW:".

- *Bcc*: Blind carbon copy; addresses added to the SMTP delivery list but not (usually) listed in the message data, remaining invisible to other recipients.

- *Cc*: Carbon copy; Many email clients will mark email in one's inbox differently depending on whether they are in the To: or Cc: list.

- Content-Type: Information about how the message is to be displayed, usually a MIME type.

- *Precedence*: commonly with values "bulk", "junk", or "list"; used to indicate that automated "vacation" or "out of office" responses should not be returned for this mail, e.g. to prevent vacation notices from being sent to all other subscribers of a mailing list. Sendmail uses this field to affect prioritization of queued email, with "Precedence: special-delivery" messages delivered sooner. With modern high-bandwidth networks, delivery priority is less of an issue than it once was. Microsoft Exchange respects a fine-grained automatic response suppression mechanism, the *X-Auto-Response-Suppress* field.

- *References*: Message-ID of the message that this is a reply to, and the message-id of the message the previous reply was a reply to, etc.

- *Reply-To*: Address that should be used to reply to the message.

- *Sender*: Address of the actual sender acting on behalf of the author listed in the From: field (secretary, list manager, etc.).

- *Archived-At*: A direct link to the archived form of an individual email message.

Note that the *To:* field is not necessarily related to the addresses to which the message is delivered. The actual delivery list is supplied separately to the transport protocol, SMTP, which may or may not originally have been extracted from the header content. The "To:" field is similar to the addressing at the top of a conventional letter which is delivered according to the address on the outer envelope. In the same way, the "From:" field does not have to be the real sender of the email message. Some mail servers apply email authentication systems to messages being relayed. Data pertaining to server's activity is also part of the header, as defined below.

SMTP defines the *trace information* of a message, which is also saved in the header using the following two fields:

- *Received*: when an SMTP server accepts a message it inserts this trace record at the top of the header (last to first).

- *Return-Path*: when the delivery SMTP server makes the *final delivery* of a message, it inserts this field at the top of the header.

Other fields that are added on top of the header by the receiving server may be called *trace fields*, in a broader sense.

- *Authentication-Results*: when a server carries out authentication checks, it can save the results in this field for consumption by downstream agents.

- *Received-SPF*: stores results of SPF checks in more detail than Authentication-Results.

- *Auto-Submitted*: is used to mark automatically generated messages.

- *VBR-Info*: claims VBR whitelisting

Message Body

Content Encoding

Email was originally designed for 7-bit ASCII. Most email software is 8-bit clean but must assume it will communicate with 7-bit servers and mail readers. The MIME standard introduced character set specifiers and two content transfer encodings to enable transmission of non-ASCII data: quoted printable for mostly 7-bit content with a few characters outside that range and base64 for arbitrary binary data. The 8BITMIME and BINARY extensions were introduced to allow transmission of mail without the need for these encodings, but many mail transport agents still do not support them fully. In some countries, several encoding schemes coexist; as the result, by default, the message in a non-Latin alphabet language appears in non-readable form (the only exception is coincidence, when the sender and receiver use the same encoding scheme). Therefore, for international character sets, Unicode is growing in popularity.

Plain Text and HTML

Most modern graphic email clients allow the use of either plain text or HTML for the message body at the option of the user. HTML email messages often include an automatically generated plain text copy as well, for compatibility reasons. Advantages of HTML include the ability to include inline links and images, set apart previous messages in block quotes, wrap naturally on any display, use emphasis such as underlines and italics, and change font styles. Disadvantages include the increased size of the email, privacy concerns about web bugs, abuse of HTML email as a vector for phishing attacks and the spread of malicious software.

Some web-based mailing lists recommend that all posts be made in plain-text, with 72 or 80 characters per line for all the above reasons, but also because they have a significant number of readers using text-based email clients such as Mutt. Some Microsoft email clients allow rich formatting using their proprietary Rich Text Format (RTF), but this should be avoided unless the recipient is guaranteed to have a compatible email client.

Servers and Client Applications

Messages are exchanged between hosts using the Simple Mail Transfer Protocol with software programs called mail transfer agents (MTAs); and delivered to a mail store by programs called mail delivery agents (MDAs, also sometimes called local delivery agents, LDAs). Accepting a message obliges an MTA to deliver it, and when a message cannot be delivered, that MTA must send a bounce message back to the sender, indicating the problem.

The interface of an email client, Thunderbird.

Users can retrieve their messages from servers using standard protocols such as POP or IMAP, or, as is more likely in a large corporate environment, with a proprietary protocol specific to Novell Groupwise, Lotus Notes or Microsoft Exchange Servers. Programs used by users for retrieving, reading, and managing email are called mail user agents (MUAs).

Mail can be stored on the client, on the server side, or in both places. Standard formats for mailboxes include Maildir and mbox. Several prominent email clients use their own proprietary format and require conversion software to transfer email between them. Server-side storage is often in a proprietary format but since access is through a standard protocol such as IMAP, moving email from one server to another can be done with any MUA supporting the protocol.

Many current email users do not run MTA, MDA or MUA programs themselves, but use a web-based email platform, such as Gmail, Hotmail, or Yahoo! Mail, that performs the same tasks. Such webmail interfaces allow users to access their mail with any standard web browser, from any computer, rather than relying on an email client.

Filename Extensions

Upon reception of email messages, email client applications save messages in operating system files in the file system. Some clients save individual messages as separate files, while others use various database formats, often proprietary, for collective storage. A historical standard of storage is the *mbox* format. The specific format used is often indicated by special filename extensions:

eml

> Used by many email clients including Novell GroupWise, Microsoft Outlook Express, Lotus notes, Windows Mail, Mozilla Thunderbird, and Postbox. The files are plain text in MIME format, containing the email header as well as the message contents and attachments in one or more of several formats.

emlx

> Used by Apple Mail.

msg

> Used by Microsoft Office Outlook and OfficeLogic Groupware.

mbx

> Used by Opera Mail, KMail, and Apple Mail based on the mbox format.

Some applications (like Apple Mail) leave attachments encoded in messages for searching while also saving separate copies of the attachments. Others separate attachments from messages and save them in a specific directory.

URI Scheme Mailto

The URI scheme, as registered with the IANA, defines the mailto: scheme for SMTP email addresses. Though its use is not strictly defined, URLs of this form are intended to be used to open the new message window of the user's mail client when the URL is activated, with the address as defined by the URL in the *To:* field.

Types

Web-based Email

Many email providers have a web-based email client (e.g. AOL Mail, Gmail, Outlook.com and Yahoo! Mail). This allows users to log in to the email account by using any compatible web browser to send and receive their email. Mail is typically not downloaded to the client, so can't be read without a current Internet connection.

POP3 Email Services

The Post Office Protocol 3 (POP3) is a mail access protocol used by a client application to read messages from the mail server. Received messages are often deleted from the server. POP supports simple download-and-delete requirements for access to remote mailboxes (termed maildrop in the POP RFC's).

IMAP Email Servers

The Internet Message Access Protocol (IMAP) provides features to manage a mailbox from multiple devices. Small portable devices like smartphones are increasingly used to check email while travelling, and to make brief replies, larger devices with better keyboard access being used to reply at greater length. IMAP shows the headers of messages, the sender and the subject and the device needs to request to download specific messages. Usually mail is left in folders in the mail server.

MAPI Email Servers

Messaging Application Programming Interface (MAPI) is used by Microsoft Outlook to communicate to Microsoft Exchange Server - and to a range of other e-mail server products such as Axigen Mail Server, Kerio Connect, Scalix, Zimbra, HP OpenMail, IBM Lotus Notes, Zarafa, and Bynari where vendors have added MAPI support to allow their products to be accessed directly via Outlook.

Uses

Business and Organizational Use

Email has been widely accepted by business, governments and non-governmental organizations in the developed world, and it is one of the key parts of an 'e-revolution' in workplace communication (with the other key plank being widespread adoption of highspeed Internet). A 2010 study on workplace communication by Paytronics found 83% of U.S. knowledge workers felt email was critical to their success and productivity at work.

It has some key benefits to business and other organizations, including:

Facilitating Logistics

Much of the business world relies on communications between people who are not physically in the same building, area, or even country; setting up and attending an in-person meeting, telephone call, or conference call can be inconvenient, time-consuming, and costly. Email provides a method of exchanging information between two or more people with no set-up costs and that is generally far less expensive than a physical meeting or phone call.

Helping with Synchronisation

With real time communication by meetings or phone calls, participants must work on the same schedule, and each participant must spend the same amount of time in the meeting or call. Email allows asynchrony: each participant may control their schedule independently.

Reducing Cost

Sending an email is much less expensive than sending postal mail, or long distance telephone calls, telex or telegrams.

Increasing Speed

Much faster than most of the alternatives.

Creating a "Written" Record

Unlike a telephone or in-person conversation, email by its nature creates a detailed written record of the communication, the identity of the sender(s) and recipient(s) and the date and time the message was sent. In the event of a contract or legal dispute, saved emails can be used to prove that an individual was advised of certain issues, as each email has the date and time recorded on it.

Email Marketing

Email marketing via "opt-in" is often successfully used to send special sales offerings and new product information, but offering hyperlinks or generic information on consumer trends is less useful - and email sent without permission such as "opt-in" is likely to be viewed as unwelcome "email spam".

Personal Use

Desktop

Many users access their personal email from friends and family members using a desktop computer in their house or apartment.

Mobile

Email has become widely used on smartphones and Wi-Fi-enabled laptops and tablet computers. Mobile "apps" for email increase accessibility to the medium for users who are out of their home. While in the earliest years of email, users could only access email on desktop computers, in the 2010s, it is possible for users to check their email when they are away from home, whether they are across town or across the world. Alerts can also be sent to the smartphone or other device to notify them immediately of new messages. This has given email the ability to be used for more frequent communication between users and allowed them to check their email and write messages throughout the day. Today, there are an estimated 1.4 billion email users worldwide and 50 billion non-spam emails that are sent daily.

Individuals often check email on smartphones for both personal and work-related messages. It was found that US adults check their email more than they browse the web or check their Facebook accounts, making email the most popular activity for users to do on their smartphones. 78% of the respondents in the study revealed that they check their email on their phone. It was also found that 30% of consumers use only their smartphone to check their email, and 91% were likely to check their email at least once per day on their smartphone. However, the percentage of consumers using email on smartphone ranges and differs dramatically across different countries. For example, in comparison to 75% of those consumers in the US who used it, only 17% in India did.

Issues

Attachment Size Limitation

Email messages may have one or more attachments, which are additional files that are appended to the email. Typical attachments include Microsoft Word documents, pdf documents and scanned images of paper documents. In principle there is no technical restriction on the size or number of attachments, but in practice email clients, servers and Internet service providers implement various limitations on the size of files, or complete email - typically to 25MB or less. Furthermore, due to technical reasons, attachment sizes as seen by these transport systems can differ to what the user sees, which can be confusing to senders when trying to assess whether they can safely send a file by email. Where larger files need to be shared, file hosting services of various sorts are available; and generally suggested. Some large files, such as digital photos, color presentations and video or music files are too large for some email systems.

Information Overload

The ubiquity of email for knowledge workers and "white collar" employees has led to concerns that recipients face an "information overload" in dealing with increasing volumes of email. This can lead to increased stress, decreased satisfaction with work, and some observers even argue it could have a significant negative economic effect, as efforts to read the many emails could reduce productivity.

Spam

Email "spam" is the term used to describe unsolicited commercial, or bulk, email. The low cost of sending such email meant that by 2003 up to 30% of total email traffic was already spam. and was threatening the usefulness of email as a practical tool. The US CAN-SPAM Act of 2003 and similar laws elsewhere had some impact, and a number of effective anti-spam techniques now largely mitigate the impact of spam by filtering or rejecting it for most users, but the volume sent is still very high - and increasingly consists not of advertisements for products, but malicious content or links.

Malware

A range of malicious email types exist. These range from various types of email scams, including "social engineering" scams such as advance-fee scam "Nigerian letters", to phishing, email bombardment and email worms.

Email Spoofing

Email spoofing occurs when the email message header is designed to make the message appear to come from a known or trusted source. Email spam and phishing methods typically use spoofing to mislead the recipient about the true message origin. Email spoofing may be done as a prank, or as part of a criminal effort to defraud an individual or organization. An example of a potentially fraudulent email spoofing is if an individual creates an email which appears to be an invoice from a major company, and then sends it to one or more recipients. In some cases, these fraudulent emails incorporate the logo of the purported organization and even the email address may appear legitimate.

Email Bombing

Email bombing is the intentional sending of large volumes of messages to a target address. The overloading of the target email address can render it unusable and can even cause the mail server to crash.

Privacy Concerns

Today it can be important to distinguish between Internet and internal email systems. Internet email may travel and be stored on networks and computers without the sender's or the recipient's control. During the transit time it is possible that third parties read or even modify the content. Internal mail systems, in which the information never leaves the organizational network, may be more secure, although information technology personnel and others whose function may involve monitoring or managing may be accessing the email of other employees.

Email privacy, without some security precautions, can be compromised because:

- email messages are generally not encrypted.

- email messages have to go through intermediate computers before reaching their destination, meaning it is relatively easy for others to intercept and read messages.

- many Internet Service Providers (ISP) store copies of email messages on their mail servers before they are delivered. The backups of these can remain for up to several months on their server, despite deletion from the mailbox.

- the "Received:"-fields and other information in the email can often identify the sender, preventing anonymous communication.

There are cryptography applications that can serve as a remedy to one or more of the above. For example, Virtual Private Networks or the Tor anonymity network can be used to encrypt traffic from the user machine to a safer network while GPG, PGP, SMEmail, or S/MIME can be used for end-to-end message encryption, and SMTP STARTTLS or SMTP over Transport Layer Security/ Secure Sockets Layer can be used to encrypt communications for a single mail hop between the SMTP client and the SMTP server.

Additionally, many mail user agents do not protect logins and passwords, making them easy to intercept by an attacker. Encrypted authentication schemes such as SASL prevent this. Finally, attached files share many of the same hazards as those found in peer-to-peer filesharing. Attached files may contain trojans or viruses.

Flaming

Flaming occurs when a person sends a message (or many messages) with angry or antagonistic content. The term is derived from the use of the word "incendiary" to describe particularly heated email discussions. The ease and impersonality of email communications mean that the social norms that encourage civility in person or via telephone do not exist and civility may be forgotten.

Email Bankruptcy

Also known as "email fatigue", email bankruptcy is when a user ignores a large number of email messages after falling behind in reading and answering them. The reason for falling behind is often due to information overload and a general sense there is so much information that it is not possible to read it all. As a solution, people occasionally send a "boilerplate" message explaining that their email inbox is full, and that they are in the process of clearing out all the messages. Harvard University law professor Lawrence Lessig is credited with coining this term, but he may only have popularized it.

Tracking of Sent Mail

The original SMTP mail service provides limited mechanisms for tracking a transmitted message, and none for verifying that it has been delivered or read. It requires that each mail server must either deliver it onward or return a failure notice (bounce message), but both software bugs and system failures can cause messages to be lost. To remedy this, the IETF introduced Delivery Status Notifications (delivery receipts) and Message Disposition Notifications (return receipts); however, these are not universally deployed in production. (A complete Message Tracking mechanism was also defined, but it never gained traction.

Many ISPs now deliberately disable non-delivery reports (NDRs) and delivery receipts due to the activities of spammers:

- Delivery Reports can be used to verify whether an address exists and if so, this indicates to a spammer that it is available to be spammed.

- If the spammer uses a forged sender email address (email spoofing), then the innocent email address that was used can be flooded with NDRs from the many invalid email addresses the spammer may have attempted to mail. These NDRs then constitute spam from the ISP to the innocent user.

In the absence of standard methods, a range of system based around the use of web bugs have been developed. However, these are often seen as underhand or raising privacy concerns, and only work with e-mail clients that support rendering of HTML. Many mail clients now default to not showing "web content". Webmail providers can also disrupt web bugs by pre-caching images.

U.S. Government

The U.S. state and federal governments have been involved in electronic messaging and the development of email in several different ways. Starting in 1977, the U.S. Postal Service (USPS) recognized that electronic messaging and electronic transactions posed a significant threat to First Class mail volumes and revenue. The USPS explored an electronic messaging initiative in 1977 and later disbanded it. Twenty years later, in 1997, when email volume overtook postal mail volume, the USPS was again urged to embrace email, and the USPS declined to provide email as a service. The USPS initiated an experimental email service known as E-COM. E-COM provided a method for the simple exchange of text messages. In 2011, shortly after the USPS reported its state of financial bankruptcy, the USPS Office of Inspector General (OIG) began exploring the possibilities of generating revenue through email servicing. Electronic messages were transmitted to a post office, printed out, and delivered as hard copy. To take advantage of the service, an individual had to transmit at least 200 messages. The delivery time of the messages was the same as First Class mail and cost 26 cents. Both the Postal Regulatory Commission and the Federal Communications Commission opposed E-COM. The FCC concluded that E-COM constituted common carriage under its jurisdiction and the USPS would have to file a tariff. Three years after initiating the service, USPS canceled E-COM and attempted to sell it off.

The early ARPANET dealt with multiple email clients that had various, and at times incompatible, formats. For example, in the Multics, the "@" sign meant "kill line" and anything before the "@" sign was ignored, so Multics users had to use a command-line option to specify the destination system. The Department of Defense DARPA desired to have uniformity and interoperability for email and therefore funded efforts to drive towards unified inter-operable standards. This led to David Crocker, John Vittal, Kenneth Pogran, and Austin Henderson publishing RFC 733, "Standard for the Format of ARPA Network Text Message" (November 21, 1977), a subset of which provided a stable base for common use on the ARPANET, but which was not fully effective, and in 1979, a meeting was held at BBN to resolve incompatibility issues. Jon Postel recounted the meeting in RFC 808, "Summary of Computer Mail Services Meeting Held at BBN on 10 January 1979" (March 1, 1982), which includes an appendix listing the varying email systems at the time. This, in turn, led to the release of David Crocker's RFC 822, "Standard for the Format of ARPA Internet Text Messages" (August 13, 1982). RFC 822 is a small adaptation of RFC 733's details, notably enhancing the host portion, to use Domain Names, that were being developed at the same time.

The National Science Foundation took over operations of the ARPANET and Internet from the Department of Defense, and initiated NSFNet, a new backbone for the network. A part of the NSFNet AUP forbade commercial traffic. In 1988, Vint Cerf arranged for an interconnection of MCI Mail with NSFNET on an experimental basis. The following year Compuserve email interconnected with NSFNET. Within a few years the commercial traffic restriction was removed from NSFNETs AUP, and NSFNET was privatised. In the late 1990s, the Federal Trade Commission grew concerned with fraud transpiring in email, and initiated a series of procedures on spam, fraud, and phishing. In 2004, FTC jurisdiction over spam was codified into law in the form of the CAN SPAM Act. Several other U.S. federal agencies have also exercised jurisdiction including the Department of Justice and the Secret Service. NASA has provided email capabilities to astronauts aboard the Space Shuttle and International Space Station since 1991 when a Macintosh Portable was used aboard Space Shuttle mission STS-43 to send the first email via AppleLink. Today astronauts aboard the International Space Station have email capabilities via the wireless networking throughout the station and are connected to the ground at 10 Mbit/s Earth to station and 3 Mbit/s station to Earth, comparable to home DSL connection speeds.

Internet Access

Internet access is the process that enables individuals and organisations to connect to the Internet using computer terminals, computers, mobile devices, sometimes via computer networks. Once connected to the Internet, users can access Internet services, such as email and the World Wide Web. Internet service providers (ISPs) offer Internet access through various technologies that offer a wide range of data signaling rates (speeds).

Consumer use of the Internet first became popular through dial-up Internet access in the 1990s. By the first decade of the 21st century, many consumers in developed nations used faster, broadband Internet access technologies. By 2014 this was almost ubiquitous worldwide, with a global average connection speed exceeding 4 Mbit/s.

History

The Internet developed from the ARPANET, which was funded by the US government to support projects within the government and at universities and research laboratories in the US – but grew over time to include most of the world's large universities and the research arms of many technology companies. Use by a wider audience only came in 1995 when restrictions on the use of the Internet to carry commercial traffic were lifted.

In the early to mid-1980s, most Internet access was from personal computers and workstations directly connected to local area networks or from dial-up connections using modems and analog telephone lines. LANs typically operated at 10 Mbit/s, while modem data-rates grew from 1200 bit/s in the early 1980s, to 56 kbit/s by the late 1990s. Initially, dial-up connections were made from terminals or computers running terminal emulation software to terminal servers on LANs. These dial-up connections did not support end-to-end use of the Internet protocols and only provided terminal to host connections. The introduction of network access servers supporting the Serial Line Internet Protocol (SLIP) and later the point-to-point protocol (PPP) extended the Internet protocols and made the full range of Internet services available to dial-up users; although slower, due to the lower data rates available using dial-up.

Broadband Internet access, often shortened to just broadband, is simply defined as "Internet access that is always on, and faster than the traditional dial-up access" and so covers a wide range of technologies. Broadband connections are typically made using a computer's built in Ethernet networking capabilities, or by using a NIC expansion card.

Most broadband services provide a continuous "always on" connection; there is no dial-in process required, and it does not interfere with voice use of phone lines. Broadband provides improved access to Internet services such as:

- Faster world wide web browsing

- Faster downloading of documents, photographs, videos, and other large files

- Telephony, radio, television, and videoconferencing

- Virtual private networks and remote system administration

- Online gaming, especially massively multiplayer online role-playing games which are interaction-intensive

In the 1990s, the National Information Infrastructure initiative in the U.S. made broadband Internet access a public policy issue. In 2000, most Internet access to homes was provided using dial-up, while many businesses and schools were using broadband connections. In 2000 there were just under 150 million dial-up subscriptions in the 34 OECD countries and fewer than 20 million broadband subscriptions. By 2004, broadband had grown and dial-up had declined so that the number of subscriptions were roughly equal at 130 million each. In 2010, in the OECD countries, over 90% of the Internet access subscriptions used broadband, broadband had grown to more than 300 million subscriptions, and dial-up subscriptions had declined to fewer than 30 million.

The broadband technologies in widest use are ADSL and cable Internet access. Newer technologies include VDSL and optical fibre extended closer to the subscriber in both telephone and cable plants. Fibre-optic communication, while only recently being used in premises and to the curb schemes, has played a crucial role in enabling broadband Internet access by making transmission of information at very high data rates over longer distances much more cost-effective than copper wire technology.

In areas not served by ADSL or cable, some community organizations and local governments are installing Wi-Fi networks. Wireless and satellite Internet are often used in rural, undeveloped, or other hard to serve areas where wired Internet is not readily available.

Newer technologies being deployed for fixed (stationary) and mobile broadband access include WiMAX, LTE, and fixed wireless, e.g., Motorola Canopy.

Starting in roughly 2006, mobile broadband access is increasingly available at the consumer level using "3G" and "4G" technologies such as HSPA, EV-DO, HSPA+, and LTE.

Availability

In addition to access from home, school, and the workplace Internet access may be available from public places such as libraries and Internet cafes, where computers with Internet connections are available. Some libraries provide stations for physically connecting users' laptops to local area networks (LANs).

Wireless Internet access points are available in public places such as airport halls, in some cases just for brief use while standing. Some access points may also provide coin-operated computers. Various terms are used, such as "public Internet kiosk", "public access terminal", and "Web pay-phone". Many hotels also have public terminals, usually fee based.

Coffee shops, shopping malls, and other venues increasingly offer wireless access to computer networks, referred to as hotspots, for users who bring their own wireless-enabled devices such as a laptop or PDA. These services may be free to all, free to customers only, or fee-based. A Wi-Fi hotspot need not be limited to a confined location since multiple ones combined can cover a whole campus or park, or even an entire city can be enabled.

Additionally, Mobile broadband access allows smart phones and other digital devices to connect to the Internet from any location from which a mobile phone call can be made, subject to the capabilities of that mobile network.

Speed

The bit rates for dial-up modems range from as little as 110 bit/s in the late 1950s, to a maximum of from 33 to 64 kbit/s (V.90 and V.92) in the late 1990s. Dial-up connections generally require the dedicated use of a telephone line. Data compression can boost the effective bit rate for a dial-up modem connection to from 220 (V.42bis) to 320 (V.44) kbit/s. However, the effectiveness of data compression is quite variable, depending on the type of data being sent, the condition of the telephone line, and a number of other factors. In reality, the overall data rate rarely exceeds 150 kbit/s.

Broadband technologies supply considerably higher bit rates than dial-up, generally without disrupting regular telephone use. Various minimum data rates and maximum latencies have been used in definitions of broadband, ranging from 64 kbit/s up to 4.0 Mbit/s. In 1988 the CCITT standards body defined "broadband service" as requiring transmission channels capable of supporting bit rates greater than the primary rate which ranged from about 1.5 to 2 Mbit/s. A 2006 Organization for Economic Co-operation and Development (OECD) report defined broadband as having download data transfer rates equal to or faster than 256 kbit/s. And in 2015 the U.S. Federal Communications Commission (FCC) defined "Basic Broadband" as data transmission speeds of at least 25 Mbit/s downstream (from the Internet to the user's computer) and 3 Mbit/s upstream (from the user's computer to the Internet). The trend is to raise the threshold of the broadband definition as higher data rate services become available.

The higher data rate dial-up modems and many broadband services are "asymmetric"—supporting much higher data rates for download (toward the user) than for upload (toward the Internet).

Data rates, including those given in this article, are usually defined and advertised in terms of the maximum or peak download rate. In practice, these maximum data rates are not always reliably available to the customer. Actual end-to-end data rates can be lower due to a number of factors. In late June 2016, internet connection speeds averaged about 6 Mbit/s globally. Physical link quality can vary with distance and for wireless access with terrain, weather, building construction, antenna placement, and interference from other radio sources. Network bottlenecks may exist at points anywhere on the path from the end-user to the remote server or service being used and not just on the first or last link providing Internet access to the end-user.

Network Congestion

Users may share access over a common network infrastructure. Since most users do not use their full connection capacity all of the time, this aggregation strategy (known as contended service) usually works well and users can burst to their full data rate at least for brief periods. However, peer-to-peer (P2P) file sharing and high-quality streaming video can require high data-rates for extended periods, which violates these assumptions and can cause a service to become oversubscribed, resulting in congestion and poor performance. The TCP protocol includes flow-control mechanisms that automatically throttle back on the bandwidth being used during periods of network congestion. This is fair in the sense that all users that experience congestion receive less bandwidth, but it can be frustrating for customers and a major problem for ISPs. In some cases the amount of bandwidth actually available may fall below the threshold required to support a particular service such as video conferencing or streaming live video–effectively making the service unavailable.

When traffic is particularly heavy, an ISP can deliberately throttle back the bandwidth available to classes of users or for particular services. This is known as traffic shaping and careful use can ensure a better quality of service for time critical services even on extremely busy networks. However, overuse can lead to concerns about fairness and network neutrality or even charges of censorship, when some types of traffic are severely or completely blocked.

Outages

An Internet blackout or outage can be caused by local signaling interruptions. Disruptions of submarine communications cables may cause blackouts or slowdowns to large areas, such as in the 2008 submarine cable disruption. Less-developed countries are more vulnerable due to a small number of high-capacity links. Land cables are also vulnerable, as in 2011 when a woman digging for scrap metal severed most connectivity for the nation of Armenia. Internet blackouts affecting almost entire countries can be achieved by governments as a form of Internet censorship, as in the blockage of the Internet in Egypt, whereby approximately 93% of networks were without access in 2011 in an attempt to stop mobilization for anti-government protests.

On April 25, 1997, due to a combination of human error and software bug, an incorrect routing table at MAI Network Service (a Virginia Internet Service Provider) propagated across backbone routers and caused major disruption to Internet traffic for a few hours.

Technologies

When the Internet is accessed using a modem, digital data is converted to analog for transmission over analog networks such as the telephone and cable networks. A computer or other device accessing the Internet would either be connected directly to a modem that communicates with an Internet service provider (ISP) or the modem's Internet connection would be shared via a Local Area Network (LAN) which provides access in a limited area such as a home, school, computer laboratory, or office building.

Although a connection to a LAN may provide very high data-rates within the LAN, actual Internet access speed is limited by the upstream link to the ISP. LANs may be wired or wireless. Ethernet over twisted pair cabling and Wi-Fi are the two most common technologies used to build LANs today, but ARCNET, Token Ring, Localtalk, FDDI, and other technologies were used in the past.

Ethernet is the name of the IEEE 802.3 standard for physical LAN communication and Wi-Fi is a trade name for a wireless local area network (WLAN) that uses one of the IEEE 802.11 standards. Ethernet cables are interconnected via switches & routers. Wi-Fi networks are built using one or more wireless antenna called access points.

Many "modems" provide the additional functionality to host a LAN so most Internet access today is through a LAN, often a very small LAN with just one or two devices attached. And while LANs are an important form of Internet access, this raises the question of how and at what data rate the LAN itself is connected to the rest of the global Internet. The technologies described below are used to make these connections.

Hardwired Broadband Access

The term broadband includes a broad range of technologies, all of which provide higher data rate access to the Internet. The following technologies use wires or cables in contrast to wireless broadband described later.

Dial-up Access

Dial-up Internet access uses a modem and a phone call placed over the public switched telephone network (PSTN) to connect to a pool of modems operated by an ISP. The modem converts a computer's digital signal into an analog signal that travels over a phone line's local loop until it reaches a telephone company's switching facilities or central office (CO) where it is switched to another phone line that connects to another modem at the remote end of the connection.

Operating on a single channel, a dial-up connection monopolizes the phone line and is one of the slowest methods of accessing the Internet. Dial-up is often the only form of Internet access available in rural areas as it requires no new infrastructure beyond the already existing telephone network, to connect to the Internet. Typically, dial-up connections do not exceed a speed of 56 kbit/s, as they are primarily made using modems that operate at a maximum data rate of 56 kbit/s downstream (towards the end user) and 34 or 48 kbit/s upstream (toward the global Internet).

Multilink Dial-up

Multilink dial-up provides increased bandwidth by channel bonding multiple dial-up connections and accessing them as a single data channel. It requires two or more modems, phone lines, and dial-up accounts, as well as an ISP that supports multilinking – and of course any line and data charges are also doubled. This inverse multiplexing option was briefly popular with some high-end users before ISDN, DSL and other technologies became available. Diamond and other vendors created special modems to support multilinking.

Integrated Services Digital Network

Integrated Services Digital Network (ISDN) is a switched telephone service capable of transporting voice and digital data, is one of the oldest Internet access methods. ISDN has been used for voice, video conferencing, and broadband data applications. ISDN was very popular in Europe, but less common in North America. Its use peaked in the late 1990s before the availability of DSL and cable modem technologies.

Basic rate ISDN, known as ISDN-BRI, has two 64 kbit/s "bearer" or "B" channels. These channels can be used separately for voice or data calls or bonded together to provide a 128 kbit/s service. Multiple ISDN-BRI lines can be bonded together to provide data rates above 128 kbit/s. Primary rate ISDN, known as ISDN-PRI, has 23 bearer channels (64 kbit/s each) for a combined data rate of 1.5 Mbit/s (US standard). An ISDN E1 (European standard) line has 30 bearer channels and a combined data rate of 1.9 Mbit/s.

Leased Lines

Leased lines are dedicated lines used primarily by ISPs, business, and other large enterprises to connect LANs and campus networks to the Internet using the existing infrastructure of the public telephone network or other providers. Delivered using wire, optical fiber, and radio, leased lines are used to provide Internet access directly as well as the building blocks from which several other forms of Internet access are created.

T-carrier technology dates to 1957 and provides data rates that range from 56 and 64 kbit/s (DS0) to 1.5 Mbit/s (DS1 or T1), to 45 Mbit/s (DS3 or T3). A T1 line carries 24 voice or data channels (24 DS0s), so customers may use some channels for data and others for voice traffic or use all 24 channels for clear channel data. A DS3 (T3) line carries 28 DS1 (T1) channels. Fractional T1 lines are also available in multiples of a DS0 to provide data rates between 56 and 1,500 kbit/s. T-carrier lines require special termination equipment that may be separate from or integrated into a router or switch and which may be purchased or leased from an ISP. In Japan the equivalent standard is J1/J3. In Europe, a slightly different standard, E-carrier, provides 32 user channels (64 kbit/s) on an E1 (2.0 Mbit/s) and 512 user channels or 16 E1s on an E3 (34.4 Mbit/s).

Synchronous Optical Networking (SONET, in the U.S. and Canada) and Synchronous Digital Hierarchy (SDH, in the rest of the world) are the standard multiplexing protocols used to carry high-data-rate digital bit-streams over optical fiber using lasers or highly coherent light from light-emitting diodes (LEDs). At lower transmission rates data can also be transferred via an electrical interface. The basic unit of framing is an OC-3c (optical) or STS-3c (electrical) which carries

155.520 Mbit/s. Thus an OC-3c will carry three OC-1 (51.84 Mbit/s) payloads each of which has enough capacity to include a full DS3. Higher data rates are delivered in OC-3c multiples of four providing OC-12c (622.080 Mbit/s), OC-48c (2.488 Gbit/s), OC-192c (9.953 Gbit/s), and OC-768c (39.813 Gbit/s). The "c" at the end of the OC labels stands for "concatenated" and indicates a single data stream rather than several multiplexed data streams.

The 1, 10, 40, and 100 gigabit Ethernet (GbE, 10 GbE, 40/100 GbE) IEEE standards (802.3) allow digital data to be delivered over copper wiring at distances to 100 m and over optical fiber at distances to 40 km.

Cable Internet Access

Cable Internet provides access using a cable modem on hybrid fiber coaxial wiring originally developed to carry television signals. Either fiber-optic or coaxial copper cable may connect a node to a customer's location at a connection known as a cable drop. In a cable modem termination system, all nodes for cable subscribers in a neighborhood connect to a cable company's central office, known as the "head end." The cable company then connects to the Internet using a variety of means – usually fiber optic cable or digital satellite and microwave transmissions. Like DSL, broadband cable provides a continuous connection with an ISP.

Downstream, the direction toward the user, bit rates can be as much as 400 Mbit/s for business connections, and 250 Mbit/s for residential service in some countries. Upstream traffic, originating at the user, ranges from 384 kbit/s to more than 20 Mbit/s. Broadband cable access tends to service fewer business customers because existing television cable networks tend to service residential buildings and commercial buildings do not always include wiring for coaxial cable networks. In addition, because broadband cable subscribers share the same local line, communications may be intercepted by neighboring subscribers. Cable networks regularly provide encryption schemes for data traveling to and from customers, but these schemes may be thwarted.

Digital Subscriber Line (DSL, ADSL, SDSL, and VDSL)

Digital Subscriber Line (DSL) service provides a connection to the Internet through the telephone network. Unlike dial-up, DSL can operate using a single phone line without preventing normal use of the telephone line for voice phone calls. DSL uses the high frequencies, while the low (audible) frequencies of the line are left free for regular telephone communication. These frequency bands are subsequently separated by filters installed at the customer's premises.

DSL originally stood for "digital subscriber loop". In telecommunications marketing, the term digital subscriber line is widely understood to mean Asymmetric Digital Subscriber Line (ADSL), the most commonly installed variety of DSL. The data throughput of consumer DSL services typically ranges from 256 kbit/s to 20 Mbit/s in the direction to the customer (downstream), depending on DSL technology, line conditions, and service-level implementation. In ADSL, the data throughput in the upstream direction, (i.e. in the direction to the service provider) is lower than that in the downstream direction (i.e. to the customer), hence the designation of asymmetric. With a symmetric digital subscriber line (SDSL), the downstream and upstream data rates are equal.

Very-high-bit-rate digital subscriber line (VDSL or VHDSL, ITU G.993.1) is a digital subscriber line (DSL) standard approved in 2001 that provides data rates up to 52 Mbit/s downstream and 16 Mbit/s upstream over copper wires and up to 85 Mbit/s down- and upstream on coaxial cable. VDSL is capable of supporting applications such as high-definition television, as well as telephone services (voice over IP) and general Internet access, over a single physical connection.

VDSL2 (ITU-T G.993.2) is a second-generation version and an enhancement of VDSL. Approved in February 2006, it is able to provide data rates exceeding 100 Mbit/s simultaneously in both the upstream and downstream directions. However, the maximum data rate is achieved at a range of about 300 meters and performance degrades as distance and loop attenuation increases.

DSL Rings

DSL Rings (DSLR) or Bonded DSL Rings is a ring topology that uses DSL technology over existing copper telephone wires to provide data rates of up to 400 Mbit/s.

Fiber to the Home

Fiber-to-the-home (FTTH) is one member of the Fiber-to-the-x (FTTx) family that includes Fiber-to-the-building or basement (FTTB), Fiber-to-the-premises (FTTP), Fiber-to-the-desk (FTTD), Fiber-to-the-curb (FTTC), and Fiber-to-the-node (FTTN). These methods all bring data closer to the end user on optical fibers. The differences between the methods have mostly to do with just how close to the end user the delivery on fiber comes. All of these delivery methods are similar to hybrid fiber-coaxial (HFC) systems used to provide cable Internet access.

The use of optical fiber offers much higher data rates over relatively longer distances. Most high-capacity Internet and cable television backbones already use fiber optic technology, with data switched to other technologies (DSL, cable, POTS) for final delivery to customers.

Australia began rolling out its National Broadband Network across the country using fiber-optic cables to 93 percent of Australian homes, schools, and businesses. The project was abandoned by the subsequent LNP government, in favour of a hybrid FTTN design, which turned out to be more expensive and introduced delays. Similar efforts are underway in Italy, Canada, India, and many other countries.

Power-line Internet

Power-line Internet, also known as Broadband over power lines (BPL), carries Internet data on a conductor that is also used for electric power transmission. Because of the extensive power line infrastructure already in place, this technology can provide people in rural and low population areas access to the Internet with little cost in terms of new transmission equipment, cables, or wires. Data rates are asymmetric and generally range from 256 kbit/s to 2.7 Mbit/s.

Because these systems use parts of the radio spectrum allocated to other over-the-air communication services, interference between the services is a limiting factor in the introduction of power-line Internet systems. The IEEE P1901 standard specifies that all power-line protocols must detect existing usage and avoid interfering with it.

Power-line Internet has developed faster in Europe than in the U.S. due to a historical difference in power system design philosophies. Data signals cannot pass through the step-down transformers used and so a repeater must be installed on each transformer. In the U.S. a transformer serves a small cluster of from one to a few houses. In Europe, it is more common for a somewhat larger transformer to service larger clusters of from 10 to 100 houses. Thus a typical U.S. city requires an order of magnitude more repeaters than in a comparable European city.

ATM and Frame Relay

Asynchronous Transfer Mode (ATM) and Frame Relay are wide-area networking standards that can be used to provide Internet access directly or as building blocks of other access technologies. For example, many DSL implementations use an ATM layer over the low-level bitstream layer to enable a number of different technologies over the same link. Customer LANs are typically connected to an ATM switch or a Frame Relay node using leased lines at a wide range of data rates.

While still widely used, with the advent of Ethernet over optical fiber, MPLS, VPNs and broadband services such as cable modem and DSL, ATM and Frame Relay no longer play the prominent role they once did.

Wireless Broadband Access

Wireless broadband is used to provide both fixed and mobile Internet access with the following technologies.

Satellite Broadband

Satellite Internet access via VSAT in Ghana

Satellite Internet access provides fixed, portable, and mobile Internet access. Data rates range from 2 kbit/s to 1 Gbit/s downstream and from 2 kbit/s to 10 Mbit/s upstream. In the northern hemisphere, satellite antenna dishes require a clear line of sight to the southern sky, due to the equatorial position of all geostationary satellites. In the southern hemisphere, this situation is reversed, and dishes are pointed north. Service can be adversely affected by moisture, rain, and snow (known as rain fade). The system requires a carefully aimed directional antenna.

Satellites in geostationary Earth orbit (GEO) operate in a fixed position 35,786 km (22,236 miles) above the Earth's equator. At the speed of light (about 300,000 km/s or 186,000 miles per second), it takes a quarter of a second for a radio signal to travel from the Earth to the satellite and back. When other switching and routing delays are added and the delays are doubled to allow for a full round-trip transmission, the total delay can be 0.75 to 1.25 seconds. This latency is large when compared to other forms of Internet access with typical latencies that range from 0.015 to 0.2 seconds. Long latencies negatively affect some applications that require real-time response, particularly online games, voice over IP, and remote control devices. TCP tuning and TCP acceleration techniques can mitigate some of these problems. GEO satellites do not cover the Earth's polar regions. HughesNet, Exede, AT&T and Dish Network have GEO systems.

Satellites in low Earth orbit (LEO, below 2000 km or 1243 miles) and medium Earth orbit (MEO, between 2000 and 35,786 km or 1,243 and 22,236 miles) are less common, operate at lower altitudes, and are not fixed in their position above the Earth. Lower altitudes allow lower latencies and make real-time interactive Internet applications more feasible. LEO systems include Globalstar and Iridium. The O3b Satellite Constellation is a proposed MEO system with a latency of 125 ms. COMMStellation™ is a LEO system, scheduled for launch in 2015, that is expected to have a latency of just 7 ms.

Mobile Broadband

Service mark for GSMA

Mobile broadband is the marketing term for wireless Internet access delivered through mobile phone towers to computers, mobile phones (called "cell phones" in North America and South Africa, and "hand phones" in Asia), and other digital devices using portable modems. Some mobile services allow more than one device to be connected to the Internet using a single cellular connection using a process called tethering. The modem may be built into laptop computers, tablets, mobile phones, and other devices, added to some devices using PC cards, USB modems, and USB sticks or dongles, or separate wireless modems can be used.

New mobile phone technology and infrastructure is introduced periodically and generally involves a change in the fundamental nature of the service, non-backwards-compatible transmission technology, higher peak data rates, new frequency bands, wider channel frequency bandwidth in Hertz becomes available. These transitions are referred to as generations. The first mobile data services became available during the second generation (2G).

The download (to the user) and upload (to the Internet) data rates given above are peak or maximum rates and end users will typically experience lower data rates.

WiMAX was originally developed to deliver fixed wireless service with wireless mobility added in 2005. CDPD, CDMA2000 EV-DO, and MBWA are no longer being actively developed.

In 2011, 90% of the world's population lived in areas with 2G coverage, while 45% lived in areas with 2G and 3G coverage.

WiMAX

Worldwide Interoperability for Microwave Access (WiMAX) is a set of interoperable implementations of the IEEE 802.16 family of wireless-network standards certified by the WiMAX Forum. WiMAX enables "the delivery of last mile wireless broadband access as an alternative to cable and DSL". The original IEEE 802.16 standard, now called "Fixed WiMAX", was published in 2001 and provided 30 to 40 megabit-per-second data rates. Mobility support was added in 2005. A 2011 update provides data rates up to 1 Gbit/s for fixed stations. WiMax offers a metropolitan area network with a signal radius of about 50 km (30 miles), far surpassing the 30-metre (100-foot) wireless range of a conventional Wi-Fi local area network (LAN). WiMAX signals also penetrate building walls much more effectively than Wi-Fi.

Wireless ISP

Wi-Fi logo

Wireless Internet service providers (WISPs) operate independently of mobile phone operators. WISPs typically employ low-cost IEEE 802.11 Wi-Fi radio systems to link up remote locations over great distances (Long-range Wi-Fi), but may use other higher-power radio communications systems as well.

Traditional 802.11b is an unlicensed omnidirectional service designed to span between 100 and 150 m (300 to 500 ft). By focusing the radio signal using a directional antenna 802.11b can operate reliably over a distance of many km(miles), although the technology's line-of-sight requirements hamper connectivity in areas with hilly or heavily foliated terrain. In addition, compared to hard-wired connectivity, there are security risks (unless robust security protocols are enabled); data rates are significantly slower (2 to 50 times slower); and the network can be less stable, due to interference from other wireless devices and networks, weather and line-of-sight problems.

Deploying multiple adjacent Wi-Fi access points is sometimes used to create city-wide wireless networks. Some are by commercial WISPs but grassroots efforts have also led to wireless community networks. Rural wireless-ISP installations are typically not commercial in nature and are instead a patchwork of systems built up by hobbyists mounting antennas on radio masts and towers, agricultural storage silos, very tall trees, or whatever other tall objects are available. There are a number of companies that provide this service.

Proprietary technologies like Motorola Canopy & Expedience can be used by a WISP to offer wireless access to rural and other markets that are hard to reach using Wi-Fi or WiMAX.

Local Multipoint Distribution Service

Local Multipoint Distribution Service (LMDS) is a broadband wireless access technology that uses microwave signals operating between 26 GHz and 29 GHz. Originally designed for digital television transmission (DTV), it is conceived as a fixed wireless, point-to-multipoint technology for utilization in the last mile. Data rates range from 64 kbit/s to 155 Mbit/s. Distance is typically limited to about 1.5 miles (2.4 km), but links of up to 5 miles (8 km) from the base station are possible in some circumstances.

LMDS has been surpassed in both technological and commercial potential by the LTE and WiMAX standards.

Pricing and Spending

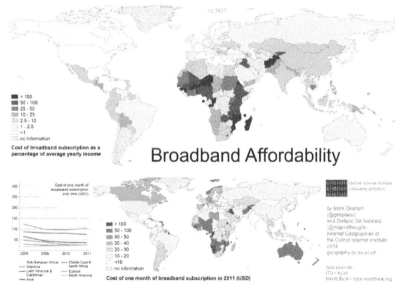

Broadband affordability in 2011
This map presents an overview of broadband affordability, as the relationship between average yearly income per capita and the cost of a broadband subscription (data referring to 2011).
Source: Information Geographies at the Oxford Internet Institute.

Internet access is limited by the relation between pricing and available resources to spend. Regarding the latter, it is estimated that 40% of the world's population has less than US$20 per year available to spend on information and communications technology (ICT). In Mexico, the poorest 30% of the society counts with an estimated US$35 per year (US$3 per month) and in Brazil, the poorest 22% of the population counts with merely US$9 per year to spend on ICT (US$0.75 per month). From Latin America it is known that the borderline between ICT as a necessity good and ICT as a luxury good is roughly around the "magical number" of US$10 per person per month, or US$120 per year. This is the amount of ICT spending people esteem to be a basic necessity. Current Internet access prices exceed the available resources by large in many countries.

Dial-up users pay the costs for making local or long distance phone calls, usually pay a monthly subscription fee, and may be subject to additional per minute or traffic based charges, and connect time limits by their ISP. Though less common today than in the past, some dial-up access is offered for "free" in return for watching banner ads as part of the dial-up service. NetZero, BlueLight, Juno, Freenet (NZ), and Free-nets are examples of services providing free access. Some Wireless community networks continue the tradition of providing free Internet access.

Fixed broadband Internet access is often sold under an "unlimited" or flat rate pricing model, with price determined by the maximum data rate chosen by the customer, rather than a per minute or traffic based charge. Per minute and traffic based charges and traffic caps are common for mobile broadband Internet access.

Internet services like Facebook, Wikipedia and Google have built special programs to partner with mobile network operators (MNO) to introduce *zero-rating* the cost for their data volumes as a means to provide their service more broadly into developing markets.

With increased consumer demand for streaming content such as video on demand and peer-to-peer file sharing, demand for bandwidth has increased rapidly and for some ISPs the flat rate pricing model may become unsustainable. However, with fixed costs estimated to represent 80–90% of the cost of providing broadband service, the marginal cost to carry additional traffic is low. Most ISPs do not disclose their costs, but the cost to transmit a gigabyte of data in 2011 was estimated to be about $0.03.

Some ISPs estimate that a small number of their users consume a disproportionate portion of the total bandwidth. In response some ISPs are considering, are experimenting with, or have implemented combinations of traffic based pricing, time of day or "peak" and "off peak" pricing, and bandwidth or traffic caps. Others claim that because the marginal cost of extra bandwidth is very small with 80 to 90 percent of the costs fixed regardless of usage level, that such steps are unnecessary or motivated by concerns other than the cost of delivering bandwidth to the end user.

In Canada, Rogers Hi-Speed Internet and Bell Canada have imposed bandwidth caps. In 2008 Time Warner began experimenting with usage-based pricing in Beaumont, Texas. In 2009 an effort by Time Warner to expand usage-based pricing into the Rochester, New York area met with public resistance, however, and was abandoned. On August 1, 2012 in Nashville, Tennessee and on October 1, 2012 in Tucson, Arizona Comcast began tests that impose data caps on area residents. In Nashville exceeding the 300 Gbyte cap mandates a temporary purchase of 50 Gbytes of additional data.

Digital Divide

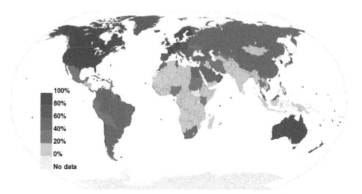

Internet users in 2012 as a percentage of a country's population

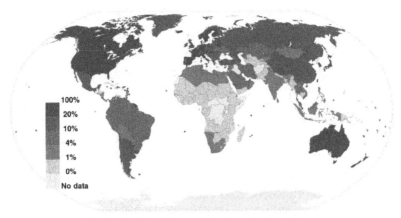

Fixed broadband Internet subscriptions in 2012
as a percentage of a country's population

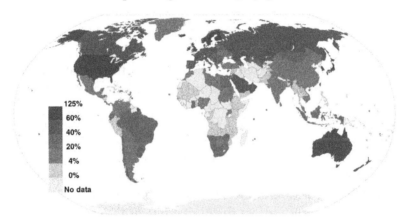

Mobile broadband Internet subscriptions in 2012
as a percentage of a country's population

Gini coefficients for telecommunication capacity *(in kbps)* per individual worldwide (incl. 172 countries)

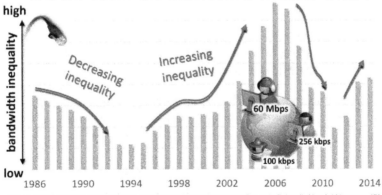

Hilbert, M. (2016). The bad news is that the digital access divide is here to stay: Domestically installed bandwidths among 172 countries for 1986–2014. *Telecommunications Policy*. www.martinhilbert.net/the-bad-news-is-that-the-digital-access-divide-is-here-to-stay/

The digital divide measured in terms of bandwidth is not closing, but fluctuating up and down. Gini coefficients for telecommunication capacity (in kbit/s) among individuals worldwide

Despite its tremendous growth, Internet access is not distributed equally within or between countries. The digital divide refers to "the gap between people with effective access to information and communications technology (ICT), and those with very limited or no access". The gap between people with Internet access and those without is one of many aspects of the digital divide. Whether someone has access to the Internet can depend greatly on financial status, geographical location as well as government policies. "Low-income, rural, and minority populations have received special scrutiny as the technological "have-nots."

Government policies play a tremendous role in bringing Internet access to or limiting access for underserved groups, regions, and countries. For example, in Pakistan, which is pursuing an aggressive IT policy aimed at boosting its drive for economic modernization, the number of Internet users grew from 133,900 (0.1% of the population) in 2000 to 31 million (17.6% of the population) in 2011. In countries such as North Korea and Cuba there is relatively little access to the Internet due to the governments' fear of political instability that might accompany the benefits of access to the global Internet. The U.S. trade embargo is another barrier limiting Internet access in Cuba.

Access to computers is a dominant factor in determining the level of Internet access. In 2011, in developing countries, 25% of households had a computer and 20% had Internet access, while in developed countries the figures were 74% of households had a computer and 71% had Internet access. When buying computers was legalized in Cuba in 2007, the private ownership of computers soared (there were 630,000 computers available on the island in 2008, a 23% increase over 2007).

Internet access has changed the way in which many people think and has become an integral part of peoples economic, political, and social lives. The United Nations has recognized that providing Internet access to more people in the world will allow them to take advantage of the "political, social, economic, educational, and career opportunities" available over the Internet. Several of the 67 principles adopted at the World Summit on the Information Society convened by the United Nations in Geneva in 2003, directly address the digital divide. To promote economic development and a reduction of the digital divide, national broadband plans have been and are being developed to increase the availability of affordable high-speed Internet access throughout the world.

Growth in Number of Users

Access to the Internet grew from an estimated 10 million people in 1993, to almost 40 million in 1995, to 670 million in 2002, and to 2.7 billion in 2013. With market saturation, growth in the number of Internet users is slowing in industrialized countries, but continues in Asia, Africa, Latin America, the Caribbean, and the Middle East.

There were roughly 0.6 billion fixed broadband subscribers and almost 1.2 billion mobile broadband subscribers in 2011. In developed countries people frequently use both fixed and mobile broadband networks. In developing countries mobile broadband is often the only access method available.

Bandwidth Divide

Traditionally the divide has been measured in terms of the existing numbers of subscriptions and digital devices ("have and have-not of subscriptions"). Recent studies have measured the digital divide not in terms of technological devices, but in terms of the existing bandwidth per individual (in kbit/s per capita). As shown in the Figure on the side, the digital divide in kbit/s is not monotonically decreasing, but re-opens up with each new innovation. For example, "the massive diffusion of narrow-band Internet and mobile phones during the late 1990s" increased digital inequality, as well as "the initial introduction of broadband DSL and cable modems during 2003–2004 increased levels of inequality". This is because a new kind of connectivity is never introduced instantaneously and uniformly to society as a whole at once, but diffuses slowly through social networks. As shown by the Figure, during the mid-2000s, communication capacity was more unequally distributed than during the late 1980s, when only fixed-line phones existed. The most recent increase in digital equality stems from the massive diffusion of the latest digital innovations (i.e. fixed and mobile broadband infrastructures, e.g. 3G and fiber optics FTTH). As shown in the Figure, Internet access in terms of bandwidth is more unequally distributed in 2014 as it was in the mid-1990s.

In the United States

In the United States, billions of dollars have been invested in efforts to narrow the digital divide and bring Internet access to more people in low-income and rural areas of the United States. Internet availability varies widely state by state in the U.S. In 2011 for example, 87.1% of all New Hampshire residents lived in a household where Internet was available, ranking first in the nation. Meanwhile, 61.4% of all Mississippi residents lived in a household where Internet was available, ranking last in the nation. The Obama administration has continued this commitment to narrowing the digital divide through the use of stimulus funding. The National Center for Education Statistics reported that 98% of all U.S. classroom computers had Internet access in 2008 with roughly one computer with Internet access available for every three students. The percentage and ratio of students to computers was the same for rural schools (98% and 1 computer for every 2.9 students).

Rural Access

One of the great challenges for Internet access in general and for broadband access in particular is to provide service to potential customers in areas of low population density, such as to farmers, ranchers, and small towns. In cities where the population density is high, it is easier for a service provider to recover equipment costs, but each rural customer may require expensive equipment to get connected. While 66% of Americans had an Internet connection in 2010, that figure was only 50% in rural areas, according to the Pew Internet & American Life Project. Virgin Media advertised over 100 towns across the United Kingdom "from Cwmbran to Clydebank" that have access to their 100 Mbit/s service.

Wireless Internet Service Provider (WISPs) are rapidly becoming a popular broadband option for rural areas. The technology's line-of-sight requirements may hamper connectivity in some areas with hilly and heavily foliated terrain. However, the Tegola project, a successful pilot in remote Scotland, demonstrates that wireless can be a viable option.

The Broadband for Rural Nova Scotia initiative is the first program in North America to guarantee access to "100% of civic addresses" in a region. It is based on Motorola Canopy technology. As of November 2011, under 1000 households have reported access problems. Deployment of a new cell network by one Canopy provider (Eastlink) was expected to provide the alternative of 3G/4G service, possibly at a special unmetered rate, for areas harder to serve by Canopy.

A rural broadband initiative in New Zealand is a joint project between Vodafone and Chorus, with

Chorus providing the fibre infrastructure and Vodafone providing wireless broadband, supported by the fibre backhaul.

Access as a Civil or Human Right

The actions, statements, opinions, and recommendations outlined below have led to the suggestion that Internet access itself is or should become a civil or perhaps a human right.

Several countries have adopted laws requiring the state to work to ensure that Internet access is broadly available and/or preventing the state from unreasonably restricting an individual's access to information and the Internet:

- Costa Rica: A 30 July 2010 ruling by the Supreme Court of Costa Rica stated: "Without fear of equivocation, it can be said that these technologies [information technology and communication] have impacted the way humans communicate, facilitating the connection between people and institutions worldwide and eliminating barriers of space and time. At this time, access to these technologies becomes a basic tool to facilitate the exercise of fundamental rights and democratic participation (e-democracy) and citizen control, education, freedom of thought and expression, access to information and public services online, the right to communicate with government electronically and administrative transparency, among others. This includes the fundamental right of access to these technologies, in particular, the right of access to the Internet or World Wide Web."

- Estonia: In 2000, the parliament launched a massive program to expand access to the countryside. The Internet, the government argues, is essential for life in the 21st century.

- Finland: By July 2010, every person in Finland was to have access to a one-megabit per second broadband connection, according to the Ministry of Transport and Communications. And by 2015, access to a 100 Mbit/s connection.

- France: In June 2009, the Constitutional Council, France's highest court, declared access to the Internet to be a basic human right in a strongly-worded decision that struck down portions of the HADOPI law, a law that would have tracked abusers and without judicial review automatically cut off network access to those who continued to download illicit material after two warnings

- Greece: Article 5A of the Constitution of Greece states that all persons has a right to participate in the Information Society and that the state has an obligation to facilitate the production, exchange, diffusion, and access to electronically transmitted information.

- Spain: Starting in 2011, Telefónica, the former state monopoly that holds the country's "universal service" contract, has to guarantee to offer "reasonably" priced broadband of at least one megabyte per second throughout Spain.

In December 2003, the World Summit on the Information Society (WSIS) was convened under the auspice of the United Nations. After lengthy negotiations between governments, businesses and civil society representatives the WSIS Declaration of Principles was adopted reaffirming the importance of the Information Society to maintaining and strengthening human rights:

1. We, the representatives of the peoples of the world, assembled in Geneva from 10–12 December 2003 for the first phase of the World Summit on the Information Society, declare our common desire and commitment to build a people-centred, inclusive and development-oriented Information Society, where everyone can create, access, utilize and share information and knowledge, enabling individuals, communities and peoples to achieve their full potential in promoting their sustainable development and improving their quality of life, premised on the purposes and principles of the Charter of the United Nations and respecting fully and upholding the Universal Declaration of Human Rights.

3. We reaffirm the universality, indivisibility, interdependence and interrelation of all human rights and fundamental freedoms, including the right to development, as enshrined in the Vienna Declaration. We also reaffirm that democracy, sustainable development, and respect for human rights and fundamental freedoms as well as good governance at all levels are interdependent and mutually reinforcing. We further resolve to strengthen the rule of law in international as in national affairs.

The WSIS Declaration of Principles makes specific reference to the importance of the right to freedom of expression in the "Information Society" in stating:

4. We reaffirm, as an essential foundation of the Information Society, and as outlined in Article 19 of the Universal Declaration of Human Rights, that everyone has the right to freedom of opinion and expression; that this right includes freedom to hold opinions without interference and to seek, receive and impart information and ideas through any media and regardless of frontiers. Communication is a fundamental social process, a basic human need and the foundation of all social organisation. It is central to the Information Society. Everyone, everywhere should have the opportunity to participate and no one should be excluded from the benefits of the Information Society offers."

A poll of 27,973 adults in 26 countries, including 14,306 Internet users, conducted for the BBC World Service between 30 November 2009 and 7 February 2010 found that almost four in five Internet users and non-users around the world felt that access to the Internet was a fundamental right. 50% strongly agreed, 29% somewhat agreed, 9% somewhat disagreed, 6% strongly disagreed, and 6% gave no opinion.

The 88 recommendations made by the Special Rapporteur on the promotion and protection of the right to freedom of opinion and expression in a May 2011 report to the Human Rights Council of the United Nations General Assembly include several that bear on the question of the right to Internet access:

67. Unlike any other medium, the Internet enables individuals to seek, receive and impart information and ideas of all kinds instantaneously and inexpensively across national borders. By vastly expanding the capacity of individuals to enjoy their right to freedom of opinion and expression, which is an "enabler" of other human rights, the Internet boosts economic, social and political development, and contributes to the progress of humankind as a whole. In this regard, the Special Rapporteur encourages other Special Procedures mandate holders to engage on the issue of the Internet with respect to their particular mandates.

78. While blocking and filtering measures deny users access to specific content on the Internet, States have also taken measures to cut off access to the Internet entirely. The Special Rapporteur considers cutting off users from Internet access, regardless of the justification provided, including on the grounds of violating intellectual property rights law, to be disproportionate and thus a violation of article 19, paragraph 3, of the International Covenant on Civil and Political Rights.

79. The Special Rapporteur calls upon all States to ensure that Internet access is maintained at all times, including during times of political unrest.

85. Given that the Internet has become an indispensable tool for realizing a range of human rights, combating inequality, and accelerating development and human progress, ensuring universal access to the Internet should be a priority for all States. Each State should thus develop a concrete and effective policy, in consultation with individuals from all sections of society, including the private sector and relevant Government ministries, to make the Internet widely available, accessible and affordable to all segments of population.

Network Neutrality

Network neutrality (also net neutrality, Internet neutrality, or net equality) is the principle that Internet service providers and governments should treat all data on the Internet equally, not discriminating or charging differentially by user, content, site, platform, application, type of attached equipment, or mode of communication. Advocates of net neutrality have raised concerns about the ability of broadband providers to use their last mile infrastructure to block Internet applications and content (e.g. websites, services, and protocols), and even to block out competitors. Opponents claim net neutrality regulations would deter investment into improving broadband infrastructure and try to fix something that isn't broken.

Natural Disasters and Access

Natural disasters disrupt internet access in profound ways. This is important—not only for telecommunication companies who own the networks and the businesses who use them, but for emergency crew and displaced citizens as well. The situation is worsened when hospitals or other buildings necessary to disaster response lose their connection. Knowledge gained from studying past internet disruptions by natural disasters could be put to use in planning or recovery. Additionally, because of both natural and man-made disasters, studies in network resiliency are now being conducted to prevent large-scale outages.

One way natural disasters impact internet connection is by damaging end sub-networks (subnets), making them unreachable. A study on local networks after Hurricane Katrina found that 26% of subnets within the storm coverage were unreachable. At Hurricane Katrina's peak intensity, almost 35% of networks in Mississippi were without power, while around 14% of Louisiana's networks were disrupted. Of those unreachable subnets, 73% were disrupted for four weeks or longer and 57% were at "network edges where important emergency organizations such as hospitals and government agencies are mostly located". Extensive infrastructure damage and inaccessible areas were two explanations for the long delay in returning service. The company Cisco has revealed a Network Emergency Response Vehicle (NERV), a truck that makes portable communications possible for emergency responders despite traditional networks being disrupted.

A second way natural disasters destroy internet connectivity is by severing submarine cables—fiber-optic cables placed on the ocean floor that provide international internet connection. The 2006 undersea earthquake near Taiwan (Richter scale 7.2) cut six out of seven international cables connected to that country and caused a tsunami that wiped out one of its cable and landing stations. The impact slowed or disabled internet connection for five days within the Asia-Pacific region as well as between the region and the United States and Europe.

With the rise in popularity of cloud computing, concern has grown over access to cloud-hosted data in the event of a natural disaster. Amazon Web Services (AWS) has been in the news for major network outages in April 2011 and June 2012. AWS, like other major cloud hosting companies, prepares for typical outages and large-scale natural disasters with backup power as well as backup data centers in other locations. AWS divides the globe into five regions and then splits each region into availability zones. A data center in one availability zone should be backed up by a data center in a different availability zone. Theoretically, a natural disaster would not affect more than one availability zone. This theory plays out as long as human error is not added to the mix. The June 2012 major storm only disabled the primary data center, but human error disabled the secondary and tertiary backups, affecting companies such as Netflix, Pinterest, Reddit, and Instagram.

Bluetooth

Bluetooth is a wireless technology standard for exchanging data over short distances (using short-wavelength UHF radio waves in the ISM band from 2.4 to 2.485 GHz) from fixed and mobile devices, and building personal area networks (PANs). Invented by telecom vendor Ericsson in 1994, it was originally conceived as a wireless alternative to RS-232 data cables. It can connect several devices, overcoming problems of synchronization.

Bluetooth is managed by the Bluetooth Special Interest Group (SIG), which has more than 25,000 member companies in the areas of telecommunication, computing, networking, and consumer electronics. The IEEE standardized Bluetooth as IEEE 802.15.1, but no longer maintains the standard. The Bluetooth SIG oversees development of the specification, manages the qualification program, and protects the trademarks. A manufacturer must meet Bluetooth SIG standards to market it as a Bluetooth device. A network of patents apply to the technology, which are licensed to individual qualifying devices.

Origin

The development of the "short-link" radio technology, later named Bluetooth, was initiated in 1989 by Dr. Nils Rydbeck, CTO at Ericsson Mobile in Lund, and Dr. Johan Ullman. The purpose was to develop wireless headsets, according to two inventions by Johan Ullman, SE 8902098-6, issued 1989-06-12 and SE 9202239, issued 1992-07-24. Nils Rydbeck tasked Tord Wingren with specifying and Jaap Haartsen and Sven Mattisson with developing. Both were working for Ericsson in Lund, Sweden. The specification is based on frequency-hopping spread spectrum technology.

Name and Logo

Etymology of the Name

The name "Bluetooth" is an Anglicised version of the Scandinavian *Blåtand/Blåtann* (Old Norse *blátnn*), the epithet of the tenth-century king Harald Bluetooth who united dissonant Danish tribes into a single kingdom and, according to legend, introduced Christianity as well. The idea of this name was proposed in 1997 by Jim Kardach who developed a system that would allow mobile phones to communicate with computers. At the time of this proposal he was reading Frans G. Bengtsson's historical novel *The Long Ships* about Vikings and King Harald Bluetooth. The implication is that Bluetooth does the same with communications protocols, uniting them into one universal standard.

Logo

The Bluetooth logo is a bind rune merging the Younger Futhark runes ᚼ (Hagall) (ᚼ) and ᛒ (Bjarkan) (ᛒ), Harald's initials.

Implementation

Bluetooth operates at frequencies between 2402 and 2480 MHz, or 2400 and 2483.5 MHz including guard bands 2 MHz wide at the bottom end and 3.5 MHz wide at the top. This is in the globally unlicensed (but not unregulated) Industrial, Scientific and Medical (ISM) 2.4 GHz short-range radio frequency band. Bluetooth uses a radio technology called frequency-hopping spread spectrum. Bluetooth divides transmitted data into packets, and transmits each packet on one of 79 designated Bluetooth channels. Each channel has a bandwidth of 1 MHz. It usually performs 800 hops per second, with Adaptive Frequency-Hopping (AFH) enabled. Bluetooth low energy uses 2 MHz spacing, which accommodates 40 channels.

Originally, Gaussian frequency-shift keying (GFSK) modulation was the only modulation scheme available. Since the introduction of Bluetooth 2.0+EDR, $\pi/4$-DQPSK (Differential Quadrature Phase Shift Keying) and 8DPSK modulation may also be used between compatible devices. Devices functioning with GFSK are said to be operating in basic rate (BR) mode where an instantaneous data rate of 1 Mbit/s is possible. The term Enhanced Data Rate (EDR) is used to describe $\pi/4$-DPSK and 8DPSK schemes, each giving 2 and 3 Mbit/s respectively. The combination of these (BR and EDR) modes in Bluetooth radio technology is classified as a "BR/EDR radio".

Bluetooth is a packet-based protocol with a master-slave structure. One master may communicate with up to seven slaves in a piconet. All devices share the master's clock. Packet exchange is based on the basic clock, defined by the master, which ticks at 312.5 µs intervals. Two clock ticks make up a slot of 625 µs, and two slots make up a slot pair of 1250 µs. In the simple case of single-slot packets the master transmits in even slots and receives in odd slots. The slave, conversely, receives in even slots and transmits in odd slots. Packets may be 1, 3 or 5 slots long, but in all cases the master's transmission begins in even slots and the slave's in odd slots.

The above is valid for "classic" BT. Bluetooth Low Energy, introduced in the 4.0 specification, uses the same spectrum but somewhat differently.

Communication and Connection

A master Bluetooth device can communicate with a maximum of seven devices in a piconet (an ad-hoc computer network using Bluetooth technology), though not all devices reach this maximum. The devices can switch roles, by agreement, and the slave can become the master (for example, a headset initiating a connection to a phone necessarily begins as master—as initiator of the connection—but may subsequently operate as slave).

The Bluetooth Core Specification provides for the connection of two or more piconets to form a scatternet, in which certain devices simultaneously play the master role in one piconet and the slave role in another.

At any given time, data can be transferred between the master and one other device (except for the little-used broadcast mode.) The master chooses which slave device to address; typically, it switches rapidly from one device to another in a round-robin fashion. Since it is the master that chooses which slave to address, whereas a slave is (in theory) supposed to listen in each receive slot, being a master is a lighter burden than being a slave. Being a master of seven slaves is possible; being a slave of more than one master is only possible on more advanced devices which confirm with Bluetooth 4.1 onwards. The specification is vague as to required behavior in scatternets.

Uses

Class	Max. permitted power		Typ. range (m)
	(mW)	(dBm)	
1	100	20	~100
2	2.5	4	~10
3	1	0	~1
4	0.5	-3	~0.5

Bluetooth is a standard wire-replacement communications protocol primarily designed for low-power consumption, with a short range based on low-cost transceiver microchips in each device. Because the devices use a radio (broadcast) communications system, they do not have to be in visual line of sight of each other, however a *quasi optical* wireless path must be viable. Range is power-class-dependent, but effective ranges vary in practice; the table on the right.

Officially Class 3 radios have a range of up to 1 metre (3 ft), Class 2, most commonly found in mobile devices, 10 metres (33 ft), and Class 1, primarily for industrial use cases,100 metres (300 ft). Bluetooth Marketing qualifies that Class 1 range is in most cases 20–30 metres (66–98 ft), and Class 2 range 5–10 metres (16–33 ft).

Bluetooth Version	Maximum Speed	Maximum Range
3.0	25 Mbit/s	
4.0	25 Mbit/s	200 feet(60.96 m)
5.0	50 Mbit/s	800 feet(243.84 m)

The effective range varies due to propagation conditions, material coverage, production sample variations, antenna configurations and battery conditions. Most Bluetooth applications are for indoor conditions, where attenuation of walls and signal fading due to signal reflections make the range far lower than specified line-of-sight ranges of the Bluetooth products. Most Bluetooth applications are battery powered Class 2 devices, with little difference in range whether the other end of the link is a Class 1 or Class 2 device as the lower powered device tends to set the range limit. In some cases the effective range of the data link can be extended when a Class 2 device is connecting to a Class 1 transceiver with both higher sensitivity and transmission power than a typical Class 2 device. Mostly, however, the Class 1 devices have a similar sensitivity to Class 2 devices. Connecting two Class 1 devices with both high sensitivity and high power can allow ranges far in excess of the typical 100m, depending on the throughput required by the application. Some such devices allow open field ranges of up to 1 km and beyond between two similar devices without exceeding legal emission limits.

The Bluetooth Core Specification mandates a range of not less than 10 metres (33 ft), but there is no upper limit on actual range. Manufacturers' implementations can be tuned to provide the range needed for each case.

Bluetooth Profiles

To use Bluetooth wireless technology, a device must be able to interpret certain Bluetooth profiles, which are definitions of possible applications and specify general behaviours that Bluetooth-enabled devices use to communicate with other Bluetooth devices. These profiles include settings to parametrize and to control the communication from start. Adherence to profiles saves the time for transmitting the parameters anew before the bi-directional link becomes effective. There are a wide range of Bluetooth profiles that describe many different types of applications or use cases for devices.

List of Applications

A typical Bluetooth mobile phone headset.

- Wireless control of and communication between a mobile phone and a handsfree headset. This was one of the earliest applications to become popular.

- Wireless control of and communication between a mobile phone and a Bluetooth compatible car stereo system.

- Wireless control of and communication with iOS and Android device phones, tablets and portable wireless speakers.

- Wireless Bluetooth headset and Intercom. Idiomatically, a headset is sometimes called "a Bluetooth".

- Wireless streaming of audio to headphones with or without communication capabilities.

- Wireless streaming of data collected by Bluetooth-enabled fitness devices to phone or PC.

- Wireless networking between PCs in a confined space and where little bandwidth is required.

- Wireless communication with PC input and output devices, the most common being the mouse, keyboard and printer.

- Transfer of files, contact details, calendar appointments, and reminders between devices with OBEX.

- Replacement of previous wired RS-232 serial communications in test equipment, GPS receivers, medical equipment, bar code scanners, and traffic control devices.

- For controls where infrared was often used.

- For low bandwidth applications where higher USB bandwidth is not required and cable-free connection desired.

- Sending small advertisements from Bluetooth-enabled advertising hoardings to other, discoverable, Bluetooth devices.

- Wireless bridge between two Industrial Ethernet (*e.g.*, PROFINET) networks.

- Seventh and eighth generation game consoles such as Nintendo's Wii, and Sony's PlayStation 3 use Bluetooth for their respective wireless controllers.

- Dial-up internet access on personal computers or PDAs using a data-capable mobile phone as a wireless modem.

- Short range transmission of health sensor data from medical devices to mobile phone, set-top box or dedicated telehealth devices.

- Allowing a DECT phone to ring and answer calls on behalf of a nearby mobile phone.

- Real-time location systems (RTLS), are used to track and identify the location of objects in real-time using "Nodes" or "tags" attached to, or embedded in the objects tracked, and "Readers" that receive and process the wireless signals from these tags to determine their locations.

- Personal security application on mobile phones for prevention of theft or loss of items. The protected item has a Bluetooth marker (*e.g.*, a tag) that is in constant communication with the phone. If the connection is broken (the marker is out of range of the phone) then an alarm is raised. This can also be used as a man overboard alarm. A product using this technology has been available since 2009.

- Calgary, Alberta, Canada's Roads Traffic division uses data collected from travelers' Bluetooth devices to predict travel times and road congestion for motorists.

- Wireless transmission of audio (a more reliable alternative to FM transmitters)

Bluetooth vs. Wi-Fi (IEEE 802.11)

Bluetooth and Wi-Fi (the brand name for products using IEEE 802.11 standards) have some similar applications: setting up networks, printing, or transferring files. Wi-Fi is intended as a replacement for high speed cabling for general local area network access in work areas or home. This category of applications is sometimes called wireless local area networks (WLAN). Bluetooth was intended for portable equipment and its applications. The category of applications is outlined as the wireless personal area network (WPAN). Bluetooth is a replacement for cabling in a variety of personally carried applications in any setting, and also works for fixed location applications such as smart energy functionality in the home (thermostats, etc.).

Wi-Fi and Bluetooth are to some extent complementary in their applications and usage. Wi-Fi is usually access point-centered, with an asymmetrical client-server connection with all traffic routed through the access point, while Bluetooth is usually symmetrical, between two Bluetooth devices. Bluetooth serves well in simple applications where two devices need to connect with minimal configuration like a button press, as in headsets and remote controls, while Wi-Fi suits better in applications where some degree of client configuration is possible and high speeds are required, especially for network access through an access node. However, Bluetooth access points do exist and ad-hoc connections are possible with Wi-Fi though not as simply as with Bluetooth. Wi-Fi Direct was recently developed to add a more Bluetooth-like ad-hoc functionality to Wi-Fi.

Devices

A Bluetooth USB dongle with a 100 m range.

Bluetooth exists in many products, such as telephones, tablets, media players, robotics systems, handheld, laptops and console gaming equipment, and some high definition headsets, modems, and watches. The technology is useful when transferring information between two or more devices that are near each other in low-bandwidth situations. Bluetooth is commonly used to transfer sound data with telephones (i.e., with a Bluetooth headset) or byte data with hand-held computers (transferring files).

Bluetooth protocols simplify the discovery and setup of services between devices. Bluetooth devices can advertise all of the services they provide. This makes using services easier, because more of the security, network address and permission configuration can be automated than with many other network types.

Computer Requirements

A typical Bluetooth USB dongle.

An internal notebook Bluetooth card (14×36×4 mm).

A personal computer that does not have embedded Bluetooth can use a Bluetooth adapter that enables the PC to communicate with Bluetooth devices. While some desktop computers and most recent laptops come with a built-in Bluetooth radio, others require an external adapter, typically in the form of a small USB "dongle."

Unlike its predecessor, IrDA, which requires a separate adapter for each device, Bluetooth lets multiple devices communicate with a computer over a single adapter.

Operating System Implementation

For Microsoft platforms, Windows XP Service Pack 2 and SP3 releases work natively with Bluetooth v1.1, v2.0 and v2.0+EDR. Previous versions required users to install their Bluetooth adapter's own drivers, which were not directly supported by Microsoft. Microsoft's own Bluetooth dongles (packaged with their Bluetooth computer devices) have no external drivers and thus require at least Windows XP Service Pack 2. Windows Vista RTM/SP1 with the Feature Pack for Wireless or Windows Vista SP2 work with Bluetooth v2.1+EDR. Windows 7 works with Bluetooth v2.1+EDR and Extended Inquiry Response (EIR).

The Windows XP and Windows Vista/Windows 7 Bluetooth stacks support the following Blue-tooth profiles natively: PAN, SPP, DUN, HID, HCRP. The Windows XP stack can be replaced by a third party stack that supports more profiles or newer Bluetooth versions. The Windows Vista/Windows 7 Bluetooth stack supports vendor-supplied additional profiles without requiring that the Microsoft stack be replaced.

Apple products have worked with Bluetooth since Mac OS X v10.2, which was released in 2002.

Linux has two popular Bluetooth stacks, BlueZ and Affix. The BlueZ stack is included with most Linux kernels and was originally developed by Qualcomm. The Affix stack was developed by Nokia.

FreeBSD features Bluetooth since its v5.0 release.

NetBSD features Bluetooth since its v4.0 release. Its Bluetooth stack has been ported to OpenBSD as well.

Specifications and Features

The specifications were formalized by the Bluetooth Special Interest Group (SIG). The SIG was for-mally announced on 20 May 1998. Today it has a membership of over 30,000 companies world-wide. It was established by Ericsson, IBM, Intel, Toshiba and Nokia, and later joined by many other companies.

All versions of the Bluetooth standards support downward compatibility. That lets the latest stan-dard cover all older versions.

The Bluetooth Core Specification Working Group (CSWG) produces mainly 4 kinds of specifica-tions

- The Bluetooth Core Specification, release cycle is typically a few years in between

- Core Specification Addendum (CSA), release cycle can be as tight as a few times per year

- Core Specification Supplements (CSS), can be released very quickly

- Errata

Bluetooth v1.0 and v1.0B

Versions 1.0 and 1.0B had many problems and manufacturers had difficulty making their products interoperable. Versions 1.0 and 1.0B also included mandatory Bluetooth hardware device address (BD_ADDR) transmission in the Connecting process (rendering anonymity impossible at the pro-tocol level), which was a major setback for certain services planned for use in Bluetooth environ-ments.

Bluetooth v1.1

- Ratified as IEEE Standard 802.15.1–2002

- Many errors found in the v1.0B specifications were fixed.

- Added possibility of non-encrypted channels.

- Received Signal Strength Indicator (RSSI).

Bluetooth v1.2

Major enhancements include the following:

- Faster Connection and Discovery

- *Adaptive frequency-hopping spread spectrum (AFH)*, which improves resistance to radio frequency interference by avoiding the use of crowded frequencies in the hopping sequence.

- Higher transmission speeds in practice, up to 721 kbit/s, than in v1.1.

- Extended Synchronous Connections (eSCO), which improve voice quality of audio links by allowing retransmissions of corrupted packets, and may optionally increase audio latency to provide better concurrent data transfer.

- Host Controller Interface (HCI) operation with three-wire UART.

- Ratified as IEEE Standard 802.15.1–2005

- Introduced Flow Control and Retransmission Modes for L2CAP.

Bluetooth v2.0 + EDR

This version of the Bluetooth Core Specification was released in 2004. The main difference is the introduction of an Enhanced Data Rate (EDR) for faster data transfer. The nominal rate of EDR is about 3 Mbit/s, although the practical data transfer rate is 2.1 Mbit/s. EDR uses a combination of GFSK and Phase Shift Keying modulation (PSK) with two variants, π/4-DQPSK and 8DPSK. EDR can provide a lower power consumption through a reduced duty cycle.

The specification is published as *Bluetooth v2.0 + EDR*, which implies that EDR is an optional feature. Aside from EDR, the v2.0 specification contains other minor improvements, and products may claim compliance to "Bluetooth v2.0" without supporting the higher data rate. At least one commercial device states "Bluetooth v2.0 without EDR" on its data sheet.

Bluetooth v2.1 + EDR

Bluetooth Core Specification Version 2.1 + EDR was adopted by the Bluetooth SIG on 26 July 2007.

The headline feature of v2.1 is secure simple pairing (SSP): this improves the pairing experience for Bluetooth devices, while increasing the use and strength of security. The section on Pairing below for more details.

Version 2.1 allows various other improvements, including "Extended inquiry response" (EIR), which provides more information during the inquiry procedure to allow better filtering of devices before connection; and sniff subrating, which reduces the power consumption in low-power mode.

Bluetooth v3.0 + HS

Version 3.0 + HS of the Bluetooth Core Specification was adopted by the Bluetooth SIG on 21 April 2009. Bluetooth v3.0 + HS provides theoretical data transfer speeds of up to 24 Mbit/s, though not over the Bluetooth link itself. Instead, the Bluetooth link is used for negotiation and establishment, and the high data rate traffic is carried over a colocated 802.11 link.

The main new feature is AMP (Alternative MAC/PHY), the addition of 802.11 as a high speed transport. The High-Speed part of the specification is not mandatory, and hence only devices that display the "+HS" logo actually support Bluetooth over 802.11 high-speed data transfer. A Bluetooth v3.0 device without the "+HS" suffix is only required to support features introduced in Core Specification Version 3.0 or earlier Core Specification Addendum 1.

L2CAP Enhanced Modes

Enhanced Retransmission Mode (ERTM) implements reliable L2CAP channel, while Streaming Mode (SM) implements unreliable channel with no retransmission or flow control. Introduced in Core Specification Addendum 1.

Alternative MAC/PHY

Enables the use of alternative MAC and PHYs for transporting Bluetooth profile data. The Bluetooth radio is still used for device discovery, initial connection and profile configuration. However, when large quantities of data must be sent, the high speed alternative MAC PHY 802.11 (typically associated with Wi-Fi) transports the data. This means that Bluetooth uses proven low power connection models when the system is idle, and the faster radio when it must send large quantities of data. AMP links require enhanced L2CAP modes.

Unicast Connectionless Data

Permits sending service data without establishing an explicit L2CAP channel. It is intended for use by applications that require low latency between user action and reconnection/transmission of data. This is only appropriate for small amounts of data.

Enhanced Power Control

Updates the power control feature to remove the open loop power control, and also to clarify ambiguities in power control introduced by the new modulation schemes added for EDR. Enhanced power control removes the ambiguities by specifying the behaviour that is expected. The feature also adds closed loop power control, meaning RSSI filtering can start as the response is received. Additionally, a "go straight to maximum power" request has been introduced. This is expected to deal with the headset link loss issue typically observed when a user puts their phone into a pocket on the opposite side to the headset.

Ultra-wideband

The high speed (AMP) feature of Bluetooth v3.0 was originally intended for UWB, but the WiMedia Alliance, the body responsible for the flavor of UWB intended for Bluetooth, announced in March 2009 that it was disbanding, and ultimately UWB was omitted from the Core v3.0 specification.

On 16 March 2009, the WiMedia Alliance announced it was entering into technology transfer agreements for the WiMedia Ultra-wideband (UWB) specifications. WiMedia has transferred all current and future specifications, including work on future high speed and power optimized implementations, to the Bluetooth Special Interest Group (SIG), Wireless USB Promoter Group and the USB Implementers Forum. After successful completion of the technology transfer, marketing, and related administrative items, the WiMedia Alliance ceased operations.

In October 2009 the Bluetooth Special Interest Group suspended development of UWB as part of the alternative MAC/PHY, Bluetooth v3.0 + HS solution. A small, but significant, number of former WiMedia members had not and would not sign up to the necessary agreements for the IP transfer. The Bluetooth SIG is now in the process of evaluating other options for its longer term roadmap.

Bluetooth v4.0

The Bluetooth SIG completed the Bluetooth Core Specification version 4.0 (called Bluetooth Smart) and has been adopted as of 30 June 2010. It includes *Classic Bluetooth, Bluetooth high speed* and *Bluetooth low energy* protocols. Bluetooth high speed is based on Wi-Fi, and Classic Bluetooth consists of legacy Bluetooth protocols.

Bluetooth low energy, previously known as Wibree, is a subset of Bluetooth v4.0 with an entirely new protocol stack for rapid build-up of simple links. As an alternative to the Bluetooth standard protocols that were introduced in Bluetooth v1.0 to v3.0, it is aimed at very low power applications running off a coin cell. Chip designs allow for two types of implementation, dual-mode, single-mode and enhanced past versions. The provisional names *Wibree* and *Bluetooth ULP* (Ultra Low Power) were abandoned and the BLE name was used for a while. In late 2011, new logos "Bluetooth Smart Ready" for hosts and "Bluetooth Smart" for sensors were introduced as the general-public face of BLE.

- In a single-mode implementation, only the low energy protocol stack is implemented. ST-Microelectronics, AMICCOM, CSR, Nordic Semiconductor and Texas Instruments have released single mode Bluetooth low energy solutions.

- In a dual-mode implementation, Bluetooth Smart functionality is integrated into an existing Classic Bluetooth controller. As of March 2011, the following semiconductor companies have announced the availability of chips meeting the standard: Qualcomm-Atheros, CSR, Broadcom and Texas Instruments. The compliant architecture shares all of Classic Bluetooth's existing radio and functionality resulting in a negligible cost increase compared to Classic Bluetooth.

Cost-reduced single-mode chips, which enable highly integrated and compact devices, feature a lightweight Link Layer providing ultra-low power idle mode operation, simple device discovery, and reliable point-to-multipoint data transfer with advanced power-save and secure encrypted connections at the lowest possible cost.

General improvements in version 4.0 include the changes necessary to facilitate BLE modes, as well the Generic Attribute Profile (GATT) and Security Manager (SM) services with AES Encryption.

Core Specification Addendum 2 was unveiled in December 2011; it contains improvements to the audio Host Controller Interface and to the High Speed (802.11) Protocol Adaptation Layer.

Core Specification Addendum 3 revision 2 has an adoption date of 24 July 2012.

Core Specification Addendum 4 has an adoption date of 12 February 2013.

Bluetooth v4.1

The Bluetooth SIG announced formal adoption of the Bluetooth v4.1 specification on 4 December 2013. This specification is an incremental software update to Bluetooth Specification v4.0, and not a hardware update. The update incorporates Bluetooth Core Specification Addenda (CSA 1, 2, 3 & 4) and adds new features that improve consumer usability. These include increased co-existence support for LTE, bulk data exchange rates—and aid developer innovation by allowing devices to support multiple roles simultaneously.

New features of this specification include:

- Mobile Wireless Service Coexistence Signaling

- Train Nudging and Generalized Interlaced Scanning

- Low Duty Cycle Directed Advertising

- L2CAP Connection Oriented and Dedicated Channels with Credit Based Flow Control

- Dual Mode and Topology

- LE Link Layer Topology

- 802.11n PAL

- Audio Architecture Updates for Wide Band Speech

- Fast Data Advertising Interval

- Limited Discovery Time

Notice that some features were already available in a Core Specification Addendum (CSA) before the release of v4.1.

Bluetooth v4.2

Bluetooth v4.2 was released on December 2, 2014. It Introduces some key features for IoT. Some features, such as Data Length Extension, require a hardware update. But some older Bluetooth hardware may receive some Bluetooth v4.2 features, such as privacy updates via firmware.

The major areas of improvement are:

- LE Data Packet Length Extension

- LE Secure Connections

- Link Layer Privacy

- Link Layer Extended Scanner Filter Policies

- IP connectivity for Bluetooth Smart devices to become available soon after the introduction of BT v4.2 via the new Internet Protocol Support Profile (IPSP).

- IPSP adds an IPv6 connection option for Bluetooth Smart, to support connected home and other IoT implementations.

Bluetooth v5

Bluetooth 5 was announced in June 2016. It will quadruple the range, double the speed, and provide an eight-fold increase in data broadcasting capacity of low energy Bluetooth connections, in addition to adding functionality for connectionless services like location-relevant information and navigation.

It is mainly focused on Internet of Things emerging technology. The release of products is scheduled for late 2016 to early 2017.

Technical Information

Bluetooth Protocol Stack

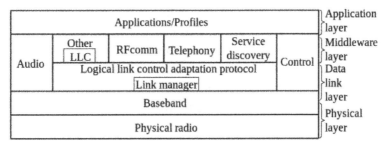

Bluetooth Protocol Stack

Bluetooth is defined as a layer protocol architecture consisting of core protocols, cable replacement protocols, telephony control protocols, and adopted protocols. Mandatory protocols for all Bluetooth stacks are: LMP, L2CAP and SDP. In addition, devices that communicate with Bluetooth almost universally can use these protocols: HCI and RFCOMM.

LMP

The *Link Management Protocol* (LMP) is used for set-up and control of the radio link between two devices. Implemented on the controller.

L2CAP

The *Logical Link Control and Adaptation Protocol* (L2CAP) is used to multiplex multiple logical connections between two devices using different higher level protocols. Provides segmentation and reassembly of on-air packets.

In *Basic* mode, L2CAP provides packets with a payload configurable up to 64 kB, with 672 bytes as the default MTU, and 48 bytes as the minimum mandatory supported MTU.

In *Retransmission and Flow Control* modes, L2CAP can be configured either for isochronous data or reliable data per channel by performing retransmissions and CRC checks.

Bluetooth Core Specification Addendum 1 adds two additional L2CAP modes to the core specification. These modes effectively deprecate original Retransmission and Flow Control modes:

- Enhanced Retransmission Mode (ERTM): This mode is an improved version of the original retransmission mode. This mode provides a reliable L2CAP channel.

- Streaming Mode (SM): This is a very simple mode, with no retransmission or flow control. This mode provides an unreliable L2CAP channel.

Reliability in any of these modes is optionally and/or additionally guaranteed by the lower layer Bluetooth BDR/EDR air interface by configuring the number of retransmissions and flush timeout (time after which the radio flushes packets). In-order sequencing is guaranteed by the lower layer.

Only L2CAP channels configured in ERTM or SM may be operated over AMP logical links.

SDP

The *Service Discovery Protocol* (SDP) allows a device to discover services offered by other devices, and their associated parameters. For example, when you use a mobile phone with a Bluetooth headset, the phone uses SDP to determine which Bluetooth profiles the headset can use (Headset Profile, Hands Free Profile, Advanced Audio Distribution Profile (A2DP) etc.) and the protocol multiplexer settings needed for the phone to connect to the headset using each of them. Each service is identified by a Universally Unique Identifier (UUID), with official services (Bluetooth profiles) assigned a short form UUID (16 bits rather than the full 128).

RFCOMM

Radio Frequency Communications (RFCOMM) is a cable replacement protocol used to generate a virtual serial data stream. RFCOMM provides for binary data transport and emulates EIA-232 (formerly RS-232) control signals over the Bluetooth baseband layer, i.e. it is a serial port emulation.

RFCOMM provides a simple reliable data stream to the user, similar to TCP. It is used directly by many telephony related profiles as a carrier for AT commands, as well as being a transport layer for OBEX over Bluetooth.

Many Bluetooth applications use RFCOMM because of its widespread support and publicly available API on most operating systems. Additionally, applications that used a serial port to communicate can be quickly ported to use RFCOMM.

BNEP

The *Bluetooth Network Encapsulation Protocol* (BNEP) is used for transferring another protocol stack's data via an L2CAP channel. Its main purpose is the transmission of IP packets in the Personal Area Networking Profile. BNEP performs a similar function to SNAP in Wireless LAN.

AVCTP

The *Audio/Video Control Transport Protocol* (AVCTP) is used by the remote control profile to transfer AV/C commands over an L2CAP channel. The music control buttons on a stereo headset use this protocol to control the music player.

AVDTP

The *Audio/Video Distribution Transport Protocol* (AVDTP) is used by the advanced audio distribution profile to stream music to stereo headsets over an L2CAP channel intended for video distribution profile in the Bluetooth transmission.

TCS

The *Telephony Control Protocol – Binary* (TCS BIN) is the bit-oriented protocol that defines the call control signaling for the establishment of voice and data calls between Bluetooth devices. Additionally, "TCS BIN defines mobility management procedures for handling groups of Bluetooth TCS devices."

TCS-BIN is only used by the cordless telephony profile, which failed to attract implementers. As such it is only of historical interest.

Adopted Protocols

Adopted protocols are defined by other standards-making organizations and incorporated into Bluetooth's protocol stack, allowing Bluetooth to code protocols only when necessary. The adopted protocols include:

- Point-to-Point Protocol (PPP): Internet standard protocol for transporting IP datagrams over a point-to-point link.

- TCP/IP/UDP: Foundation Protocols for TCP/IP protocol suite

- Object Exchange Protocol (OBEX): Session-layer protocol for the exchange of objects, providing a model for object and operation representation

- Wireless Application Environment/Wireless Application Protocol (WAE/WAP): WAE specifies an application framework for wireless devices and WAP is an open standard to provide mobile users access to telephony and information services.

Baseband Error Correction

Depending on packet type, individual packets may be protected by error correction, either 1/3 rate forward error correction (FEC) or 2/3 rate. In addition, packets with CRC will be retransmitted until acknowledged by automatic repeat request (ARQ).

Setting up Connections

Any Bluetooth device in *discoverable mode* transmits the following information on demand:

- Device name

- Device class

- List of services

- Technical information (for example: device features, manufacturer, Bluetooth specification used, clock offset)

Any device may perform an inquiry to find other devices to connect to, and any device can be configured to respond to such inquiries. However, if the device trying to connect knows the address of the device, it always responds to direct connection requests and transmits the information shown in the list above if requested. Use of a device's services may require pairing or acceptance by its owner, but the connection itself can be initiated by any device and held until it goes out of range. Some devices can be connected to only one device at a time, and connecting to them prevents them from connecting to other devices and appearing in inquiries until they disconnect from the other device.

Every device has a unique 48-bit address. However, these addresses are generally not shown in inquiries. Instead, friendly Bluetooth names are used, which can be set by the user. This name appears when another user scans for devices and in lists of paired devices.

Most cellular phones have the Bluetooth name set to the manufacturer and model of the phone by default. Most cellular phones and laptops show only the Bluetooth names and special programs are required to get additional information about remote devices. This can be confusing as, for example, there could be several cellular phones in range named T610.

Pairing and Bonding

Motivation

Many services offered over Bluetooth can expose private data or let a connecting party control the Bluetooth device. Security reasons make it necessary to recognize specific devices, and thus enable control over which devices can connect to a given Bluetooth device. At the same time, it is useful for Bluetooth devices to be able to establish a connection without user intervention (for example, as soon as in range).

To resolve this conflict, Bluetooth uses a process called *bonding*, and a bond is generated through a process called *pairing*. The pairing process is triggered either by a specific request from a user to generate a bond (for example, the user explicitly requests to "Add a Bluetooth device"), or it is triggered automatically when connecting to a service where (for the first time) the identity of a device is required for security purposes. These two cases are referred to as dedicated bonding and general bonding respectively.

Pairing often involves some level of user interaction. This user interaction confirms the identity of the devices. When pairing successfully completes, a bond forms between the two devices, enabling those two devices to connect to each other in the future without repeating the pairing process to confirm device identities. When desired, the user can remove the bonding relationship.

Implementation

During pairing, the two devices establish a relationship by creating a shared secret known as a *link key*. If both devices store the same link key, they are said to be *paired* or *bonded*. A device that wants to communicate only with a bonded device can cryptographically authenticate the identity of the other device, ensuring it is the same device it previously paired with. Once a link key is generated, an authenticated Asynchronous Connection-Less (ACL) link between the devices may be encrypted to protect exchanged data against eavesdropping. Users can delete link keys from either device, which removes the bond between the devices—so it is possible for one device to have a stored link key for a device it is no longer paired with.

Bluetooth services generally require either encryption or authentication and as such require pairing before they let a remote device connect. Some services, such as the Object Push Profile, elect not to explicitly require authentication or encryption so that pairing does not interfere with the user experience associated with the service use-cases.

Pairing Mechanisms

Pairing mechanisms changed significantly with the introduction of Secure Simple Pairing in Bluetooth v2.1. The following summarizes the pairing mechanisms:

- *Legacy pairing*: This is the only method available in Bluetooth v2.0 and before. Each device must enter a PIN code; pairing is only successful if both devices enter the same PIN code. Any 16-byte UTF-8 string may be used as a PIN code; however, not all devices may be capable of entering all possible PIN codes.

    o *Limited input devices*: The obvious example of this class of device is a Bluetooth Hands-free headset, which generally have few inputs. These devices usually have a *fixed PIN*, for example "0000" or "1234", that are hard-coded into the device.

    o *Numeric input devices*: Mobile phones are classic examples of these devices. They allow a user to enter a numeric value up to 16 digits in length.

    o *Alpha-numeric input devices*: PCs and smartphones are examples of these devices. They allow a user to enter full UTF-8 text as a PIN code. If pairing with a less capable device the user must be aware of the input limitations on the other device, there is no mechanism available for a capable device to determine how it should limit the available input a user may use.

- *Secure Simple Pairing* (SSP): This is required by Bluetooth v2.1, although a Bluetooth v2.1 device may only use legacy pairing to interoperate with a v2.0 or earlier device. Secure Simple Pairing uses a form of public key cryptography, and some types can help protect against man in the middle, or MITM attacks. SSP has the following authentication mechanisms:

o   *Just works*: As the name implies, this method just works, with no user interaction. However, a device may prompt the user to confirm the pairing process. This method is typically used by headsets with very limited IO capabilities, and is more secure than the fixed PIN mechanism this limited set of devices uses for legacy pairing. This method provides no man-in-the-middle (MITM) protection.

o   *Numeric comparison*: If both devices have a display, and at least one can accept a binary yes/no user input, they may use Numeric Comparison. This method displays a 6-digit numeric code on each device. The user should compare the numbers to ensure they are identical. If the comparison succeeds, the user(s) should confirm pairing on the device(s) that can accept an input. This method provides MITM protection, assuming the user confirms on both devices and actually performs the comparison properly.

o   *Passkey Entry*: This method may be used between a device with a display and a device with numeric keypad entry (such as a keyboard), or two devices with numeric keypad entry. In the first case, the display is used to show a 6-digit numeric code to the user, who then enters the code on the keypad. In the second case, the user of each device enters the same 6-digit number. Both of these cases provide MITM protection.

o   *Out of band* (OOB): This method uses an external means of communication, such as Near Field Communication (NFC) to exchange some information used in the pairing process. Pairing is completed using the Bluetooth radio, but requires information from the OOB mechanism. This provides only the level of MITM protection that is present in the OOB mechanism.

SSP is considered simple for the following reasons:

- In most cases, it does not require a user to generate a passkey.

- For use-cases not requiring MITM protection, user interaction can be eliminated.

- For *numeric comparison*, MITM protection can be achieved with a simple equality comparison by the user.

- Using OOB with NFC enables pairing when devices simply get close, rather than requiring a lengthy discovery process.

Security Concerns

Prior to Bluetooth v2.1, encryption is not required and can be turned off at any time. Moreover, the encryption key is only good for approximately 23.5 hours; using a single encryption key longer than this time allows simple XOR attacks to retrieve the encryption key.

- Turning off encryption is required for several normal operations, so it is problematic to detect if encryption is disabled for a valid reason or for a security attack.

Bluetooth v2.1 addresses this in the following ways:

- Encryption is required for all non-SDP (Service Discovery Protocol) connections

- A new Encryption Pause and Resume feature is used for all normal operations that require that encryption be disabled. This enables easy identification of normal operation from security attacks.

- The encryption key must be refreshed before it expires.

Link keys may be stored on the device file system, not on the Bluetooth chip itself. Many Bluetooth chip manufacturers let link keys be stored on the device—however, if the device is removable, this means that the link key moves with the device.

Security

Overview

Bluetooth implements confidentiality, authentication and key derivation with custom algorithms based on the SAFER+ block cipher. Bluetooth key generation is generally based on a Bluetooth PIN, which must be entered into both devices. This procedure might be modified if one of the devices has a fixed PIN (e.g., for headsets or similar devices with a restricted user interface). During pairing, an initialization key or master key is generated, using the E22 algorithm. The E0 stream cipher is used for encrypting packets, granting confidentiality, and is based on a shared cryptographic secret, namely a previously generated link key or master key. Those keys, used for subsequent encryption of data sent via the air interface, rely on the Bluetooth PIN, which has been entered into one or both devices.

An overview of Bluetooth vulnerabilities exploits was published in 2007 by Andreas Becker.

In September 2008, the National Institute of Standards and Technology (NIST) published a Guide to Bluetooth Security as a reference for organizations. It describes Bluetooth security capabilities and how to secure Bluetooth technologies effectively. While Bluetooth has its benefits, it is susceptible to denial-of-service attacks, eavesdropping, man-in-the-middle attacks, message modification, and resource misappropriation. Users and organizations must evaluate their acceptable level of risk and incorporate security into the lifecycle of Bluetooth devices. To help mitigate risks, included in the NIST document are security checklists with guidelines and recommendations for creating and maintaining secure Bluetooth piconets, headsets, and smart card readers.

Bluetooth v2.1 – finalized in 2007 with consumer devices first appearing in 2009 – makes significant changes to Bluetooth's security, including pairing. The pairing mechanisms section for more about these changes.

Bluejacking

Bluejacking is the sending of either a picture or a message from one user to an unsuspecting user through *Bluetooth* wireless technology. Common applications include short messages, *e.g.*, "You've just been bluejacked!". Bluejacking does not involve the removal or alteration of any data from the device. Bluejacking can also involve taking control of a mobile device wirelessly and phoning a premium rate line, owned by the bluejacker. Security advances have alleviated this issue.

History of Security Concerns

2001–2004

In 2001, Jakobsson and Wetzel from Bell Laboratories discovered flaws in the Bluetooth pairing protocol and also pointed to vulnerabilities in the encryption scheme. In 2003, Ben and Adam Laurie from A.L. Digital Ltd. discovered that serious flaws in some poor implementations of Bluetooth security may lead to disclosure of personal data. In a subsequent experiment, Martin Herfurt from the trifinite.group was able to do a field-trial at the CeBIT fairgrounds, showing the importance of the problem to the world. A new attack called BlueBug was used for this experiment. In 2004 the first purported virus using Bluetooth to spread itself among mobile phones appeared on the Symbian OS. The virus was first described by Kaspersky Lab and requires users to confirm the installation of unknown software before it can propagate. The virus was written as a proof-of-concept by a group of virus writers known as "29A" and sent to anti-virus groups. Thus, it should be regarded as a potential (but not real) security threat to Bluetooth technology or Symbian OS since the virus has never spread outside of this system. In August 2004, a world-record-setting experiment (see also Bluetooth sniping) showed that the range of Class 2 Bluetooth radios could be extended to 1.78 km (1.11 mi) with directional antennas and signal amplifiers. This poses a potential security threat because it enables attackers to access vulnerable Bluetooth devices from a distance beyond expectation. The attacker must also be able to receive information from the victim to set up a connection. No attack can be made against a Bluetooth device unless the attacker knows its Bluetooth address and which channels to transmit on, although these can be deduced within a few minutes if the device is in use.

2005

In January 2005, a mobile malware worm known as Lasco.A began targeting mobile phones using Symbian OS (Series 60 platform) using Bluetooth enabled devices to replicate itself and spread to other devices. The worm is self-installing and begins once the mobile user approves the transfer of the file (velasco.sis) from another device. Once installed, the worm begins looking for other Bluetooth enabled devices to infect. Additionally, the worm infects other .SIS files on the device, allowing replication to another device through use of removable media (Secure Digital, Compact Flash, etc.). The worm can render the mobile device unstable.

In April 2005, Cambridge University security researchers published results of their actual implementation of passive attacks against the PIN-based pairing between commercial Bluetooth devices. They confirmed that attacks are practicably fast, and the Bluetooth symmetric key establishment method is vulnerable. To rectify this vulnerability, they designed an implementation that showed that stronger, asymmetric key establishment is feasible for certain classes of devices, such as mobile phones.

In June 2005, Yaniv Shaked and Avishai Wool published a paper describing both passive and active methods for obtaining the PIN for a Bluetooth link. The passive attack allows a suitably equipped attacker to eavesdrop on communications and spoof, if the attacker was present at the time of initial pairing. The active method makes use of a specially constructed message that must be inserted at a specific point in the protocol, to make the master and slave repeat the pairing process. After that, the first method can be used to crack the PIN. This attack's major weakness is that it requires the user of the devices under attack to re-enter the PIN during the attack when the device prompts them to. Also, this active attack probably requires custom hardware, since most commercially available Bluetooth devices are not capable of the timing necessary.

In August 2005, police in Cambridgeshire, England, issued warnings about thieves using Bluetooth enabled phones to track other devices left in cars. Police are advising users to ensure that any mobile networking connections are de-activated if laptops and other devices are left in this way.

2006

In April 2006, researchers from Secure Network and F-Secure published a report that warns of the large number of devices left in a visible state, and issued statistics on the spread of various Bluetooth services and the ease of spread of an eventual Bluetooth worm.

2007

In October 2007, at the Luxemburgish Hack.lu Security Conference, Kevin Finistere and Thierry Zoller demonstrated and released a remote root shell via Bluetooth on Mac OS X v10.3.9 and v10.4. They also demonstrated the first Bluetooth PIN and Linkkeys cracker, which is based on the research of Wool and Shaked.

Mitigation

Options to mitigate against Bluetooth security attacks include:

- Enable Bluetooth only when required

- Enable Bluetooth discovery only when necessary, and disable discovery when finished

- Do not enter link keys or PINs when unexpectedly prompted to do so

- Remove paired devices when not in use

- Regularly update firmware on Bluetooth-enabled devices

Health Concerns

Bluetooth uses the microwave radio frequency spectrum in the 2.402 GHz to 2.480 GHz range, which is harmless non-ionizing radiation that cannot cause cancer. Maximum power output from a Bluetooth radio is 100 mW for class 1, 2.5 mW for class 2, and 1 mW for class 3 devices. Even the maximum power output of class 1 is a lower level than the lowest powered mobile phones. UMTS & W-CDMA outputs 250 mW, GSM1800/1900 outputs 1000 mW, and GSM850/900 outputs 2000 mW.

Interference Caused by USB 3.0

USB 3.0 devices, ports and cables have been proven to interfere with Bluetooth devices due to the electronic noise they release falling over the same operating band as Bluetooth. The close proximity of Bluetooth and USB 3.0 devices can result in a drop in throughput or complete connection loss of the Bluetooth device/s connected to a computer.

Various strategies can be applied to resolve the problem, ranging from simple solutions such as increasing the distance of USB 3.0 devices from any Bluetooth devices or purchasing better shielded USB cables. Other solutions include applying additional shielding to the internal USB components of a computer.

Bluetooth Award Programs

The Bluetooth Innovation World Cup, a marketing initiative of the Bluetooth Special Interest Group (SIG), was an international competition that encouraged the development of innovations for applications leveraging Bluetooth technology in sports, fitness and health care products. The aim of the competition was to stimulate new markets.

The Bluetooth Innovation World Cup morphed into the Bluetooth Breakthrough Awards in 2013. The Breakthrough Awards Bluetooth program highlights the most innovative products and applications available today, prototypes coming soon, and student-led projects in the making.

Mobile App

A mobile app is a software application designed to run on mobile devices such as smartphones and tablet computers. Most such devices are sold with several apps bundled as pre-installed software, such as a web browser, email client, calendar, mapping program, and an app for buying music or other media or more apps. Some pre-installed apps can be removed by an ordinary uninstall process, thus leaving more storage space for desired ones. Where the software does not allow this, some devices can be rooted to eliminate the undesired apps.

Native mobile apps often stand in contrast to applications that run on desktop computers, and with web applications which run in mobile web browsers rather than directly on the mobile device.

Overview

Apps that are not preinstalled are usually available through distribution platforms called app stores. They began appearing in 2008 and are typically operated by the owner of the mobile operating system, such as the Apple App Store, Google Play, Windows Phone Store, and BlackBerry App World. Some apps are free, while others must be bought. Usually, they are downloaded from the platform to a target device, but sometimes they can be downloaded to laptops or desktop computers. For apps with a price, generally a percentage, 20-30%, goes to the distribution provider (such as iTunes), and the rest goes to the producer of the app. The same app can therefore cost a different price depending on the mobile platform.

The term "app" is a shortening of the term "application software". It has become very popular, and in 2010 was listed as "Word of the Year" by the American Dialect Society. In 2009, technology columnist David Pogue said that newer smartphones could be nicknamed "app phones" to distinguish them from earlier less-sophisticated smartphones.

Mobile apps were originally offered for general productivity and information retrieval, including email, calendar, contacts, stock market and weather information. However, public demand and the availability of developer tools drove rapid expansion into other categories, such as those handled by desktop application software packages. As with other software, the explosion in number and variety of apps made discovery a challenge, which in turn led to the creation of a wide range of review, recommendation, and curation sources, including blogs, magazines, and dedicated online app-discovery services. In 2014 government regulatory agencies began trying to regulate and curate apps, particularly medical apps. Some companies offer apps as an alternative method to deliver content with certain advantages over an official website.

The official US Army iPhone app presents the service's technology news, updates and media in a single place

Usage of mobile apps has become increasingly prevalent across mobile phone users. A May 2012 comScore study reported that during the previous quarter, more mobile subscribers used apps than browsed the web on their devices: 51.1% vs. 49.8% respectively. Researchers found that usage of mobile apps strongly correlates with user context and depends on user's location and time of the day. Mobile apps are playing an ever increasing role within healthcare and when designed and integrated correctly can yield many benefits.

Market research firm Gartner predicted that 102 billion apps would be downloaded in 2013 (91% of them free), which would generate $26 billion in the US, up 44.4% on 2012's US$18 billion. By Q2 2015, the Google Play and Apple stores alone generated $5 billion. An analyst report estimates that the app economy creates revenues of more than €10 billion per year within the European Union, while over 529,000 jobs have been created in 28 EU states due to the growth of the app market.

Development

Developers at work

Developing apps for mobile devices requires considering the constraints and features of these devices. Mobile devices run on battery and have less powerful processors than personal computers and also have more features such as location detection and cameras. Developers also have to consider a wide array of screen sizes, hardware specifications and configurations because of intense competition in mobile software and changes within each of the platforms (although these issues can be overcome with mobile device detection).

Mobile application development requires use of specialized integrated development environments. Mobile apps are first tested within the development environment using emulators and later subjected to field testing. Emulators provide an inexpensive way to test applications on mobile phones to which developers may not have physical access.

Mobile user interface (UI) Design is also essential. Mobile UI considers constraints and contexts, screen, input and mobility as outlines for design. The user is often the focus of interaction with their device, and the interface entails components of both hardware and software. User input allows for the users to manipulate a system, and device's output allows the system to indicate the effects of the users' manipulation. Mobile UI design constraints include limited attention and form factors, such as a mobile device's screen size for a user's hand. Mobile UI contexts signal cues from user activity, such as location and scheduling that can be shown from user interactions within a mobile application. Overall, mobile UI design's goal is primarily for an understandable, user-friendly interface.

Mobile UIs, or front-ends, rely on mobile back-ends to support access to enterprise systems. The mobile back-end facilitates data routing, security, authentication, authorization, working off-line, and service orchestration. This functionality is supported by a mix of middleware components including mobile app servers, Mobile Backend as a service (MBaaS), and SOA infrastructure.

Conversational interfaces display the computer interface and present interactions through text instead of graphic elements. They emulate conversations with real humans. There are two main types of conversational interfaces: voice assistants (like the Amazon Echo) and chatbots.

Conversational interfaces are growing particularly practical as users are starting to feel overwhelmed with mobile apps (a term known as "app fatigue").

David Limp, Amazon's senior vice president of devices, says in an interview with Bloomberg, "We believe the next big platform is voice."

Distribution

The two biggest app stores are Google Play for Android and App Store for iOS.

Google Play

Google Play (formerly known as the Android Market) is an international online software store developed by Google for Android devices. It opened in October 2008. In July 2013, the number of apps downloaded via the Google Play Store surpassed 50 billion, of the over 1 million apps available. As of February 2015, According to Statista the number of apps available exceeded 1.4 million.

App Store

Apple's App Store for iOS was not the first app distribution service, but it ignited the mobile revolution and was opened on July 10, 2008, and as of January 2011, reported over 10 billion downloads. The original AppStore was first demonstrated to Steve Jobs in 1993 by Jesse Tayler at NeXTWorld Expo As of June 6, 2011, there were 425,000 apps available, which had been downloaded by 200 million iOS users. During Apple's 2012 Worldwide Developers Conference, Apple CEO Tim Cook announced that the App Store has 650,000 available apps to download as well as 30 billion apps downloaded from the app store until that date. From an alternative perspective, figures seen in July 2013 by the BBC from tracking service Adeven indicate over two-thirds of apps in the store are "zombies", barely ever installed by consumers.

Others

- Amazon Appstore is an alternative application store for the Android operating system. It was opened in March 2011 and as of June 2015, the app store has nearly 334,000 apps. The Amazon Appstore's Android Apps can also be installed and run on BlackBerry 10 devices.

- BlackBerry World is the application store for BlackBerry 10 and BlackBerry OS devices. It opened in April 2009 as BlackBerry App World.

- Ovi (Nokia) for Nokia phones was launched internationally in May 2009. In May 2011, Nokia announced plans to rebrand its Ovi product line under the Nokia brand and Ovi Store was renamed Nokia Store in October 2011. Nokia Store will no longer allow developers to publish new apps or app updates for its legacy Symbian and MeeGo operating systems from January 2014.

- Windows Phone Store was introduced by Microsoft for its Windows Phone platform, which was launched in October 2010. As of October 2012, it has over 120,000 apps available.

- Windows Store was introduced by Microsoft for its Windows 8 and Windows RT platforms. While it can also carry listings for traditional desktop programs certified for compatibility with Windows 8, it is primarily used to distribute "Windows Store apps"—which are primarily built for use on tablets and other touch-based devices (but can still be used with a keyboard and mouse, and on desktop computers and laptops).

- Samsung Apps was introduced in September 2009. As of October 2011, Samsung Apps reached 10 million downloads. The store is available in 125 countries and it offers apps for Windows Mobile, Android and Bada platforms.

- The Electronic AppWrapper was the first electronic distribution service to collectively provide encryption and purchasing electronically

- F-Droid — Free and open Source Android app repository.

- There are numerous other independent app stores for Android devices.

Enterprise Management

Mobile application management (MAM) describes software and services responsible for provisioning and controlling access to internally developed and commercially available mobile apps used in business settings. The strategy is meant to off-set the security risk of a Bring Your Own Device (BYOD) work strategy. When an employee brings a personal device into an enterprise setting, mobile application management enables the corporate IT staff to transfer required applications, control access to business data, and remove locally cached business data from the device if it is lost, or when its owner no longer works with the company. Containerization is an alternate BYOD security solution. Rather than controlling an employees entire device, containerization apps create isolated and secure pockets separate from all personal data. Company control of the device only extends to that separate container.

App Wrapping vs. Native App Management

Especially when employees "bring your own device", mobile apps can be a significant security risk for businesses, because they transfer unprotected sensitive data to the Internet without knowledge and consent of the users. Reports of stolen corporate data show how quickly corporate and personal data can fall into the wrong hands. Data theft is not just the loss of confidential information, but makes companies vulnerable to attack and blackmail.

Professional mobile application management helps companies protect their data. One option for securing corporate data is app wrapping. But there also are some disadvantages like copyright infringement or the loss of warranty rights. Functionality, productivity and user experience are particularly limited under app wrapping. The policies of a wrapped app can't be changed. If required, it must be recreated from scratch, adding cost.

Alternatively, it is possible to offer native apps securely through enterprise mobility management without limiting the native user experience. This enables more flexible IT management as apps can be easily implemented and policies adjusted at any time.

Mobile Cloud Computing

Mobile Cloud Computing (MCC) is the combination of cloud computing, mobile computing and wireless networks to bring rich computational resources to mobile users, network operators, as well as cloud computing providers. The ultimate goal of MCC is to enable execution of rich mobile applications on a plethora of mobile devices, with a rich user experience. MCC provides business opportunities for mobile network operators as well as cloud providers. More comprehensively, MCC can be defined as "a rich mobile computing technology that leverages unified elastic resources of varied clouds and network technologies toward unrestricted functionality, storage, and mobility to serve a multitude of mobile devices anywhere, anytime through the channel of Ethernet or Internet regardless of heterogeneous environments and platforms based on the pay-as-you-use principle."

Architecture

MCC uses computational augmentation approaches (computations are executed remotely instead of on the device) by which resource-constraint mobile devices can utilize computational resources of varied cloud-based resources. In MCC, there are four types of cloud-based resources, namely distant immobile clouds, proximate immobile computing entities, proximate mobile computing entities, and hybrid (combination of the other three model). Giant clouds such as Amazon EC2 are in the distant immobile groups whereas cloudlet or surrogates are member of proximate immobile computing entities. Smartphones, tablets, handheld devices, and wearable computing devices are part of the third group of cloud-based resources which is proximate mobile computing entities.

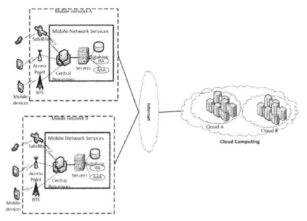

Mobile cloud architecture

Vodafone, Orange and Verizon have started to offer cloud computing services for companies.

Frameworks

MobiByte

MobiByte is a mobile cloud application development framework that enables mobile devices to offload computations to the cloud to achieve performance enhancement, energy efficiency, and application execution support.

Challenges

In the MCC landscape, an amalgam of mobile computing, cloud computing, and communication networks (to augment smartphones) creates several complex challenges such as Mobile Computation Offloading, Seamless Connectivity, Long WAN Latency, Mobility Management, Context-Processing, Energy Constraint, Vendor/data Lock-in, Security and Privacy, Elasticity that hinder MCC success and adoption.

Open Research Issues

Although significant research and development in MCC is available in the literature, efforts in the following domains is still lacking:

- Architectural issues: A reference architecture for heterogeneous MCC environment is a crucial requirement for unleashing the power of mobile computing towards unrestricted ubiquitous computing.

- Energy-efficient transmission: MCC requires frequent transmissions between cloud platform and mobile devices, due to the stochastic nature of wireless networks, the transmission protocol should be carefully designed.

- Context-awareness issues: Context-aware and socially-aware computing are inseparable traits of contemporary handheld computers. To achieve the vision of mobile computing among heterogeneous converged networks and computing devices, designing resource-efficient environment-aware applications is an essential need.

- Live VM migration issues: Executing resource-intensive mobile application via Virtual Machine (VM) migration-based application offloading involves encapsulation of application in VM instance and migrating it to the cloud, which is a challenging task due to additional overhead of deploying and managing VM on mobile devices.

- Mobile communication congestion issues: Mobile data traffic is tremendously hiking by ever increasing mobile user demands for exploiting cloud resources which impact on mobile network operators and demand future efforts to enable smooth communication between mobile and cloud endpoints.

- Trust, security, and privacy issues: Trust is an essential factor for the success of the burgeoning MCC paradigm.

MCC Research Groups

Several academic and industrial research groups in MCC have been emerging since last few years. Some of the MCC research groups in academia with large number of researchers and publications include:

- MobCC lab, Faculty of Computer Science and Information Technology, University Malaya. The lab was established in 2010 under the High Impact Research Grant, Ministry of Higher Education, Malaysia. It has 17 researchers and has track of 22 published articles in international conference and peer reviewed CS journals.

- ICCLAB, Zürich University of Applied Sciences has a segment working on MCC. The InIT Cloud Computing Lab is a research lab within the Institute of Applied Information Technology (InIT) of Zürich University of Applied Sciences (ZHAW). It covers topic areas across the entire cloud computing technology stack.

- Mobile & Cloud Lab, Institute of Computer Science, University of Tartu. Mobile & Cloud Lab conducts research and teaching in the mobile computing and cloud computing domains. The research topics of the group include cloud computing, mobile application development, mobile cloud, mobile web services and migrating scientific computing and enterprise applications to the cloud.

- SmartLab, Data Management Systems Laboratory, Department of Computer Science, University of Cyprus. SmartLab is a first-of-a-kind open cloud of smartphones that enables a new line of systems-oriented mobile computing research.

Mobile Operating System

A mobile operating system (or mobile OS) is an operating system for smartphones, tablets, PDAs, or other mobile devices. While computers such as typical laptops are mobile, the operating systems usually used on them are not considered mobile ones, as they were originally designed for desktop computers that historically did not have or need specific "mobile" features. This distinction is becoming blurred in some newer operating systems that are hybrids made for both uses.

Mobile operating systems combine features of a personal computer operating system with other features useful for mobile or handheld use; usually including, and most of the following considered essential in modern mobile systems; a touchscreen, cellular, Bluetooth, Wi-Fi, GPS mobile navigation, camera, video camera, speech recognition, voice recorder, music player, near field communication and infrared blaster.

Mobile devices with mobile communications capabilities (e.g. smartphones) contain two mobile operating systems – the main user-facing software platform is supplemented by a second low-level proprietary real-time operating system which operates the radio and other hardware. Research has shown that these low-level systems may contain a range of security vulnerabilities permitting malicious base stations to gain high levels of control over the mobile device.

Timeline

Mobile operating system milestones mirror the development of mobile phones and smartphones:

- 1973–1993 Mobile phones use embedded systems to control operation.

- 1994 The first smartphone, the IBM Simon, has a touchscreen, email and PDA features.

- 1996 Palm Pilot 1000 personal digital assistant is introduced with the Palm OS mobile operating system.

- 1996 First Windows CE Handheld PC devices are introduced.

- 1999 Nokia S40 OS is officially introduced along with the Nokia 7110.

- 2000 Symbian becomes the first modern mobile OS on a smartphone with the launch of the Ericsson R380.

- 2001 The Kyocera 6035 is the first smartphone with Palm OS.

- 2002 Microsoft's first Windows CE (Pocket PC) smartphones are introduced.

- 2002 BlackBerry releases its first smartphone.

- 2005 Nokia introduces Maemo OS on the first internet tablet N770.

- 2007 Apple iPhone with iOS is introduced as an iPhone, "mobile phone" and "internet communicator."

- 2007 Open Handset Alliance (OHA) formed by Google, HTC, Sony, Dell, Intel, Motorola, Samsung, LG, etc.

- 2008 OHA releases Android (based on Linux Kernel) 1.0 with the HTC Dream (T-Mobile G1) as the first Android phone.

- 2009 Palm introduces webOS with the Palm Pre. By 2012 webOS devices were no longer sold.

- 2009 Samsung announces the Bada OS with the introduction of the Samsung S8500.

- 2010 Windows Phone OS phones are released but are not compatible with the previous Windows Mobile OS.

- 2010 MIUI are release by Xiaomi Inc which based on Google's Android Open Source Project(AOSP).

- 2011 MeeGo the first mobile Linux, combining Maemo and Moblin, is introduced with the Nokia N9, a collaboration of Nokia, Intel and Linux Foundation

- 2011 Samsung, Intel and the Linux Foundation announced, in September 2011, that their efforts will shift from Bada, MeeGo to Tizen during 2011 and 2012.

- 2011 the Mer project was announced, in October 2011, centered around an ultra-portable Linux + HTML5/QML/JavaScript core for building products with, derived from the MeeGo codebase.

- 2012 Mozilla announced in July 2012 that the project previously known as "Boot to Gecko"(which was built on top of Android Linux kernel and using Android drivers, however it doesn't use any Java-like code of Android) was now Firefox OS and had several handset OEMs on board.

- 2013 Canonical announced Ubuntu Touch, a version of the Linux distribution expressly designed for smartphones. The OS is built on the Android Linux kernel, using Android drivers, but does not use any of the Java-like code of Android.

- 2013 BlackBerry releases their new operating system for smartphones, BlackBerry 10.

- 2013 Google releases Android KitKat 4.4.

- 2014 Microsoft releases Windows Phone 8.1 in February 2014.

- 2014 Xiaomi releases MIUI v6 in August 2014.

- 2014 Apple releases iOS 8 in September 2014.

- 2014 BlackBerry release BlackBerry 10.3 with integration with the Amazon Appstore in September 2014.

- 2014 Google releases Android 5.0 "Lollipop" in November 2014.

- 2015 Microsoft releases Windows 10 Mobile in November 2015.

- 2015 Google releases Android 5.1 "Lollipop" in February 2015.

- 2015 Apple releases iOS 9 in September 2015.

- 2015 Google releases Android 6.0 "Marshmallow" in September 2015.

- 2016 Apple announced iOS 10 in June 2016.

- 2016 Google released Android 7.0 "Nougat" in August 2016.

- 2016 Apple releases iOS 10 in September 2016.

Current Software Platforms

Note that these operating systems often run on top of baseband or other real time operating systems that handle hardware aspects of the phone.

Android

Google Android Marshmallow OS

Android (based on the Linux kernel) is from Google Inc. Besides having the largest installed base worldwide on smartphones, it is also the most popular operating system for general purpose computers (a category that includes desktop computers as well as mobile devices), even though Android is not a popular operating system for regular ("desktop") PCs. Although the Android operating system is free and open-source software, in actual devices, much of the software bundled with it (including Google apps and vendor-installed software) is proprietary and closed source.

Android's releases prior to 2.0 (1.0, 1.5, 1.6) were used exclusively on mobile phones. Android 2.x releases were mostly used for mobile phones but also some tablets. Android 3.0 was a tablet-oriented release and does not officially run on mobile phones. The current Android version is 7.0.

Android's releases are named after sweets or dessert items (except for the first and second releases):

- 1.0 : (API Level 1)

- 1.1 – Alpha : (API Level 2)

- 1.2 – Beta

- 1.5 – Cupcake : (API Level 3)

- 1.6 – Donut : (API Level 4)

- 2.0 – Eclair : (API Level 5)

- 2.0.1 – Eclair : (API Level 6)

- 2.1 – Eclair : (API Level 7)

- 2.2.x – Frozen Yogurt ("Froyo") : (API Level 8)

- 2.3 – Gingerbread (Minor UI Tweak): (API Level 9)

- 2.3.3 – Gingerbread: (API Level 10)

- 3.0 – Honeycomb (Major UI revamp): (API Level 11)

- 3.1 – Honeycomb: (API Level 12)

- 3.2 – Honeycomb: (API Level 13)

- 4.0 – Ice Cream Sandwich (Minor UI Tweak): (API Level 14)

- 4.0.3 – Ice Cream Sandwich: (API Level 15)

- 4.1 – Jelly Bean: (API Level 16)

- 4.2 – Jelly Bean: (API Level 17)

- 4.3 – Jelly Bean: (API Level 18)

- 4.4.4 – KitKat: (API Level 19)

- 5.0, 5.0.1, 5.0.2 – Lollipop (Major UI revamp) : (API Level 21)

- 5.1, 5.1.1 – Lollipop : (API Level 22)

- 6.0 & 6.0.1 – Marshmallow: (API Level 23)

- 7.0 - Nougat (API Level 24)

AOKP

AOKP, short for Android Open Kang Project, is a custom ROM which is based on the Android Open Source Project (AOSP). Similar to CyanogenMod, AOKP allows Android users who can no longer obtain update support from their manufacturer to continue updating their OS version to the latest one based on official release from Google AOSP and heavy theme customization together with customizable system functions.

Current AOKP version list:

- AOKP (Based on Android "Ice Cream Sandwich" 4.0.x)

- AOKP (Based on Android "Jelly Bean" 4.1.x – 4.3.x)

- AOKP (Based on Android "KitKat" 4.4.x)

ColorOS

ColorOS is based on the open source Android Open Source Project (AOSP) and develop by OPPO Electronics Corp. Currently, ColorOS are officially release together with every OPPO devices and OPPO had release an official ColorOS ROM for Oneplus One.

Current ColorOS version list:

- ColorOS 1.0 (Based on Android "Jelly Bean" 4.1.x – 4.3.x) (Initial release)

- ColorOS 2.0 (Based on Android "KitKat" 4.4.x) (Minor UI upgrade)

- ColorOS 2.1 (Based on Android "Lollipop" 5.0.x – 5.1.x) (Minor UI upgrade)

CyanogenMod

CyanogenMod is based on the open source Android Open Source Project (AOSP). It is a custom ROM that was co-developed by the CyanogenMod community; therefore, the OS does not include any proprietary apps unless the user installs them. Due to its open source nature, CyanogenMod allows Android users who can no longer obtain update support from their manufacturer to continue updating their OS version to the latest one based on official release from Google AOSP and heavy theme customization. The current version of the OS is CyanogenMod 13 which is based on Android Marshmallow.

Current CyanogenMod version list:

- CyanogenMod 3 (Based on Android "Cupcake" 1.5.x, initial release)

- CyanogenMod 4 (Based on Android "Cupcake" and "Donut" 1.5.x and 1.6.x)

- CyanogenMod 5 (Based on Android "Eclair" 2.0/2.1)

- CyanogenMod 6 (Based on Android "Froyo" 2.2.x)

- CyanogenMod 7 (Based on Android "Gingerbread" 2.3.x)

- CyanogenMod 9 (Based on Android "Ice Cream Sandwich" 4.0.x, major UI revamp)

- CyanogenMod 10 (Based on Android "Jelly Bean" 4.1.x – 4.3.x)

- CyanogenMod 11 (Based on Android "KitKat" 4.4.x)

- CyanogenMod 12 (Based on Android "Lollipop" 5.0.x – 5.1.x, major UI revamp)

- CyanogenMod 13 (Based on Android "Marshmallow" 6.0.x)

- CyanogenMod 14 (Based on Android "Nougat" 7.0.x, still not officially launched)

Cyanogen OS

Cyanogen OS is based on CyanogenMod and maintained by Cyanogen Inc, however it includes proprietary apps and it is only available for commercial uses.

Current Cyanogen OS version list:

- Cyanogen OS 11s (Based on Android "KitKat" 4.4.x, initial release)
- Cyanogen OS 12 (Based on Android "Lollipop" 5.0.x, major UI revamp)
- Cyanogen OS 12.1(Based on android "Lollipop" 5.1.x)
- Cyanogen OS 13 (Based on Android "Marshmallow" 6.0.x)

EMUI

EMUI (which stands for *Emotion User Interface*) is a ROM/OS that is developed by Huawei Technologies Co. Ltd. and is based on Google's Android Open Source Project (AOSP). EMUI is preinstalled on most Huawei Smartphone devices and its subsidiaries the Honor series.

Current EMUI version list:

- EMUI 1.x (Based on Android "Ice Cream Sandwich" and "Jelly Bean" 4.0.x and 4.1.x – 4.3.x)(Initial release)
- EMUI 2.x (Based on Android "Ice Cream Sandwich", "Jelly Bean" and "KitKat" 4.0.x, 4.1.x – 4.3.x and 4.4.x)(Minor UI tweak)
- EMUI 3.x (Based on Android "KitKat" and "Lollipop" 4.4.x and 5.0.x – 5.1.x)(Minor UI tweak)
- EMUI 4.x (Based on Android "Marshmallow" 6.x)
- EMUI 5.x (Based on Android "Nougat" 7.x)

Fire OS

Fire OS is an operating system launched by Amazon and is based on Google's Android Open Source Project (AOSP). Currently only a few devices have Fire OS installed, like Fire Phone, the Kindle Fire series and Amazon's Fire TV. Although the OS was built on top on Google's AOSP, it does not pre-install Google apps and ship with custom Amazon services.

Current Fire OS version list:

- Fire OS 3.0.x (Based on Android "Jelly Bean" 4.2.2, official release as Fire OS)
- Fire OS 4.x.x (Based on Android "Jelly Bean" and "KitKat" 4.2.2 and 4.4.x, major UI revamp to match the Amazon's Fire Phone)
- Fire OS 5.x.x (Based on Android "Lollipop" 5.0.x – 5.1.x)

Flyme OS

Flyme OS is an operating system developed by Meizu Technology Co., Ltd., an open source OS based on Google Android Open Source Project (AOSP). Flyme OS is mainly installed on Meizu Smartphones such as the MX's series, however it also has official ROM support for a few Android devices.

Current Flyme OS version list:

- Flyme OS 1.x.x (Based on Android "Ice Cream Sandwich" 4.0.3, initial release)

- Flyme OS 2.x.x (Based on Android "Jelly Bean" 4.1.x – 4.2.x)

- Flyme OS 3.x.x (Based on Android "Jelly Bean" 4.3.x)

- Flyme OS 4.x.x (Based on Android "KitKat" 4.4.x)

- Flyme OS 5.x.x (Based on Android "Lollipop" 5.0.x – 5.1.x)

HTC Sense

HTC Sense is a software suite developed by HTC, used primarily on the company's Android-based devices. Serving as a successor to HTC's TouchFLO 3D software for Windows Mobile, Sense modifies many aspects of the Android user experience, incorporating additional features (such as an altered home screen and keyboard), additional widgets, re-designed applications, and additional HTC-developed applications. The first device with Sense, the HTC Hero, was released in 2009.

- HTC Sense 1.x (Based on Android "Eclair" 2.0/2.1, Initial release)

- HTC Sense 2.x (Based on Android "Eclair", "Froyo" and "Gingerbread" 2.0/2.1, 2.2.x and 2.3.x, redesign UI)

- HTC Sense 3.x (Based on Android "Gingerbread" 2.3.x, redesign UI)

- HTC Sense 4.x (Based on Android "Ice Cream Sandwich" and "Jelly Bean" 4.0.x and 4.1.x, redesign UI)

- HTC Sense 5.x (Based on Android "Jelly Bean" 4.1.x – 4.3.x, redesign UI)

- HTC Sense 6.x (Based on Android "KitKat" 4.4.x, redesign UI)

- HTC Sense 7.x (Based on Android "Lollipop" 5.0.x, redesign UI)

- HTC Sense 8.x (Based on Android "Marshmallow" 6.0.x, redesign UI)

MIUI

MIUI (which stands for *Mi User Interface*), an operating system developed by a Chinese electronic company Xiaomi Tech, is a mobile operating system which based on Google Android Open Source Project (AOSP). MIUI is found in Xiaomi smartphones such as the Mi and Redmi Series, however it also has official ROM support for a few Android devices. Although MIUI is based on AOSP, which is Open Source, it consists of closed source and proprietary software of its own.

Current MIUI version list:

- MIUI V1 (Based on Android "Froyo" 2.2.x, Initial release)

- MIUI V2 (Based on Android "Froyo" 2.2.x, redesign UI)

- MIUI V3 (Based on Android "Gingerbread" 2.3.x, redesign UI)

- MIUI V4 (Based on Android "Ice Cream Sandwich" and "Jelly Bean" 4.0.x and 4.1.x, redesign UI)

- MIUI V5 (Based on Android "Jelly Bean" and "KitKat" 4.1.x – 4.3.x and 4.4.x, redesign UI)

- MIUI V6 (Based on Android "KitKat" and "Lollipop" 4.4.x and 5.0.x, redesign UI)

- MIUI V7 (Based on Android "KitKat" and "Lollipop" and "Marshmallow" 4.4.x and 5.0.x and 6.0.x)

- MIUI V8 (Based on Android "Marshmallow" 6.0.x)

Nokia X Platform

The Nokia X platform was developed by Nokia Corporation and later on maintained by Microsoft Mobile. It is a project which is based on the open source Android Open Source Project (AOSP). It removes all the Google Services and Apps and replaces them with Nokia and Microsoft apps. Its overall UI mimics the Windows Phone UI.

Current Nokia X platform version list:

- Nokia X platform 1.x (Based on Android "Jelly Bean" 4.1.x)(Initial release)

- Nokia X platform 2.x (Based on Android "Jelly Bean" 4.3.x)(Minor UI tweak)

LG UX

LG UX (formally known as Optimus UI) is a front-end touch interface developed by LG Electronics with partners, featuring a full touch user interface. It is sometimes incorrectly identified as an operating system. LG UX is used internally by LG for sophisticated feature phones and tablet computers, and is not available for licensing by external parties.

Optimus UI 2 which based on Android 4.1.2 has been released on the Optimus K II and the Optimus Neo 3. It features a more refined user interface as compared to the previous version which based on Android 4.1.1, would include together which new functionality such as voice shutter and quick memo.

Current LG UX version list:

- Optimus UI 1.x (Based on Android "Gingerbread" 2.3.x) (Initial release)

- Optimus UI 2.x (Based on Android "Ice Cream Sandwich" and "Jelly Bean" 4.0.x and 4.1.x – 4.3.x, redesign UI)

- LG UX 3.x (Based on Android "KitKat" and "Lollipop" 4.4.x and 5.0.x, redesign UI)

- LG UX 4.x (Based on Android "Lollipop" 5.1.x and "Marshmallow" 5.1.x and 6.0.x, redesign UI)

OxygenOS

OxygenOS is based on the open source Android Open Source Project (AOSP) and is developed by OnePlus to replace Cyanogen OS on OnePlus devices such as the OnePlus One, and it is preinstalled on the OnePlus 2, OnePlus 3, and OnePlus X. As stated by Oneplus, OxygenOS is focused on stabilization and maintaining of "stock" like those found on Nexus devices. It consists of mainly Google apps and minor UI customization to maintain the sleekness of "pure" Android.

Current OxygenOS version list:

- Oxygen OS 1.0.X (Based on Android "Lollipop" 5.0.x) (Initial release)

- Oxygen OS 2.0.X (Based on Android "Lollipop" 5.1.x) (Overall maintenance update)

- Oxygen OS 3.0.X (Based on Android "Marshmallow" 6.0.0) (Upgrade Android main version)

- Oxygen OS 3.1.X (Based on Android "Marshmallow" 6.0.1) (Maintenance update)

TouchWiz

TouchWiz is a front-end touch interface developed by Samsung Electronics with partners, featuring a full touch user interface. It is sometimes incorrectly identified as an independent operating system. TouchWiz is used internally by Samsung for smartphones, feature phones and tablet computers, and is not available for licensing by external parties. The Android version of TouchWiz also comes with Samsung-made apps preloaded (except starting with the Galaxy S6 which have removed all Samsung pre-loaded apps installed, leaving one with Galaxy Apps, to save storage space and initially due to the removal of MicroSD).

Current TouchWiz version list:

- TouchWiz 3.0 & 3.0 Lite (Based on Android "Eclair" and "Froyo" 2.0/2.1 and 2.2.x) (Initial release)

- TouchWiz 4.0 (Based on Android "Gingerbread" and "Ice Cream Sandwich" 2.3.x and 4.0.x) (Redesign UI)

- TouchWiz Nature UX "1.0" and Lite (Based on Android "Ice Cream Sandwich" and "Jelly Bean" 4.0.x and 4.1.x) (Redesign UI)

- TouchWiz Nature UX 2.x (Based on Android "Jelly Bean" and "KitKat" 4.2.x – 4.3.x and 4.4.x) (Redesign UI)

- TouchWiz Nature UX 3.x (Based on Android "KitKat" and "Lollipop" 4.4.x and 5.0.x) (Redesign UI)

- TouchWiz Nature UX 5.x (Based on Android "Lollipop" 5.0.x – 5.1.x) (Redesign UI)

ZenUI

ZenUI is a front-end touch interface developed by ASUS with partners, featuring a full touch user interface. ZenUI is used by Asus for its Android phones and tablet computers, and is not available for licensing by external parties. ZenUI also comes preloaded with Asus-made apps like ZenLink (PC Link, Share Link, Party Link & Remote Link).

iOS

iOS (previously known as iPhone OS) is from Apple Inc. It has the second largest installed base worldwide on smartphones, but the largest profits, due to aggressive price competition between Android-based manufacturers. It is closed source and proprietary and built on open source Darwin core OS. The Apple iPhone, iPod Touch, iPad and second-generation Apple TV all use iOS, which is derived from OS X.

Native third party applications were not officially supported until the release of iOS 2.0 on July 11, 2008. Before this, "jailbreaking" allowed third party applications to be installed, and this method is still available.

Currently all iOS devices are developed by Apple and manufactured by Foxconn or another of Apple's partners.

As of 2014, the global market share of iOS was 15.4%.

Current iOS version list:

- iPhone OS 1.x

- iPhone OS 2.x

- iPhone OS 3.x

- iOS 4.x

- iOS 5.x

- iOS 6.x

- iOS 7.x (Major UI revamp)

- iOS 8.x

- iOS 9.x

- iOS 10.x

Windows 10 Mobile

Windows 10 Mobile (formerly called Windows Phone) is from Microsoft. It is closed source and proprietary. It has the third largest installed base on smartphones behind Android and iOS.

Unveiled on February 15, 2010, Windows Phone includes a user interface inspired by Microsoft's

"Metro Design Language". It is integrated with Microsoft services such as OneDrive and Office, Xbox Music, Xbox Video, Xbox Live games and Bing, but also integrates with many other non-Microsoft services such as Facebook and Google accounts. Windows Phone devices are made primarily by Microsoft Mobile/Nokia, and also by HTC and Samsung.

On 21 January 2015, Microsoft announced that the Windows Phone brand will be phased out and replaced with Windows 10 Mobile, bringing tighter integration and unification with its PC counterpart Windows 10, and provide a platform for smartphones as well as tablets with screen size under 8 inches.

As of 2016, Windows 10 Mobile global market share dropped below 0.6%.

Current Windows Phone version list:

- Windows Phone 7

- Windows Phone 7.5

- Windows Phone 7.8 (Minor UI tweak)

- Windows Phone 8 (GDR1, GDR2 & GDR3) & (Minor UI tweak)

- Windows Phone 8.1 (GDR1 & GDR2) & (Minor UI tweak)

- Windows 10 Mobile

BlackBerry 10

BlackBerry 10 (based on the QNX OS) is from BlackBerry. As a smart phone OS, it is closed source and proprietary, and only runs on phones and tablets manufactured by Blackberry.

Once one of the dominant platforms in the world, its global market share was reduced to 0.4% by the end of 2014.

Current BlackBerry 10 version list:

- BlackBerry 10.0

- BlackBerry 10.1

- BlackBerry 10.2

- BlackBerry 10.3 (Major UI revamp)

- BlackBerry 10.4 (Developing, expected to release by 2016)

Firefox OS

Firefox OS is from Mozilla. It is open source and is released under the Mozilla Public License. It is built on the Android Linux kernel and uses Android drivers, but doesn't use any Java-like code of Android.

According to Ars Technica, "Mozilla says that B2G is motivated by a desire to demonstrate that the standards-based open Web has the potential to be a competitive alternative to the existing single-vendor application development stacks offered by the dominant mobile operating systems."

Current Firefox OS version list:

- 1.0.x

- 1.1.x

- 1.2.x

- 1.3.x

- 1.4.x

- 1.5.x

- 2.0.0

- 2.1.0

- 2.2.0

- 2.5.0

Sailfish OS

Sailfish OS is from Jolla. It is partly open source and adopts GPL (core and middleware), however the user interface is closed source.

After Nokia abandoned in 2011 the MeeGo project most of the MeeGo team left Nokia, and established Jolla as a company to use MeeGo and Mer business opportunities. Thanks to MER standard it can be launched on any hardware with kernel compatible with MER. In 2012 Linux Sailfish OS based on MeeGo and using middleware of MER core stack distribution has been launched for public use. The first device, the Jolla smartphone, was unveiled on 20 May 2013. In 2015 has been launched Jolla Tablet and the BRICS countries has declared it officially supported OS there. Jolla started licensing Sailfish OS 2.0 for 3rd parties. Sold already devices are updateable to Sailfish 2.0 without limitations.

Each Sailfish OS version releases are named after Finnish lakes:

- 1.0.0.5 – Update – (Kaajanlampi)

- 1.0.1.1x – Update 1 (Laadunjärvi)

- 1.0.2.5 – Update 2 (Maadajävri)

- 1.0.3.8 – Update 3 (Naamankajärvi)

- 1.0.4.20 – Update 4 (Ohijärvi)

- 1.0.5.1x – Update 5 (Paarlamp)

- 1.0.7.16 – Update 7 (Saapunki)

- 1.0.8.19 – Update 8 (Tahkalampi)

- 1.1.0.3x – Update 9 (Uitukka)

- 1.1.1.2x – Update 10 (Vaarainjärvi)

- 1.1.2.1x – Update 11 (Yliaavanlampi)

- 1.1.4.28 – Update 13 (Äijänpäivänjärvi)

- 1.1.6.27 – Update 15 (Aaslakkajärvi)

- 1.1.7.24 – Update 16 (Björnträsket)

- 1.1.9.28 – Update 17 pre-transition to Sailfish OS 2.0 (Eineheminlampi) (Major UI revamp)

- 2.0.0.10 – Update 18 complete-transition to Sailfish 2.0 (Saimaa) (Minor UI and functionality improvement)

Tizen

Tizen is hosted by the Linux Foundation and support from the Tizen Association, guided by a Technical Steering Group composed of Intel and Samsung.

Tizen is an operating system for devices including smartphones, tablets, in-vehicle infotainment (IVI) devices, and smart TVs. It is an open source system (however the SDK was closed source and proprietary) that aims to offer a consistent user experience across devices. Tizen's main components are the Linux kernel and the WebKit runtime. According to Intel, Tizen "combines the best of LiMo and MeeGo." HTML5 apps are emphasized, with MeeGo encouraging its members to transition to Tizen, stating that the "future belongs to HTML5-based applications, outside of a relatively small percentage of apps, and we are firmly convinced that our investment needs to shift toward HTML5." Tizen will be targeted at a variety of platforms such as handsets, touch pc, smart TVs and in-vehicle entertainment. On May 17, 2013, Tizen released version 2.1, code-named Nectarine.

Currently Tizen are the fourth largest Mobile OS in term of market share. Tizen has the second-largest market share in the budget segment of smartphones in India as of Q4 2015.

Current Tizen version list:

- 1.0 (Larkspur)

- 2.0 (Magnolia)

- 2.1 (Nectarine)

- 2.2.x

- 2.3.x

- 2.4.x (minor UI tweaks)

- 3.0 (Under development)

Ubuntu Touch OS

Ubuntu Touch OS is from Canonical Ltd.. It is open source and uses the GPL license. The OS is built on the Android Linux kernel, using Android drivers, but does not use any of the Java-like code of Android.

Current Ubuntu Touch version list:

- Preview Version (Initial release)

- OTA 2.x

- OTA 3.x

- OTA 4.x

- OTA 5.x

- OTA 6.x

- OTA 7.x

- OTA 8.x

- OTA 9.x

- OTA 10.x

- OTA 11.x

- OTA 12.x

- OTA 13.x

H5OS

H5OS is from Acadine Technologies. The OS is based on Firefox OS.

Current H5OS version list:

- v1.0 (Initial release)

Discontinued software platforms

Bada

Bada platform (stylized as bada; Korean: 바다) was an operating system for mobile devices such as smartphones and tablet computers. It was developed by Samsung Electronics. Its name is derived from "바다 (bada)", meaning "ocean" or "sea" in Korean. It ranges from mid- to high-end smart-phones. To foster adoption of Bada OS, since 2011 Samsung reportedly has considered releasing the source code under an open-source license, and expanding device support to include Smart TVs. Samsung announced in June 2012 intentions to merge Bada into the Tizen project,but would meanwhile use its own Bada operating system, in parallel with Google Android OS and Microsoft Windows Phone, for its smartphones. All Bada-powered devices are branded under the Wave name, but not all of Samsung's Android-powered devices are branded under the name Galaxy. On 25 February 2013, Samsung announced that it will stop developing Bada, moving development to Tizen instead.Bug reporting was finally terminated in April 2014.

Symbian

The Symbian platform was developed by Nokia for certain models of smartphones. It is proprietary software. The operating system was discontinued in 2012, although a slimmed-down version for basic phones was still developed until July 2014. Microsoft officially shelved the platform in favor of Windows Phone after the acquisition of Nokia.

Windows Mobile

Windows Mobile is a discontinued operating system from Microsoft that it replaced with Windows Phone. It is closed source and proprietary.

The Windows CE operating system and Windows Mobile middleware was widely spread in Asia (which mostly uses Android now). The two improved variants of this operating system, Windows Mobile 6 Professional (for touch screen devices) and Windows Mobile 6 Standard, were unveiled in February 2007. It was criticized for having a user interface which is not optimized for touch input by fingers; instead, it is more usable with a stylus. Like iOS, and most other Mobile OS, it supports both touch screen, physical and Bluetooth keyboard configurations.

Windows Mobile's market share sharply declined to just 5% in Q2 of 2010. Microsoft phased out the Windows Mobile OS to focus on Windows Phone.

Palm OS

Palm OS/Garnet OS was from Access Co. It is closed source and proprietary. webOS was introduced by Palm in January 2009 as the successor to Palm OS with Web 2.0 technologies, open architecture and multitasking capabilities.

webOS

webOS was developed by Palm, although some parts are open source. webOS is a proprietary mobile operating system running on the Linux kernel, initially developed by Palm, which launched with the Palm Pre. After being acquired by HP, two phones (the Veer and the Pre 3) and a tablet (the TouchPad) running webOS were introduced in 2011. On August 18, 2011, HP announced that webOS hardware was to be discontinued but would continue to support and update webOS software and develop the webOS ecosystem. HP released webOS as open source under the name Open webOS, and plans to update it with additional features. On February 25, 2013 HP announced the sale of WebOS to LG Electronics, who used the operating system for its "smart" or Internet-connected TVs. However HP retained patents underlying WebOS as well as cloud-based services such as the App Catalog.

Maemo

Maemo was a platform developed by Nokia for smartphones and Internet tablets. It is open source and GPL, based on Debian GNU/Linux and draws much of its GUI, frameworks and libraries from the GNOME project. It uses the Matchbox window manager and the GTK-based Hildon as its GUI and application framework.

MeeGo

MeeGo was from non-profit organization The Linux Foundation. It is open source and GPL. At the 2010 Mobile World Congress in Barcelona, Nokia and Intel both unveiled 'MeeGo', a mobile operating system that combined Moblin and Maemo to create an open-sourced experience for users across all devices. In 2011 Nokia announced that it would no longer pursue MeeGo in favor of Windows Phone. Nokia announced the Nokia N9 on June 21, 2011 at the Nokia Connection event in Singapore. LG announced its support for the platform.

LiMo

LiMo was from the LiMo Foundation. LiMo Foundation launched LiMo 4 on February 14, 2011. LiMo 4 delivers middleware and application functionality, including a flexible user interface, extended widget libraries, 3D window effects, advanced multimedia, social networking and location-based service frameworks, sensor frameworks, multi-tasking and multi-touch capabilities. In addition, support for scalable screen resolution and consistent APIs means that the platform can deliver a consistent user experience across multiple device types and form factors.

Market Share

In 2006, Android, iOS and Windows Phone did not exist and just 64 million smartphones were sold. In 2015 Q1, global market share was 80.7% for Android, 15.4% for iOS, 2.8% for Windows Phone, 0.6% for Blackberry and remaining 0.5% for all other platforms. In 2016 Q1, more than a billion smartphones were sold and global market share was 84.1% for Android, 14.8% for iOS, 0.7% for Windows Phone, 0.2% for Blackberry and remaining 0.2% for all other platforms.

World-Wide Share or Shipments

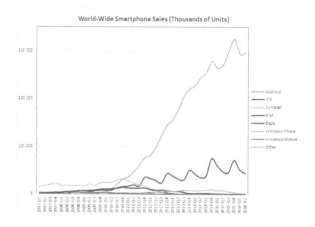

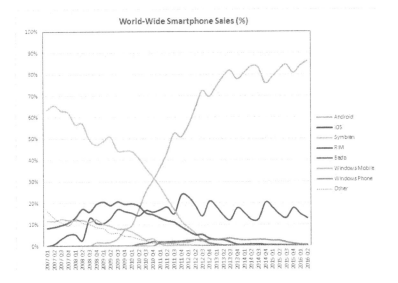

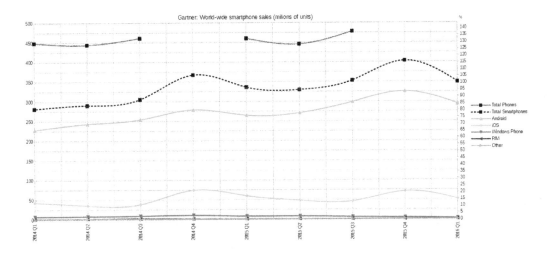

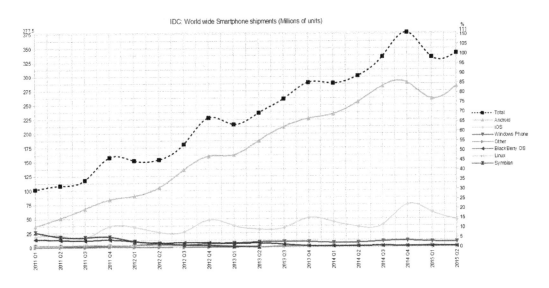

Market Share by Country or Region

Mobile Internet Traffic Share

As of November 2013, mobile data usage showed 55.17% of mobile data traffic to be from iOS, 33.89% from Android, 4.49% from Java ME (Nokia S40), 4.12% from Symbian, 1.65% from Windows Phone and 1% from BlackBerry. Many mobile browsers such as Internet Explorer Mobile, Firefox for Mobile, and Google Chrome can be switched to "Desktop view" by users, which identifies devices with the analogous desktop versions of those browsers. In these cases, the mobile usage would be excluded from these statistics.

Smartphone

A smartphone is a mobile phone with an advanced mobile operating system which combines features of a personal computer operating system with other features useful for mobile or handheld use. Smartphones, which are usually pocket-sized, typically combine the features of a cell phone, such as the abilities to place and receive voice calls and create and receive text messages, with those of other popular digital mobile devices like personal digital assistants (PDAs), such as an event calendar, media player, video games, GPS navigation, digital camera and digital video camera. Most smartphones can access the Internet and can run a variety of third-party software components ("apps"). They typically have a color display with a graphical user interface that covers 70% or more of the front surface. The display is often a touchscreen, which enables the user to use a virtual keyboard to type words and numbers and press onscreen icons to activate "app" features.

An array of various smartphones.

In 1999, the Japanese firm NTT DoCoMo released the first smartphones to achieve mass adoption within a country. Smartphones became widespread in the late 2000s. Most of those produced from 2012 onward have high-speed mobile broadband 4G LTE, motion sensors, and mobile payment features. In the third quarter of 2012, one billion smartphones were in use worldwide. Global smartphone sales surpassed the sales figures for regular cell phones in early 2013. As of 2013, 65% of U.S. mobile consumers own smartphones. By January 2016, smartphones held over 79% of the U.S. mobile market.

A person using a smartphone to take a digital photograph.

History

Early Years

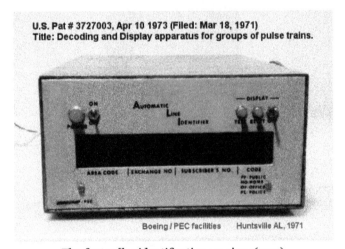

The first caller identification receiver (1971).

Devices that combined telephony and computing were first conceptualized by Nikola Tesla in 1909 and Theodore Paraskevakos in 1971 and patented in 1974, and were offered for sale beginning in 1993. Paraskevakos was the first to introduce the concepts of intelligence, data processing and visual display screens into telephones. In 1971, while he was working with Boeing in Huntsville, Alabama, Paraskevakos demonstrated a transmitter and receiver that provided additional ways to communicate with remote equipment, however it did not yet have general purpose PDA applications in a wireless device typical of smartphones. They were installed at Peoples' Telephone Company in Leesburg, Alabama and were demonstrated to several telephone companies. The original and historic working models are still in the possession of Paraskevakos.

Forerunner

The first mobile phone to incorporate PDA features was an IBM prototype developed in 1992 and demonstrated that year at the COMDEX computer industry trade show. It included PDA features and other visionary mobile applications such as maps, stock reports and news. A refined version was marketed to consumers in 1994 by BellSouth under the name Simon Personal Communicator. The Simon was the first commercially available device that could be properly referred to as a "smartphone", although it was not called that in 1994. In addition to placing and receiving cellular calls, Simon could send and receive faxes and emails and included an address book, calendar, appointment scheduler, calculator, world time clock and notepad, utilizing its touch screen display. The term "smart phone" appeared in print as early as 1995, describing AT&T's PhoneWriter Communicator.

IBM Simon and charging base (1994).

PDAs

In the mid-late 1990s, many mobile phone users carried a separate dedicated PDA device, running early versions of operating systems such as Palm OS, BlackBerry OS or Windows CE/Pocket PC. These operating systems would later evolve into mobile operating systems. In March 1996, Hewlett-Packard released the OmniGo 700LX, a modified 200LX PDA that supported a Nokia 2110-compatible phone with ROM-based software to support it. It had a 640x200 resolution CGA compatible 4-shade gray-scale LCD screen and could be used to place and receive calls, and to create and receive text messages, emails and faxes. It was also 100% DOS 5.0 compatible, allowing it to run thousands of existing software titles, including early versions of Windows.

In August 1996, Nokia released the Nokia 9000 Communicator, a digital cellular phone based on the Nokia 2110 with an integrated PDA based on the GEOS V3.0 operating system from Geoworks. The two components were attached by a hinge in what became known as a clamshell design, with the display above and a physical QWERTY keyboard below. The PDA provided e-mail; calendar, address book, calculator and notebook applications; text-based Web browsing; and could send and receive faxes. When closed, the device could be used as a digital cellular phone. In June 1999 Qualcomm released the "pdQ Smartphone", a CDMA digital PCS Smartphone with an integrated Palm PDA and Internet connectivity.

Subsequent landmark devices included:

- The Ericsson R380 (2000) by Ericsson Mobile Communications. The first device marketed as a "smartphone", it combined the functions of a mobile phone and PDA, and supported limited Web browsing with a resistive touchscreen utilizing a stylus.

- The Kyocera 6035 (early 2001) introduced by Palm, Inc. Combining a PDA with a mobile phone, it operated on the Verizon network, and supported limited Web browsing.

- Handspring's Treo 180 (2002), the first smartphone to combine the Palm OS and a GSM phone with telephony, SMS messaging and Internet access fully integrated into the OS.

Smartphones before present-day Android-, iOS- and BlackBerry-based phones typically used the Symbian operating system. Originally developed by Psion, it was the world's most widely used smartphone operating system until the last quarter of 2010.

Mass Adoption

In 1999, the Japanese firm NTT DoCoMo released the first smartphones to achieve mass adoption within a country. These phones ran on i-mode, which provided data transmission speeds up to 9.6 kbit/s. Unlike future generations of wireless services, NTT DoCoMo's i-mode used cHTML, a language which restricted some aspects of traditional HTML in favor of increasing data speed for the devices. Limited functionality, small screens and limited bandwidth allowed for phones to use the slower data speeds available. The rise of i-mode helped NTT DoCoMo accumulate an estimated 40 million subscribers by the end of 2001. It was also ranked first in market capitalization in Japan and second globally. This power would wane in the face of the rise of 3G and new phones with advanced wireless network capabilities. Outside Japan smartphones were still rare until the introduction of the Danger Hiptop in 2002, which saw moderate success in the US as the T-Mobile Sidekick. Later, in the mid-2000s, devices based on Microsoft's Windows Mobile started to gain popularity among business users in the U.S. The BlackBerry later gained mass adoption in the U.S., and American users popularized the term "CrackBerry" in 2006 due to its addictive nature. The company first released its GSM BlackBerry 6210, BlackBerry 6220, and BlackBerry 6230 devices in 2003.

Operating Systems

Symbian was the most popular smartphone OS in Europe during the middle to late 2000s. Initially, Nokia's Symbian devices were focused on business, similar to Windows Mobile and BlackBerry devices at the time. From 2006 onwards, Nokia started producing entertainment-focused smartphones, popularized by the Nseries. In Asia, with the exception of Japan, the trend was similar to that of Europe. In 2003, Motorola launched the first smartphone to use Linux, the A760 handset. While the initial release was limited to a single high-end handset only available in the Asia-Pacific region, the maker's intention was to eventually use Linux on most of its handsets, including the lower-end models. Further models to use Linux such as the Motorola Ming A1200i in 2005 and several successors to the Ming line would be unveiled through 2010. In late 2009, Motorola released the Motorola Cliq, the first of Motorola's smartphones to run the Linux-based Android operating system.

In early 2007, Apple Inc. introduced the iPhone, one of the first smartphones to use a multi-touch interface. The iPhone was notable for its use of a large touchscreen for direct finger input as its main means of interaction, instead of a stylus, keyboard, or keypad typical for smartphones at the time. In October 2008, the first phone to use Android called the HTC Dream (also known as the T-Mobile G1) was released. Android is an open-source platform founded by Andy Rubin and now owned by Google. Although Android's adoption was relatively slow at first, it started to gain widespread popularity in 2010, and in early 2012 dominated the smartphone market share worldwide, which continues to this day.

These new platforms led to the decline of earlier ones. Microsoft, for instance, started a new OS from scratch, called Windows Phone. Nokia abandoned Symbian and partnered with Microsoft to use Windows Phone on its smartphones. Windows Phone then became the third-most-popular OS. Palm's webOS was bought by Hewlett-Packard and later sold to LG Electronics for use on LG smart TVs. BlackBerry Limited, formerly known as Research In Motion, also made a new platform based on QNX, BlackBerry 10. The capacitive touchscreen also changed smartphone form factors. Before 2007, it was common for devices to have a physical numeric keypad or physical QWERTY keyboard in either a candybar or sliding form factor. However, by 2010, there were no top-selling smartphones with physical keypads.

2010s Technological Developments

In 2013, the Fairphone company launched its first "socially ethical" smartphone at the London Design Festival to address concerns regarding the sourcing of materials in the manufacturing. In late 2013, QSAlpha commenced production of a smartphone designed entirely around security, encryption and identity protection. In December 2013, the world's first curved OLED technology smartphones were introduced to the retail market with the sale of the Samsung Galaxy Round and LG G Flex models. Samsung phones with more bends and folds in the screens were expected in 2014. In 2013, water and dustproofing have made their way into mainstream high end smartphones including Sony Xperia Z, Sony Xperia Z3 and Samsung Galaxy S5. Previously, this feature was confined to special ruggedized phones designed for outdoor use.

In early 2014, smartphones were beginning to use Quad HD (2K) 2560x1440 on 5.5" screens with up to 534 PPI on devices such as the LG G3 which is a significant improvement over Apple's Retina Display. Quad HD is used in advanced televisions and computer monitors, but with 110 ppi or less on such larger displays. In 2014, Wi-Fi networks were used a lot for smartphones. As Wi-Fi became more prevalent and easier to connect to, it was predicted that Wi-Fi phone services will start to take off. In 2014, LG introduced lasers on the LG G3 to help camera focus. In 2014, some smartphones had such good digital cameras that they could be categorized as high-end point-and-shoot cameras with large sensors up to 1" with 20 megapixels and 4K video. Some can store their pictures in proprietary raw image format, but the Android (operating system) 5.0 Lollipop serves open source RAW images. By 2015, smartphones were increasingly integrated with everyday uses. For instance, credit cards, mobile payments, and mobile banking were integrated into smartphone applications and Software as a Service (SaaS) platforms. Additionally, recent technological innovations are causing the role of traditional keys to be fused into the smartphones, because a smartphone can act as a digital key and access badge for its users. In October 2015, Microsoft announced Windows Continuum, a feature that allows users to connect their devices to an external monitor via Microsoft Continuum Display Dock. HP adds a layer to the Continuum with their HP Workplace which enables user to run a Win32 app by a virtualized server. The first modular smartphone available to the public was the Fairphone 2, which was released in December 2015. Unlike most smartphones, users can remove and replace parts on this phone.

Future Possible Developments

Foldable OLED smartphones have been anticipated for years but have failed to materialize because of the relatively high failure rate when producing these screens. As well, creating a battery that can be folded is another hurdle.

Mobile Operating Systems

Android

Samsung, using Android, is the top-selling smartphone brand in 2016.

Android is a mobile operating system developed by Google Inc., and backed by an industry consortium known as the Open Handset Alliance. It is an open source platform with optional proprietary components, including a suite of flagship software for Google services, and the application and content storefront Google Play. Android was officially introduced via the release of its inaugural device, the HTC Dream (T-Mobile G1) on 20 October 2008. As an open source product, Android has also been the subject of third-party development. Development groups have used the Android source code to develop and distribute their own modified versions of the operating system, such as CyanogenMod, to add features to the OS and provide newer versions of Android to devices that no longer receive official updates from their vendor. Forked versions of Android have also been adopted by other vendors, such as Amazon.com, who used its "Fire OS" on a range of tablets and the Fire Phone. As it is a non-proprietary platform that has shipped on devices covering a wide range of market segments, Android has seen significant adoption. Gartner Research estimated that 325 million Android smartphones were sold during the fourth quarter of 2015, leading all other platforms. Samsung Electronics, who produces Android devices, was also the top smartphone vendor across all platforms in the same period of time.

iOS

iOS (formerly iPhone OS) is a proprietary mobile operating system developed by Apple Inc. primarily for its iPhone product line. The iPhone was first unveiled in January 2007. The device introduced numerous design concepts that have been adopted by modern smartphone platforms, such as the use of multi-touch gestures for navigation, eschewing physical controls such as physical keyboards in favor of those rendered by the operating system itself on its touchscreen (including the keyboard), and the use of skeumorphism—making features and controls within the user interface resemble real-world objects and concepts in order to improve their usability. In 2008, Apple introduced the App Store, a centralized storefront for purchasing new software for iPhone devices. iOS can also integrate with Apple's desktop music program iTunes to sync media to a personal computer. The dependency on a PC was removed with the introduction of iCloud on later versions of iOS, which provides synchronization of user data via internet servers between multiple devices. The iPhone line's early dominance was credited with reshaping the smartphone industry, and helping make Apple one of the world's most valuable publicly traded companies by 2011. However, the iPhone and iOS have generally been in second place in worldwide market share.

Windows Phone

Windows Phone is a series of proprietary smartphone operating systems developed by Microsoft. Its original release, Windows Phone 7, was a revamped version of the previous, Windows CE-based Windows Mobile platform; however, it was incompatible with the legacy platform. Windows Phone's user interface was designed to contrast with its competitors, utilizing a design language codenamed "Metro" which de-emphasized iconography and skeuomorphism in favor of flat, text-based designs. The platform also featured concepts such as "live tiles" on its home screen that can display dynamic content, and "Hubs"—which aggregate content from various sources and services (such as a user's local contacts, in combination with connected social networking services) into unified displays. Windows Phone also integrated with other Microsoft brands and platforms, including Bing, SkyDrive, and Xbox. Microsoft Office Mobile apps were also bundled with the operating system.

Windows Phone 8 was released in 2012; it was incompatible with existing devices, but switched to a core system based on the Windows NT platform, expanded the platform's hardware support and functionality, and added expanded enterprise-oriented functionality such as storage encryption. Windows 10 Mobile was released in late-2015; it is no longer promoted under the Windows Phone brand, as it is intended to provide greater consistency and integration with Windows 10 for PC, including cross-platform applications via Universal Windows Platform, and the ability to dock supported devices to use a desktop interface with keyboard and mouse support.

The Windows Phone series has had poor adoption in comparison to its competitors. Lack of interest in the platform also led to a decrease in third-party applications, and some vendors ended their support for Windows Phone altogether. The most prominent Windows Phone vendor was Nokia, who exclusively adopted Windows Phone as its smartphone platform in 2011 as part of a wider partnership with Microsoft. Nokia's Lumia series was the most popular line of Windows Phone devices, representing 83.3% of all Windows Phones sold in June 2013, and Microsoft acquired Nokia's mobile business for just over €5.44 billion in April 2014, forming the subsidiary Microsoft Mobile under former Nokia CEO Stephen Elop

BlackBerry

A Blackberry Classic smartphone with an integrated keyboard

In 1999, RIM released its first BlackBerry devices, providing secure real-time push-email communications on wireless devices. Services such as BlackBerry Messenger provide the integration of all communications into a single inbox. In September 2012, RIM announced that the 200 millionth BlackBerry smartphone was shipped. As of September 2014, there were around 46 million active BlackBerry service subscribers. Most recently, RIM has undergone a platform transition, changing its name to BlackBerry and making new devices on a new platform named "BlackBerry 10" and in November 2015 released an Android smartphone, the BlackBerry Priv.

Sailfish OS

The Sailfish OS is based on the Linux kernel and Mer. Additionally Sailfish OS includes a partially or completely proprietary multi-tasking user interface programmed by Jolla. This user interface differentiate Jolla smartphones from others. Sailfish OS is intended to be a system made by many of the MeeGo team, which left Nokia to form Jolla, utilizing funding from Nokia's "Bridge" program which helps establish and support start-up companies formed by ex-Nokia employees.

Tizen

Tizen is a Linux-based operating system for devices, including smartphones, tablets, in-vehicle infotainment (IVI) devices, smart TVs, laptops and smart cameras. Tizen is a project within the Linux Foundation and is governed by a Technical Steering Group (TSG) composed of Samsung and Intel among others. In April 2014, Samsung released the Samsung Gear 2 and the Gear 2 Neo, running Tizen. The Samsung Z1 is the first smartphone produced by Samsung that runs Tizen; it was released in the Indian market on January 14, 2015.

Ubuntu Touch

Ubuntu Touch (also known as Ubuntu Phone) is a mobile version of the Ubuntu operating system developed by Canonical UK Ltd and Ubuntu Community. It is designed primarily for touchscreen mobile devices such as smartphones and tablet computers.

Discontinued Operating Systems

Symbian

Symbian was originally developed by Psion as EPOC32. It was the world's most widely used smartphone operating system until Q4 2010, though the platform never gained popularity in the U.S., as it did in Europe and Asia. The first Symbian phone, the touchscreen Ericsson R380 Smartphone, was released in 2000, and was the first device marketed as a "smartphone". It combined a PDA with a mobile phone. Variants of Symbian OS began to emerge, most notably Symbian UIQ, MOAP and S60, each supported by different manufacturers. With the creation of Symbian Foundation in 2008, Symbian OS was unified under one variant under the stewardship of Nokia. In February 2011, Nokia announced that it would replace Symbian with Windows Phone as the operating system on all of its future smartphones, with the platform being abandoned over the following few years.

Windows Mobile

Windows Mobile was based on the Windows CE kernel and first appeared as the Pocket PC 2000 operating system. Throughout its lifespan, the operating system was available in both touchscreen and non-touchscreen formats. It was supplied with a suite of applications developed with the Microsoft Windows API and was designed to have features and appearance somewhat similar to desktop versions of Windows. Third parties could develop software for Windows Mobile with no restrictions imposed by Microsoft. Software applications were eventually purchasable from Windows Marketplace for Mobile during the service's brief lifespan. Windows Mobile was eventually phased out in favor of Windows Phone OS.

Bada

The Bada operating system for smartphones was announced by Samsung in November 2009. The first Bada-based phone was the Samsung Wave S8500, released in June 2010. Samsung shipped 4.5 million phones running Bada in Q2 of 2011. In 2013, Bada merged with a similar platform called Tizen.

Firefox OS

Firefox OS was demonstrated by Mozilla in February 2012. It was designed to have a complete community-based alternative system for mobile devices, using open standards and HTML5 applications. The first commercially available Firefox OS phones were ZTE Open and Alcatel One

Touch Fire. As of 2014, more companies have partnered with Mozilla including Panasonic (which is making a smart TV with Firefox OS) and Sony. In December 2015, Mozilla announced that it would phase out development of Firefox OS for smartphones, and would reposition the project to focus on other forms of Internet-connected devices.

Palm OS

In late 2001, Handspring launched the Springboard GSM phone module with limited success. In May 2002, Handspring released the Palm OS Treo 270 smartphone, that did not support Springboard, with both a touchscreen and a full keyboard. The Treo had wireless web browsing, email, calendar, a contact organizer and mobile third-party applications that could be downloaded or synced with a computer. Handspring was purchased by Palm, Inc which released the Treo 600 and continued releasing Treo devices with a few Treo devices using Windows Mobile. After buying Palm in 2011, Hewlett-Packard (HP) discontinued its webOS smartphone and tablet production.

webOS

webOS is a proprietary mobile operating system running on the Linux kernel, initially developed by Palm, which launched with the Palm Pre. After being acquired by HP, two phones (the Veer and the Pre 3) and a tablet (the TouchPad) running webOS were introduced in 2011. On August 18, 2011, HP announced that webOS hardware was to be discontinued but would continue to support and update webOS software and develop the webOS ecosystem. HP released webOS as open source under the name Open webOS, and plans to update it with additional features. On February 25, 2013 HP announced the sale of WebOS to LG Electronics, who used the operating system for its current "smart" or Internet-connected TVs, but not smartphones. In January 2014, Qualcomm has announced that it has acquired technology patents from HP, which includes all the WebOS patents.

Maemo / MeeGo

MeeGo is an operating system created from the source code of Moblin (produced by Intel) and Maemo (produced by Nokia). Before that, Nokia used Maemo on some of its smartphones and internet tablets (such as Nokia N810 and N900). MeeGo was originally envisioned to power a variety of devices from netbooks, tablets to smartphones and smart TVs. However, the only smartphones which used MeeGo was the Nokia N9 and Nokia N950 (MeeGo v1.2 Harmattan). Following Nokia's decision to move to Windows Phone OS in 2011 and to cease MeeGo development, the Linux Foundation canceled MeeGo in September 2011 in favor of the development of Tizen.

Application Stores

The introduction of Apple's App Store for the iPhone and iPod Touch in July 2008 popularized manufacturer-hosted online distribution for third-party applications (software and computer programs) focused on a single platform. There are a huge variety of apps, including video games, music products and business tools. Up until that point, smartphone application distribution depended on third-party sources providing applications for multiple platforms, such as GetJar, Handango, Handmark, and PocketGear. Following the success of the App Store, other smartphone manufacturers launched application stores, such as Google's Android Market (now Google Play Store) and

RIM's BlackBerry App World in April 2009. In February 2014, 93% of mobile developers were targeting smartphones first for mobile app development.

Display

Samsung Galaxy S6(5.1 inches), S6 Edge(5.1 inches), S6 Edge+(5.7 inches)

One of the main characteristics of smartphones is their screen. It usually fills most of the phone's front surface (about 70%); screen size usually defines the size of a smartphone. Many have an aspect ratio of 16:9; some are 4:3 or other ratios. They are measured in diagonal inches, starting from 2.45 inches. Phones with screens larger than 5.5 inches are often called "phablets". Smartphones with screens over 4.5 inches commonly are shifted while using a single hand, since most thumbs cannot reach the entire screen surface, or used in place with both hands. Liquid-crystal displays are the most common; others are IPS, LED, OLED, AMOLED and E Ink displays. In the 2010s, Braille screens, which can be used by visually impaired people are being developed. It is expected that Braille screens will use some type of microfluidics technology.

Accessories

A selection of protective cases. The small cut-out in the left-hand corner is to enable the person to use the digital camera.

As with cellphones, a range of accessories are sold for smartphones, including cases, screen protectors, power charging cables, add-on batteries, headphones, combined headphone-microphones which allow a person to use the phone without holding it to the ear, and Bluetooth-enabled powered speakers that enable users to listen to music files stored on their smartphones. Cases range from relatively inexpensive rubber or soft plastic cases which provide moderate protection from bumps and good protection from scratches to more expensive, heavy-duty cases that combine a rubber padding with a hard outer shell. Some cases have a "book"-like form, with a cover that the user opens to use the device; when the cover is closed, it protects the screen. Some "book"-like cases have additional pockets for credit cards, thus enabling people to use them as wallets. Accessories include products sold by the manufacturer of the smartphone and compatible products made by other manufacturers.

Market Share

Usage

In the third quarter of 2012, one billion smartphones were in use worldwide. Global smartphone sales surpassed the sales figures for feature phones in early 2013. As of 2013, 65 percent of U.S. mobile consumers own smartphones. The European mobile device market as of 2013 is 860 million. In China, smartphones represented more than half of all handset shipments in the second quarter of 2012 and in 2014 there were 519.7 million smartphone users, with the number estimated to grow to 700 million by 2018.

As of November 2011, 27% of all photographs were taken with camera-equipped smartphones. A study conducted in September 2012 concluded that 4 out of 5 smartphone owners use the device to shop. Another study conducted in June 2013 concluded that 56% of American adults now owned a smartphone of some kind. Android and iPhone owners account for half of the cell phone user population. Higher income adults and those under age 35 lead the way when it comes to smartphone ownership. Worldwide shipments of smartphones topped 1 billion units in 2013, up 38% from 2012's 725 million, while comprising a 55% share of the mobile phone market in 2013, up from 42% in 2012. Since 1996, smartphone shipments have had positive growth, but in Q1 2016 for the first time the shipments dropped by 3 percent year on year. The situation was caused by the maturing China market.

By Manufacturer

The LG G3, OnePlus One, Samsung Galaxy Note 4, and iPhone 6 Plus.

In 2011, Samsung had the highest shipment market share worldwide, followed by Apple. In 2013, Samsung had 31.3% market share, a slight increase from 30.3% in 2012, while Apple was at 15.3%, a decrease from 18.7% in 2012. Huawei, LG and Lenovo were at about 5% each, significantly better than 2012 figures, while others had about 40%, the same as the previous years figure. Only Apple lost market share, although their shipment volume still increased by 12.9 percent; the rest had significant increases in shipment volumes of 36 to 92 percent. In Q1 2014, Samsung had a 31% share and Apple had 16%. In Q4 2014, Apple had a 20.4% share and Samsung had 19.9%.

In January 2015 in the US, Android market share was 53.2%, Apple's iOS had a 41.3% share and Samsung's Android smartphones had 29.3%. In January 2016, the US market share for Android smartphones was 52.8%, Apple had 43.6% share, Microsoft had 2.7% and BlackBerry had 0.8 percent.

By Operating System

The market has been dominated by the Android operating system since 2010. Android's market share (measured by units shipment) rose from 33.2% in Q4 2011 to 78.1% of the market in Q4 2013. Apple's market share oscillated between 15% to 20.9% during the same period. BlackBerry's market share fell from 14.3% in Q4 2011 to 0.6% in Q4 2013. Windows Mobile market share rose from 1.5% to 3% during the same time frame. As of the end of Q3 2014, Android was the most popular operating system, with a 84.4% market share, followed by iOS with 11.7%, Windows Phone with 2.9%, BlackBerry with 0.5% and all others with 0.6%.

Issues

Smartphones have presented issues similar to those affecting other mobile telephones. As well, there are some issues which are unique to smartphones.

Battery Life

Compared to earlier non-smartphone mobile phones, smartphone battery life has generally been poor, due to the significant power requirements of their computer systems and color screens. Poor smartphone battery life has negatively affected customer satisfaction. There is also a trend towards using batteries that the user cannot replace. Smartphone users have addressed the challenge of limited battery life by purchasing additional chargers for use outside the home, at work, and in cars and by buying portable external "battery packs". External battery packs include generic models which are connected to the smartphone with a cable and custom-made models which "piggyback" onto a smartphone's case.

A high-capacity portable battery charger.

Social

A 2012 University of Southern California study found that unprotected adolescent sexual activity was more common among owners of smartphones. A study conducted by the Rensselaer Polytechnic Institute's (RPI) Lighting Research Center (LRC) concluded that smartphones, or any backlit devices, can seriously affect sleep cycles. Some persons might become psychologically attached to cellphones resulting in anxiety when separated from the devices. A "smombie" (a combination of "smartphone" and "zombie") is a walking person using a smartphone and not paying attention as they walk, possibly risking an accident in the process, an increasing social phenomenon. The issue of slow-moving smartphone users led to the temporary creation of a "mobile lane" for walking in Chongqing, China. The issue of distracted smartphone users led the city of Augsburg, Germany to embed pedestrian traffic lights in the pavement.

While Driving

A New York City driver using two handheld mobile phones at once.

Mobile phone use while driving, including talking on the phone, texting, or operating other phone features, is common but controversial. It is widely considered dangerous due to distracted driving. Being distracted while operating a motor vehicle has been shown to increase the risk of accidents. In September 2010, the US National Highway Traffic Safety Administration (NHTSA) reported that 995 people were killed by drivers distracted by cell phones. In March 2011 a US insurance company, State Farm Insurance, announced the results of a study which showed 19% of drivers surveyed accessed the Internet on a smartphone while driving. Many jurisdictions prohibit the use of mobile phones while driving. In Egypt, Israel, Japan, Portugal and Singapore, both handheld and hands-free use of a mobile phone (which uses a speakerphone) is banned. In other countries including the UK and France and in many U.S. states, only handheld phone use is banned, while hands-free use is permitted.

A 2011 study reported that over 90% of college students surveyed text (initiate, reply or read) while driving. The scientific literature on the dangers of driving while sending a text message from a mobile phone, or *texting while driving*, is limited. A simulation study at the University of Utah found a sixfold increase in distraction-related accidents when texting. Due to the increasing complexity of smartphones, this has introduced additional difficulties for law enforcement officials when attempting to distinguish one usage from another in drivers using their devices. This is more apparent in countries which ban both handheld and hands-free usage, rather than those which ban handheld use only, as officials cannot easily tell which function of the mobile phone is being used simply by looking at the driver. This can lead to drivers being stopped for using their device illegally for a phone call when, in fact, they were using the device legally, for example, when using the phone's incorporated controls for car stereo, GPS or satnav.

A sign along Bellaire Boulevard in Southside Place, Texas (Greater Houston) states that using mobile phones while driving is prohibited from 7:30 am to 9:30 am and from 2:00 pm to 4:15 pm

A 2010 study reviewed the incidence of mobile phone use while cycling and its effects on behaviour and safety. In 2013 a national survey in the US reported the number of drivers who reported using their cellphones to access the Internet while driving had risen to nearly one of four. A study conducted by the University of Illinois examined approaches for reducing inappropriate and problematic use of mobile phones, such as using mobile phones while driving.

Accidents involving a driver being distracted by talking on a mobile phone have begun to be prosecuted as negligence similar to speeding. In the United Kingdom, from 27 February 2007, motorists who are caught using a hand-held mobile phone while driving will have three penalty points added to their license in addition to the fine of £60. This increase was introduced to try to stem the increase in drivers ignoring the law. Japan prohibits all mobile phone use while driving, including use of hands-free devices. New Zealand has banned hand held cellphone use since 1 November 2009. Many states in the United States have banned texting on cell phones while driving. Illinois became the 17th American state to enforce this law. As of July 2010, 30 states had banned texting while driving, with Kentucky becoming the most recent addition on July 15.

Public Health Law Research maintains a list of distracted driving laws in the United States. This database of laws provides a comprehensive view of the provisions of laws that restrict the use of mobile communication devices while driving for all 50 states and the District of Columbia between 1992, when first law was passed through December 1, 2010. The dataset contains information on 22 dichotomous, continuous or categorical variables including, for example, activities regulated (e.g., texting versus talking, hands-free versus handheld), targeted populations, and exemptions.

Legal

A "patent war" between Samsung and Apple started when the latter claimed that the original Galaxy S Android phone copied the interface—and possibly the hardware—of Apple's iOS for the iPhone 3GS. There was also smartphone patents licensing and litigation involving Sony, Google, Apple Inc., Samsung, Microsoft, Nokia, Motorola, Huawei, and HTC, among others. The conflict is part of the wider "patent wars" between multinational technology and software corporations. To secure and increase their market share, companies granted a patent can sue to prevent competitors from using the methods the patent covers. Since 2010 the number of lawsuits, counter-suits, and trade complaints based on patents and designs in the market for smartphones, and devices based on smartphone OSes such as Android and iOS, has increased significantly. Initial suits, countersuits, rulings, licence agreements, and other major events began in 2009 as the smartphone market grew more rapidly.

Medical

With the rise in number of mobile medical apps in the market place, government regulatory agencies raised concerns on the safety of the use of such applications. These concerns were transformed into regulation initiatives worldwide with the aim of safeguarding users from untrusted medical advice.

Security

Smartphone malware is easily distributed through an insecure app store. Often malware is hidden in pirated versions of legitimate apps, which are then distributed through third-party app stores. Malware risk also comes from what's known as an "update attack", where a legitimate application is later changed to include a malware component, which users then install when they are notified that the app has been updated. As well, one out of three robberies in 2012 in the United States involved the theft of a mobile phone. An online petition has urged smartphone makers to install kill switches in their devices. In 2014, Apple's "Find my iPhone" and Google's "Android Device Manager" can disable phones that have been lost/stolen. With BlackBerry Protect in OS version 10.3.2, devices can be rendered unrecoverable to even Black-Berry's own Operating System recovery tools if incorrectly authenticated or dissociated from their account.

Sleep

Using smartphones late at night can disturb sleep, due to the brightly lit screen affecting melatonin levels and sleep cycles. In an effort to alleviate these issues, several apps that change the color temperature of a screen to a warmer hue based on the time of day to reduce the amount of blue light generated have been developed for Android, while iOS 9.3 integrated similar, system-level functionality known as "Night Shift". Amazon released a feature known as "blue shade" in their

Fire OS "Bellini" 5.0 and later. It has also been theorized that for some users, addicted use of their phones, especially before they go to bed, can result in "ego depletion".

Other Terms

"Phablet", a portmanteau of the words *phone* and *tablet*, describes smartphones with larger screens.

"Superphone" is also used by some companies to market phones with unusually large screens and other expensive features.

Mobile Content

Mobile content is any type of electronic media which is viewed or used on mobile phones, like ringtones, graphics, discount offers, games, movies, and GPS navigation. As mobile phone use has grown since the mid-1990s, the significance of the devices in everyday life has grown accordingly. Owners of mobile phones can now use their devices to make calendar appointments, send and receive text messages (SMS), listen to music, watch videos, shoot videos, redeem coupons for purchases, view office documents, get driving instructions on a map, and so forth. The use of mobile content has grown accordingly.

Camera phones not only present but produce media, for example photographs with a few million pixels, and can act as pocket video cameras.

Mobile content can also refer to text or multimedia hosted on websites, which may either be standard Internet pages, or else specific mobile pages.

Transmission

Mobile content via SMS is still the main technology for communication used to send mobile consumers messages, especially simple content such as ringtones and wallpapers. Because SMS is the main messaging technology used by young people, it is still the most effective way of reaching this target market. SMS is also ubiquitous, reaching a wider audience than any other technology available in the mobile space (MMS, bluetooth, mobile e-mail or WAP). More important than anything else, SMS is extremely easy to use, which makes adoption increase day by day.

Although SMS is an old technology that may someday be replaced by the likes of Multimedia Messaging Service (MMS) or WAP, SMS frequently gains new powers. One example is the introduction of applications whereby mobile tickets are sent to consumers via SMS, which contains a WAP-push that contains a link where a barcode is placed. This clearly substitutes MMS, which has a limited reach and still suffers from interoperability problems.

It is important to keep enhancing the consumer confidence in using SMS for mobile content applications. This means, if a consumer has ordered a new wallpaper or ringtone, this has to work properly, in a speedy and reliable way. Therefore, it is important to choose the right SMS gateway provider in order to ensure quality-of-service along the whole path of the content SMS until reaching the consumer's mobile.

Modern phones come with Bluetooth and Near field communication. This allows video to be sent from phone to phone over Bluetooth, which has the advantages that there is no data charge.

Content Types

Apps

Mobile application development, also known as mobile apps, has become a significant mobile content market since the release of the first iPhone from Apple in 2007. Prior to the release of Apple's phone product, the market for mobile applications (outside of games) had been quite limited. The bundling of the iPhone with an app store, as well as the iPhone's unique design and user interface, helped bring a large surge in mobile application use. It also enabled additional competition from other players. For example, Google's Android platform for mobile content has further increased the amount of app content available to mobile phone subscribers.

Some examples of mobile apps would be applications to manage travel schedules, buy movie tickets, preview video content, manage RSS news feeds, read digital version of popular newspapers, identify music, look at star constellations, view Wikipedia, and much more. Many televion networks have their own app to promote and present their content. iTyphoon is an example of a mobile application used to provide information about typhoons in the Philippines.

Games

Mobile games are applications that allow people to play a game on a mobile handset. The main categories of mobile games include Puzzle/Strategy, Retro/Arcade, Action/Adventure, Card/Casino, Trivia/Word, Sports/Racing, given in approximate order of their popularity.

Several studies have shown that the majority of mobile games are bought and played by women. Sixty-five percent of mobile game revenue is driven by female wireless subscribers. They are the biggest driver of revenue for the Puzzle/Strategy category; comprising 72 percent of the total share of revenue, while men made up 28 percent. Women dominate revenue generation for all mobile game categories, with the exception of Action/Adventure mobile games, in which men drive 60 percent of the revenue for that category. It's also said that teens are three times as likely as those over twenty to play cell phone games.

Images

Mobile images are used as the wallpaper to a mobile phone, and are also available as screensavers. On some handsets, images can also be set to display when a particular person calls the users. Sites like adg.ms allow users to download free content, however service operators such as Telus Mobility blocks non Telus website downloads.

Music

Mobile music is any audio file that is played on a mobile phone. Mobile music is normally formatted as an AAC (Advanced Audio Coding) file or an MP3, and comes in several different formats. Monophonic ringtones were the earliest form of ringtone, and played one tone at a time. This was improved upon with polyphonic ringtones, which played several tones at the same time so a more convincing melody could be created. The next step was to play clips of actual songs, which were dubbed Realtones. These are preferred by record labels as this evolution of the ringtone has allowed them to gain a cut of lucurative ringtone market. In short Realtones generate royalties for record labels (the master recording owners) as well as publishers (the writers), however, when Monophonic or Polyphonic ringtones are sold only publishing or "mechanical" royalties are incurred as no master recording has been exploited. Some companies promote covertones, which are ringtones that are recorded by cover bands to sound like a famous song. Recently Ringback tones have become available, which are played to the person calling the owner of the ringback tone. Voicetones are ringtones that play someone talking or shouting rather than music, and there are various of ringtones of natural and everyday sounds. Realtones are the most popular form of ringtones. As an example, they captures 76.4% of the US ringtone market in the second quarter of 2006, followed by monophonic and polyphonic ringtones at 12% and ringback tones and 11.5% – but monophonic and polyphonic ringtones are falling in popularity while ringback tones are growing. This trend is common around the globe. A recent innovation is the singtone, whereby "the user's voice is recorded singing to a popular music track and then "tuned-up" automatically to sound good. This can then be downloaded as a ringtone or sent to another user's mobile phone" said the director of Synchro Arts, the developers.

As well as mobile music there are full track downloads, which are an entire song encoded to play on a mobile phone. These can be purchased and bought over the mobile network, but data charges can make this prohibitive. The other way to get a song onto a mobile phone is by "side loading" it, which normally involves downloading the song onto a computer and then transferring it to the mobile phone via Bluetooth, infra-red or cable connections. It is possible to use a full track as a ringtone. In recent years, websites have sprung that allow users to upload audio files and customize them into ringtones using specialized applications, including Myxer, Bongotones, and Zedge.

Mobile music is becoming an integral part of the music industry as a whole. In 2005, the International Federation of Phonographic Industries (IFPI) said it expects mobile music to generate more revenues that online music before the end of that year. In the first half of 2005, the digital music market grew enough to offset the fall in the traditional music market – without including the sale of ringtones, which still makes up the majority of mobile music sales around the globe.

Video

Mobile video comes in several forms including 3GPP, MPEG-4, RTSP, and Flash Lite.

Mobishows and Cellsodes

A Mobishow or a cellsode are terms to describe a broadcast quality programme / series which has been produced, directed, edited and encoded for the mobile phone. Mobishows and Cellsodes can range from short video clips such as betting advice or the latest celebrity gossip, through to half-hour drama serials. Examples include *The Ashes* and *Mr Paparazzi Show* which both were created for mobile viewing.

Streaming

Radio

Mobile streaming radio is an application that streams on-demand audio channels or live radio stations to the mobile phone. In the U.S., mSpot was the first company to develop and commercialize streaming radio which went live in March 2005 on Sprint. Today, all major carriers offer some sort streaming radio service featuring programmed stations based on popular genres and live stations which included both music and talk.

TV

Mobile video also comes in the form of streaming TV over the mobile network, which must be a 2.5G or 3G network. This mimics a television station in that the user cannot elect to see what they wish but must watch whatever is on the channel at the time.

There is also mobile broadcast TV, which operates like a traditional television station and broadcasts the content over a different spectrum. This frees up the mobile network to handle calls and other data usage, and because of the "one-to-many" nature of mobile broadcast TV the video quality is a lot better than that streamed over the mobile networks, which is a "one-to-one" system.

The problem is that broadcast technologies don't have a natural up link, so for users to interact with the TV stream the service has to be closely integrated to the carriers mobile network. The main technologies for broadcast TV are DVB-H, Digital Multimedia Broadcasting (DMB), and MediaFLO.

Live Video

Live video can also be streamed and shared from your cell phone through applications like Qik and InstaLively. The uploaded video can be shared to your friends through emails or social networking sites. Most Live video streaming application works over the cell network or through Wi-Fi. They also require most users to have a dataplan from their cell phone carriers.

International Trends

Since the late 1990s, mobile content has become an increasingly important market worldwide. The South Koreans are the worlds Leaders in Mobile Content and 3-G mobile networks, then the Japanese, followed closely by the Europeans, are heavy users of their mobile phones and have been attaining custom mobile content for their devices for years. In fact, mobile phone use has begun to exceed the use of PCs in some countries. In the United States and Canada, mobile phone use and the accompanying use of mobile content has been slower to gain traction because of political issues and because open networks do not exist in America.

On current trends, mobile phone content will play an increasing role in the lives of millions across the globe in the years ahead, as users will depend on their mobile phones to keep in touch not only with their friends but with world news, sports scores, the latest movies and music, and more.

Mobile content is usually downloaded through WAP sites, but new methods are on the rise. In Italy, 800,000 people are registered users to Passa Parola, an application that allows users to browse a big database for mobile content and directly download it to their handsets. This tool can also be used to recommend content to others, or send content as a gift.

An increasing number of people are also beginning to use applications like Qik to upload and share their videos from their cell phone to the internet. Mobile phone software like Qik allows user to share their videos to their friends through emails, SMS, and even social networking sites like Twitter and Facebook.

References

- Hillebrand, ed. (2010). Short Message Service, the Creation of Personal Global Text Messaging. Wiley. ISBN 978-0-470-68865-6.

- Rheingold, Howard (2002) Smart Mobs: the Next Social Revolution, Perseus, Cambridge, Massachusetts, pp. xi–xxii, 157–82 ISBN 0-7382-0861-2.

- Linzmayer, Owen W. (2004). Apple confidential 2.0 : the definitive history of the world's most colorful company ([Rev. 2. ed.]. ed.). San Francisco, Calif.: No Starch Press. ISBN 1-59327-010-0.

- Telecommunications and Data Communications Handbook, Ray Horak, 2nd edition, Wiley-Interscience, 2008, 791 p., ISBN 0-470-39607-5

- Deborah Hurley, James H. Keller (1999). The First 100 Feet: Options for Internet and Broadband Access. Harvard college. ISBN 0-262-58160-4.

- Joshua Bardwell; Devin Akin (2005). Certified Wireless Network Administrator Official Study Guide (Third ed.). McGraw-Hill. p. 418. ISBN 978-0-07-225538-6.

- Security, Subbu Iyer, Director of Product Management, Bluebox. "5 things you no longer need to do for mobile security". Network World. Retrieved 16 May 2016.

- "Enterprise IT Spotlight: enterprise mobility management - 451 Research - Analyzing the Business of Enterprise IT Innovation". 451research.com. Retrieved 16 May 2016.

- The Associated Press (2016-08-22). "Google Rolling Out Latest Android System to Nexus Phones". The New York Times. ISSN 0362-4331. Retrieved 2016-08-29.

- "Gartner Says Five of Top 10 Worldwide Mobile Phone Vendors Increased Sales in Second Quarter of 2016". www.gartner.com. Retrieved 2016-08-19.

- "comScore Reports January 2016 U.S. Smartphone Subscriber Market Share". comScore. 4 March 2016. Retrieved 18 August 2016.

- "T-Mobile Announces Upcoming Availability of Motorola CLIQ with MOTOBLUR". T-Mobile Announces Upcoming Availability of Motorola CLIQ with MOTOBLUR. September 29, 2009. Retrieved May 6, 2016.

- "Gartner Says Five of Top 10 Worldwide Mobile Phone Vendors Increased Sales in Second Quarter of 2016". Gartner. Retrieved August 19, 2016.

- "Top Five Smartphone Vendors, Shipments, Market Share, and Year-Over-Year Growth, Q2 2016". International Data Corporation. Retrieved July 28, 2016.

- Hookham, Mark; Togoh, Isabel; Yeates, Alex (21 February 2016). "Walkers hit by curse of the smombie". The Sunday Times. UK. Retrieved 23 February 2016.

- Rick Noack (April 25, 2016) This city embedded traffic lights in the sidewalks so that smartphone users don't have to look up The Washington Post. Retrieved 5 May 2016.

- Mawston, Neil (July 30, 2015). "Huawei Becomes World's 3rd Largest Mobile Phone Vendor in Q2 2015".

Boston, Massachusetts: Strategy Analytics. Archived from the original on 11 August 2015. Retrieved March 20, 2016.

- PCS network launched in Baltimore-D.C. area First system in nation offers digital challenge to cellular phone industry – tribunedigital-baltimoresun. Articles.baltimoresun.com (1995-11-16). Retrieved on 2015-06-08.

- Sammy Licence; et al. (29 July 2015). "Gait Pattern Alterations during Walking, Texting and Walking and Texting during Cognitively Distractive Tasks while Negotiating Common Pedestrian Obstacles". PLOS ONE. 10: e0133281. doi:10.1371/journal.pone.0133281. Retrieved 3 August 2015.

- ICT Facts and Figures 2005, 2010, 2014, Telecommunication Development Bureau, International Telecommunication Union (ITU). Retrieved 24 May 2015.

Concerns and Challenges of Mobile Computing

The main concerns and challenges of mobile computing are mobile malware, spyware and ghost push. Mobile malware causes the system of a mobile to collapse whereas spyware is the software that collects information of people or companies without their knowledge. Spyware is usually used for malicious purposes, and is mostly used without the knowledge of the user. The aspects elucidated in this chapter are of vital importance, and provides a better understanding of mobile computing.

Mobile Malware

Mobile malware is malicious software that targets mobile phones or wireless-enabled Personal digital assistants (PDA), by causing the collapse of the system and loss or leakage of confidential information. As wireless phones and PDA networks have become more and more common and have grown in complexity, it has become increasingly difficult to ensure their safety and security against electronic attacks in the form of viruses or other malware.

History

Cell phone malware were initially demonstrated by Brazilian software engineer Marcos Velasco. He created a virus that could be used by anyone in order to educate the public of the threat.

The first known mobile virus, "Timofonica", originated in Spain and was identified by antivirus labs in Russia and Finland in June 2000. "Timofonica" sent SMS messages to GSM mobile phones that read (in Spanish) "Information for you: Telefónica is fooling you." These messages were sent through the Internet SMS gate of the MoviStar mobile operator.

In June 2004, it was discovered that a company called Ojam had engineered an anti-piracy Trojan virus in older versions of its mobile phone game, *Mosquito*. This virus sent SMS text messages to the company without the user's knowledge. Although this malware was removed from the game's more recent versions, it still exists in older, unlicensed versions, and these may still be distributed on file-sharing networks and free software download web sites.

In July 2004, computer hobbyists released a proof-of-concept mobile virus *Cabir,* that replicates and spreads itself on Bluetooth wireless networks and infects mobile phones running the Symbian OS.

In March 2005, it was reported that a computer worm called Commwarrior-A had been infecting Symbian series 60 mobile phones. This specific worm replicated itself through the phone's Multimedia Messaging Service (MMS), sending copies of itself to other phone owners listed in the phone user's address book. Although the worm is not considered harmful, experts agree that it heralded a new age of electronic attacks on mobile phones.

In August 2010, Kaspersky Lab reported a trojan designated Trojan-SMS.AndroidOS.FakePlayer.a. This was the first malicious program classified as a Trojan SMS that affects smartphones running on Google's Android operating system, and which had already infected a number of mobile devices, sending SMS messages to premium rate numbers without the owner's knowledge or consent, and accumulating huge bills.

Currently, various antivirus software companies like Trend Micro, AVG, avast!, Comodo, Kaspersky Lab, PSafe, and Softwin are working to adapt their programs to the mobile operating systems that are most at risk. Meanwhile, operating system developers try to curb the spread of infections with quality control checks on software and content offered through their digital application distribution platforms, such as Google Play or Apple's App Store. Recent studies however show that mobile antivirus programs are ineffective due to the rapid evolution of mobile malware.

Taxonomy

Four types of the most common malicious programs are known to affect mobile devices:

- Expander: Expanders target mobile meters for additional phone billing and profit

- Worm: The main objective of this stand-alone type of malware is to endlessly reproduce itself and spread to other devices. Worms may also contain harmful and misleading instructions. Mobile worms may be transmitted via text messages SMS or MMS and typically do not require user interaction for execution.

- Trojan: Unlike worms, a Trojan horse always requires user interaction to be activated. This kind of virus is usually inserted into seemingly attractive and non-malicious executable files or applications that are downloaded to the device and executed by the user. Once activated, the malware can cause serious damage by infecting and deactivating other applications or the phone itself, rendering it paralyzed after a certain period of time or a certain number of operations. Usurpation data (spyware) synchronizes with calendars, email accounts, notes, and any other source of information before it is sent to a remote server.

- Spyware: This malware poses a threat to mobile devices by collecting, using, and spreading a user's personal or sensitive information without the user's consent or knowledge. It is mostly classified into four categories: system monitors, trojans, adware, and tracking cookies.

- Ghost Push: This is a kind of malware which infects the Android OS by automatically gaining root access, downloading malicious software, converting to a system app and then losing root access which makes it virtually impossible to remove the infection with a factory reset unless the firmware is reflashed. The malware hogs all system resources making it unresponsive and drains the battery. The advertisements appear anytime either in full screen, as part of a display, or in the status bar. The unnecessary apps automatically activate and sometimes download more malicious software when connected to the internet. It is hard to detect and remove. It steals the personal data of the user from the phone.

Notable Mobile Malicious Programs

- Cabir: This malware infects mobile phones running on Symbian OS and was first identified in June 2004. When a phone is infected, the message 'Caribe' is displayed on the phone's screen and is displayed every time the phone is turned on. The worm then attempts to spread to other phones in the area using wireless Bluetooth signals, although the recipient has to confirm this manually.

- Duts: This parasitic file infector virus is the first known virus for the Pocket PC platform. It attempts to infect all EXE files that are larger than 4096 bytes in the current directory.

- Skulls: A trojan horse piece of code that targets mainly Symbian OS. Once downloaded, the virus replaces all phone desktop icons with images of a skull. It also renders all phone applications useless. This malware also tends to mass text messages containing malicious links to all contacts accessible through the device in order to spread the damage. This mass texting can also give rise to high expenses.

- Commwarrior: This malware was identified in 2005. It was the first worm to use MMS messages and can spread through Bluetooth as well. It infects devices running under OS Symbian Series 60. The executable worm file, once launched, hunts for accessible Bluetooth devices and sends the infected files under a random name to various devices.

- Gingermaster: A trojan developed for an Android platform that propagates by installing applications that incorporate a hidden malware for installation in the background. It exploits the frailty in the version Gingerbread (2.3) of the operating system to use super-user permissions by privileged escalation. It then creates a service that steals information from infected terminals (user ID, number SIM, phone number, IMEI, IMSI, screen resolution and local time) by sending it to a remote server through petitions HTTP.

- DroidKungFu: A trojan content in Android applications, which when executed, obtains root privileges and installs the file com.google. ssearch.apk, which contains a back door that allows files to be removed, open home pages to be supplied, and 'open web and download and install' application packages. This virus collects and sends to a remote server all available data on the terminal.

- Ikee: The first worm known for iOS platforms. It only works on terminals that were previously made a process of jailbreak, and spreads by trying to access other devices using the SSH protocol, first through the subnet that is connected to the device. Then, it repeats the process generating a random range and finally uses some preset ranges corresponding to the IP address of certain telephone companies. Once the computer is infected, the wallpaper is replaced by a photograph of the singer Rick Astley, a reference to the Rickroll phenomenon.

- Gunpoder : This worm file infector virus is the first known virus that officially infected the Google Play Store in few countries, including Brazil.

- Shedun: adware serving malware able to root Android devices.

- HummingBad - has infected over 10 million Android operating systems. User details are sold and adverts are tapped on without the user's knowledge thereby generating fraudulent advertising revenue.

Spyware

Spyware is software that aims to gather information about a person or organization without their knowledge and that may send such information to another entity without the consumer's consent, or that asserts control over a computer without the consumer's knowledge.

"Spyware" is mostly classified into four types: system monitors, trojans, adware, and tracking cookies. Spyware is mostly used for the purposes of tracking and storing Internet users' movements on the Web and serving up pop-up ads to Internet users.

Whenever spyware is used for malicious purposes, its presence is typically hidden from the user and can be difficult to detect. Some spyware, such as keyloggers, may be installed by the owner of a shared, corporate, or public computer intentionally in order to monitor users.

While the term *spyware* suggests software that monitors a user's computing, the functions of spyware can extend beyond simple monitoring. Spyware can collect almost any type of data, including personal information like internet surfing habits, user logins, and bank or credit account information. Spyware can also interfere with user control of a computer by installing additional software or redirecting web browsers. Some spyware can change computer settings, which can result in slow Internet connection speeds, un-authorized changes in browser settings, or changes to software settings.

Sometimes, spyware is included along with genuine software, and may come from a malicious web-site or may have been added to the intentional functionality of genuine software. In response to the emergence of spyware, a small industry has sprung up dealing in anti-spyware software. Running anti-spyware software has become a widely recognized element of computer security practices, especially for computers running Microsoft Windows. A number of jurisdictions have passed anti-spyware laws, which usually target any software that is surreptitiously installed to control a user's computer.

In German-speaking countries, spyware used or made by the government is called *govware* by computer experts (in common parlance: *Regierungstrojaner*, literally 'Government Trojan'). Govware is typically a trojan horse software used to intercept communications from the target computer. Some countries like Switzerland and Germany have a legal framework governing the use of such software. In the US, the term policeware has been used for similar purposes.

Use of the term "spyware" has eventually declined as the practice of tracking users has been pushed ever further into the mainstream by major websites and data mining companies; these generally break no known laws and compel users to be tracked, not by fraudulent practices per se, but by the default settings created for users and the language of terms-of-service agreements. As one documented example, on March 7, 2011, CBS/Cnet News reported on a Wall Street Journal analysis revealing the practice of Facebook and other websites of tracking users' browsing activity, linked to their identity, far beyond users' visit and activity within the Facebook site itself. The report stated

"Here's how it works. You go to Facebook, you log in, you spend some time there, and then ... you move on without logging out. Let's say the next site you go to is New York Times. Those buttons, without you clicking on them, have just reported back to Facebook and Twitter that you went there and also your identity within those accounts. Let's say you moved on to something like a site about depression. This one also has a tweet button, a Google widget, and those, too, can report back who you are and that you went there." The WSJ analysis was researched by Brian Kennish, founder of Disconnect, Inc.

Routes of Infection

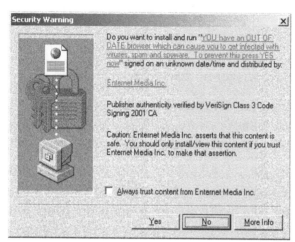

Malicious websites attempt to install spyware on readers' computers.

Spyware does not necessarily spread in the same way as a virus or worm because infected systems generally do not attempt to transmit or copy the software to other computers. Instead, spyware installs itself on a system by deceiving the user or by exploiting software vulnerabilities.

Most spyware is installed without users' knowledge, or by using deceptive tactics. Spyware may try to deceive users by bundling itself with desirable software. Other common tactics are using a Trojan horse, spy gadgets that look like normal devices but turn out to be something else, such as a USB Keylogger. These devices actually are connected to the device as memory units but are capable of recording each stroke made on the keyboard. Some spyware authors infect a system through security holes in the Web browser or in other software. When the user navigates to a Web page controlled by the spyware author, the page contains code which attacks the browser and forces the download and installation of spyware.

The installation of spyware frequently involves Internet Explorer. Its popularity and history of security issues have made it a frequent target. Its deep integration with the Windows environment make it susceptible to attack into the Windows operating system. Internet Explorer also serves as a point of attachment for spyware in the form of Browser Helper Objects, which modify the browser's behavior to add toolbars or to redirect traffic.

Effects and Behaviors

A spyware program is rarely alone on a computer: an affected machine usually has multiple infections. Users frequently notice unwanted behavior and degradation of system performance. A spyware infestation can create significant unwanted CPU activity, disk usage, and network traffic. Stability issues, such as applications freezing, failure to boot, and system-wide crashes are also common. Spyware, which interferes with networking software, commonly causes difficulty connecting to the Internet.

In some infections, the spyware is not even evident. Users assume in those situations that the performance issues relate to faulty hardware, Windows installation problems, or another infection. Some owners of badly infected systems resort to contacting technical support experts, or even buying a new computer because the existing system "has become too slow". Badly infected systems may require a clean reinstallation of all their software in order to return to full functionality.

Moreover, some types of spyware disable software firewalls and anti-virus software, and/or reduce browser security settings, which further open the system to further opportunistic infections. Some spyware disables or even removes competing spyware programs, on the grounds that more spyware-related annoyances make it even more likely that users will take action to remove the programs.

Keyloggers are sometimes part of malware packages downloaded onto computers without the owners' knowledge. Some keyloggers software is freely available on the internet while others are commercial or private applications. Most keyloggers allow not only keyboard keystrokes to be captured but also are often capable of collecting screen captures from the computer.

A typical Windows user has administrative privileges, mostly for convenience. Because of this, any program the user runs has unrestricted access to the system. As with other operating systems, Windows users are able to follow the principle of least privilege and use non-administrator accounts. Alternatively, they can also reduce the privileges of specific vulnerable Internet-facing processes such as Internet Explorer.

Since Windows Vista, by default, a computer administrator runs everything under limited user privileges. When a program requires administrative privileges, a User Account Control pop-up will prompt the user to allow or deny the action. This improves on the design used by previous versions of Windows.

Remedies and Prevention

As the spyware threat has worsened, a number of techniques have emerged to counteract it. These include programs designed to remove or block spyware, as well as various user practices which reduce the chance of getting spyware on a system.

Nonetheless, spyware remains a costly problem. When a large number of pieces of spyware have infected a Windows computer, the only remedy may involve backing up user data, and fully reinstalling the operating system. For instance, some spyware cannot be completely removed by Symantec, Microsoft, PC Tools.

Anti-spyware Programs

Many programmers and some commercial firms have released products dedicated to remove or block spyware. Programs such as PC Tools' Spyware Doctor, Lavasoft's *Ad-Aware SE* and Patrick

Kolla's *Spybot - Search & Destroy* rapidly gained popularity as tools to remove, and in some cases intercept, spyware programs. On December 16, 2004, Microsoft acquired the *GIANT AntiSpyware* software, rebranding it as *Windows AntiSpyware beta* and releasing it as a free download for Genuine Windows XP and Windows 2003 users. (In 2006 it was renamed Windows Defender).

Major anti-virus firms such as Symantec, PC Tools, McAfee and Sophos have also added anti-spyware features to their existing anti-virus products. Early on, anti-virus firms expressed reluctance to add anti-spyware functions, citing lawsuits brought by spyware authors against the authors of web sites and programs which described their products as "spyware". However, recent versions of these major firms' home and business anti-virus products do include anti-spyware functions, albeit treated differently from viruses. Symantec Anti-Virus, for instance, categorizes spyware programs as "extended threats" and now offers real-time protection against these threats.

How Anti-spyware Software Works

Anti-spyware programs can combat spyware in two ways:

1. They can provide real-time protection in a manner similar to that of anti-virus protection: they scan all incoming network data for spyware and blocks any threats it detects.

2. Anti-spyware software programs can be used solely for detection and removal of spyware software that has already been installed into the computer. This kind of anti-spyware can often be set to scan on a regular schedule.

Such programs inspect the contents of the Windows registry, operating system files, and installed programs, and remove files and entries which match a list of known spyware. Real-time protection from spyware works identically to real-time anti-virus protection: the software scans disk files at download time, and blocks the activity of components known to represent spyware. In some cases, it may also intercept attempts to install start-up items or to modify browser settings. Earlier versions of anti-spyware programs focused chiefly on detection and removal. Javacool Software's SpywareBlaster, one of the first to offer real-time protection, blocked the installation of ActiveX-based spyware.

Like most anti-virus software, many anti-spyware/adware tools require a frequently updated database of threats. As new spyware programs are released, anti-spyware developers discover and evaluate them, adding to the list of known spyware, which allows the software to detect and remove new spyware. As a result, anti-spyware software is of limited usefulness without regular updates. Updates may be installed automatically or manually.

A popular generic spyware removal tool used by those that requires a certain degree of expertise is HijackThis, which scans certain areas of the Windows OS where spyware often resides and presents a list with items to delete manually. As most of the items are legitimate windows files/registry entries it is advised for those who are less knowledgeable on this subject to post a HijackThis log on the numerous antispyware sites and let the experts decide what to delete.

If a spyware program is not blocked and manages to get itself installed, it may resist attempts to terminate or uninstall it. Some programs work in pairs: when an anti-spyware scanner (or the user) terminates one running process, the other one respawns the killed program. Likewise, some spyware will detect attempts to remove registry keys and immediately add them again. Usually, booting the infected computer in safe mode allows an anti-spyware program a better chance of removing persistent spyware. Killing the process tree may also work.

Security Practices

To detect spyware, computer users have found several practices useful in addition to installing anti-spyware programs. Many users have installed a web browser other than Internet Explorer, such as Mozilla Firefox or Google Chrome. Though no browser is completely safe, Internet Explorer was once at a greater risk for spyware infection due to its large user base as well as vulnerabilities such as ActiveX but these three major browsers are now close to equivalent when it comes to security.

Some ISPs—particularly colleges and universities—have taken a different approach to blocking spyware: they use their network firewalls and web proxies to block access to Web sites known to install spyware. On March 31, 2005, Cornell University's Information Technology department released a report detailing the behavior of one particular piece of proxy-based spyware, *Marketscore*, and the steps the university took to intercept it. Many other educational institutions have taken similar steps.

Individual users can also install firewalls from a variety of companies. These monitor the flow of information going to and from a networked computer and provide protection against spyware and malware. Some users install a large hosts file which prevents the user's computer from connecting to known spyware-related web addresses. Spyware may get installed via certain shareware programs offered for download. Downloading programs only from reputable sources can provide some protection from this source of attack.

Applications

"Stealware" and Affiliate Fraud

A few spyware vendors, notably 180 Solutions, have written what the *New York Times* has dubbed "stealware", and what spyware researcher Ben Edelman terms *affiliate fraud*, a form of click fraud. Stealware diverts the payment of affiliate marketing revenues from the legitimate affiliate to the spyware vendor.

Spyware which attacks affiliate networks places the spyware operator's affiliate tag on the user's activity – replacing any other tag, if there is one. The spyware operator is the only party that gains from this. The user has their choices thwarted, a legitimate affiliate loses revenue, networks' reputations are injured, and vendors are harmed by having to pay out affiliate revenues to an "affiliate" who is not party to a contract. Affiliate fraud is a violation of the terms of service of most affiliate marketing networks. As a result, spyware operators such as 180 Solutions have been terminated from affiliate networks including LinkShare and ShareSale. Mobile devices can also be vulnerable to chargeware, which manipulates users into illegitimate mobile charges.

Identity Theft and Fraud

In one case, spyware has been closely associated with identity theft. In August 2005, researchers from security software firm Sunbelt Software suspected the creators of the common CoolWeb-

Search spyware had used it to transmit "chat sessions, user names, passwords, bank information, etc."; however it turned out that "it actually (was) its own sophisticated criminal little trojan that's independent of CWS." This case is currently under investigation by the FBI.

The Federal Trade Commission estimates that 27.3 million Americans have been victims of identity theft, and that financial losses from identity theft totaled nearly $48 billion for businesses and financial institutions and at least $5 billion in out-of-pocket expenses for individuals.

Digital Rights Management

Some copy-protection technologies have borrowed from spyware. In 2005, Sony BMG Music Entertainment was found to be using rootkits in its XCP digital rights management technology Like spyware, not only was it difficult to detect and uninstall, it was so poorly written that most efforts to remove it could have rendered computers unable to function. Texas Attorney General Greg Abbott filed suit, and three separate class-action suits were filed. Sony BMG later provided a workaround on its website to help users remove it.

Beginning on April 25, 2006, Microsoft's Windows Genuine Advantage Notifications application was installed on most Windows PCs as a "critical security update". While the main purpose of this deliberately uninstallable application is to ensure the copy of Windows on the machine was lawfully purchased and installed, it also installs software that has been accused of "phoning home" on a daily basis, like spyware. It can be removed with the RemoveWGA tool.

Personal Relationships

Spyware has been used to monitor electronic activities of partners in intimate relationships. At least one software package, Loverspy, was specifically marketed for this purpose. Depending on local laws regarding communal/marital property, observing a partner's online activity without their consent may be illegal; the author of Loverspy and several users of the product were indicted in California in 2005 on charges of wiretapping and various computer crimes.

Browser Cookies

Anti-spyware programs often report Web advertisers' HTTP cookies, the small text files that track browsing activity, as spyware. While they are not always inherently malicious, many users object to third parties using space on their personal computers for their business purposes, and many anti-spyware programs offer to remove them.

Examples

These common spyware programs illustrate the diversity of behaviors found in these attacks. Note that as with computer viruses, researchers give names to spyware programs which may not be used by their creators. Programs may be grouped into "families" based not on shared program code, but on common behaviors, or by "following the money" of apparent financial or business connections. For instance, a number of the spyware programs distributed by Claria are collectively known as "Gator". Likewise, programs that are frequently installed together may be described as parts of the same spyware package, even if they function separately.

- CoolWebSearch, a group of programs, takes advantage of Internet Explorer vulnerabilities. The package directs traffic to advertisements on Web sites including *coolwebsearch.com*. It displays pop-up ads, rewrites search engine results, and alters the infected computer's hosts file to direct DNS lookups to these sites.

- FinFisher, sometimes called FinSpy is a high-end surveillance suite sold to law enforcement and intelligence agencies. Support services such as training and technology updates are part of the package.

- HuntBar, aka WinTools or Adware.Websearch, was installed by an ActiveX drive-by download at affiliate Web sites, or by advertisements displayed by other spyware programs—an example of how spyware can install more spyware. These programs add toolbars to IE, track aggregate browsing behavior, redirect affiliate references, and display advertisements.

- Internet Optimizer, also known as DyFuCa, redirects Internet Explorer error pages to advertising. When users follow a broken link or enter an erroneous URL, they see a page of advertisements. However, because password-protected Web sites (HTTP Basic authentication) use the same mechanism as HTTP errors, Internet Optimizer makes it impossible for the user to access password-protected sites.

- Spyware such as Look2Me hides inside system-critical processes and start up even in safe mode. With no process to terminate they are harder to detect and remove, which is a combination of both spyware and a rootkit. Rootkit technology is also seeing increasing use, as newer spyware programs also have specific countermeasures against well known anti-malware products and may prevent them from running or being installed, or even uninstall them.

- Movieland, also known as Moviepass.tv and Popcorn.net, is a movie download service that has been the subject of thousands of complaints to the Federal Trade Commission (FTC), the Washington State Attorney General's Office, the Better Business Bureau, and other agencies. Consumers complained they were held hostage by a cycle of oversized pop-up windows demanding payment of at least $29.95, claiming that they had signed up for a three-day free trial but had not cancelled before the trial period was over, and were thus obligated to pay. The FTC filed a complaint, since settled, against Movieland and eleven other defendants charging them with having "engaged in a nationwide scheme to use deception and coercion to extract payments from consumers."

- WeatherStudio has a plugin that displays a window-panel near the *bottom* of a browser window. The official website notes that it is easy to remove (uninstall) WeatherStudio from a computer, using its own uninstall-program, such as under C:\Program Files\WeatherStudio. Once WeatherStudio is removed, a browser returns to the prior display appearance, without the need to modify the browser settings.

- Zango (formerly 180 Solutions) transmits detailed information to advertisers about the Web sites which users visit. It also alters HTTP requests for affiliate advertisements linked from a Web site, so that the advertisements make unearned profit for the 180 Solutions company. It opens pop-up ads that cover over the Web sites of competing companies (as seen in their [Zango End User License Agreement]).

- Zlob trojan, or just Zlob, downloads itself to a computer via an ActiveX codec and reports information back to Control Server. Some information can be the search-history, the Websites visited, and even keystrokes. More recently, Zlob has been known to hijack routers set to defaults.

History and Development

The first recorded use of the term spyware occurred on October 16, 1995 in a Usenet post that poked fun at Microsoft's business model. *Spyware* at first denoted *software* meant for espionage purposes. However, in early 2000 the founder of Zone Labs, Gregor Freund, used the term in a press release for the ZoneAlarm Personal Firewall. Later in 2000, a parent using ZoneAlarm was alerted to the fact that "Reader Rabbit," educational software marketed to children by the Mattel toy company, was surreptitiously sending data back to Mattel. Since then, "spyware" has taken on its present sense.

According to a 2005 study by AOL and the National Cyber-Security Alliance, 61 percent of surveyed users' computers were infected with form of spyware. 92 percent of surveyed users with spyware reported that they did not know of its presence, and 91 percent reported that they had not given permission for the installation of the spyware. As of 2006, spyware has become one of the preeminent security threats to computer systems running Microsoft Windows operating systems. Computers on which Internet Explorer (IE) is the primary browser are particularly vulnerable to such attacks, not only because IE is the most widely used, but because its tight integration with Windows allows spyware access to crucial parts of the operating system.

Before Internet Explorer 6 SP2 was released as part of Windows XP Service Pack 2, the browser would automatically display an installation window for any ActiveX component that a website wanted to install. The combination of user ignorance about these changes, and the assumption by Internet Explorer that all ActiveX components are benign, helped to spread spyware significantly. Many spyware components would also make use of exploits in JavaScript, Internet Explorer and Windows to install without user knowledge or permission.

The Windows Registry contains multiple sections where modification of key values allows software to be executed automatically when the operating system boots. Spyware can exploit this design to circumvent attempts at removal. The spyware typically will link itself from each location in the registry that allows execution. Once running, the spyware will periodically check if any of these links are removed. If so, they will be automatically restored. This ensures that the spyware will execute when the operating system is booted, even if some (or most) of the registry links are removed.

Programs Distributed with Spyware

- Kazaa

- Morpheus

- WeatherBug

- WildTangent

Programs Formerly Distributed with Spyware

- AOL Instant Messenger (AOL Instant Messenger still packages Viewpoint Media Player, and WildTangent)

- DivX

- FlashGet

- magicJack

Rogue Anti-spyware Programs

Malicious programmers have released a large number of rogue (fake) anti-spyware programs, and widely distributed Web banner ads can warn users that their computers have been infected with spyware, directing them to purchase programs which do not actually remove spyware—or else, may add more spyware of their own.

The recent proliferation of fake or spoofed antivirus products that bill themselves as antispyware can be troublesome. Users may receive popups prompting them to install them to protect their computer, when it will in fact add spyware. This software is called rogue software. It is recommended that users do not install any freeware claiming to be anti-spyware unless it is verified to be legitimate. Some known offenders include:

- AntiVirus 360

- Antivirus 2009

- AntiVirus Gold

- ContraVirus

- MacSweeper

- Pest Trap

- PSGuard

- Spy Wiper

- Spydawn

- Spylocked

- Spysheriff

- SpyShredder

- Spyware Quake

- SpywareStrike

- UltimateCleaner

- WinAntiVirus Pro 2006

- Windows Police Pro

- WinFixer

- WorldAntiSpy

Fake antivirus products constitute 15 percent of all malware.

On January 26, 2006, Microsoft and the Washington state attorney general filed suit against Secure Computer for its Spyware Cleaner product.

Legal Issues

Criminal Law

Unauthorized access to a computer is illegal under computer crime laws, such as the U.S. Computer Fraud and Abuse Act, the U.K.'s Computer Misuse Act, and similar laws in other countries. Since owners of computers infected with spyware generally claim that they never authorized the installation, a *prima facie* reading would suggest that the promulgation of spyware would count as a criminal act. Law enforcement has often pursued the authors of other malware, particularly viruses. However, few spyware developers have been prosecuted, and many operate openly as strictly legitimate businesses, though some have faced lawsuits.

Spyware producers argue that, contrary to the users' claims, users do in fact give consent to installations. Spyware that comes bundled with shareware applications may be described in the legalese text of an end-user license agreement (EULA). Many users habitually ignore these purported contracts, but spyware companies such as Claria say these demonstrate that users have consented.

Despite the ubiquity of EULAs agreements, under which a single click can be taken as consent to the entire text, relatively little caselaw has resulted from their use. It has been established in most common law jurisdictions that this type of agreement can be a binding contract *in certain circumstances*. This does not, however, mean that every such agreement is a contract, or that every term in one is enforceable.

Some jurisdictions, including the U.S. states of Iowa and Washington, have passed laws criminalizing some forms of spyware. Such laws make it illegal for anyone other than the owner or operator of a computer to install software that alters Web-browser settings, monitors keystrokes, or disables computer-security software.

In the United States, lawmakers introduced a bill in 2005 entitled the Internet Spyware Prevention Act, which would imprison creators of spyware.

Administrative Sanctions

US FTC Actions

The US Federal Trade Commission has sued Internet marketing organizations under the "unfairness doctrine" to make them stop infecting consumers' PCs with spyware. In one case, that against

Seismic Entertainment Productions, the FTC accused the defendants of developing a program that seized control of PCs nationwide, infected them with spyware and other malicious software, bombarded them with a barrage of pop-up advertising for Seismic's clients, exposed the PCs to security risks, and caused them to malfunction. Seismic then offered to sell the victims an "antispyware" program to fix the computers, and stop the popups and other problems that Seismic had caused. On November 21, 2006, a settlement was entered in federal court under which a $1.75 million judgment was imposed in one case and $1.86 million in another, but the defendants were insolvent

In a second case, brought against CyberSpy Software LLC, the FTC charged that CyberSpy marketed and sold "RemoteSpy" keylogger spyware to clients who would then secretly monitor unsuspecting consumers' computers. According to the FTC, Cyberspy touted RemoteSpy as a "100% undetectable" way to "Spy on Anyone. From Anywhere." The FTC has obtained a temporary order prohibiting the defendants from selling the software and disconnecting from the Internet any of their servers that collect, store, or provide access to information that this software has gathered. The case is still in its preliminary stages. A complaint filed by the Electronic Privacy Information Center (EPIC) brought the RemoteSpy software to the FTC's attention.

Netherlands OPTA

An administrative fine, the first of its kind in Europe, has been issued by the Independent Authority of Posts and Telecommunications (OPTA) from the Netherlands. It applied fines in total value of Euro 1,000,000 for infecting 22 million computers. The spyware concerned is called DollarRevenue. The law articles that have been violated are art. 4.1 of the Decision on universal service providers and on the interests of end users; the fines have been issued based on art. 15.4 taken together with art. 15.10 of the Dutch telecommunications law.

Civil Law

Former New York State Attorney General and former Governor of New York Eliot Spitzer has pursued spyware companies for fraudulent installation of software. In a suit brought in 2005 by Spitzer, the California firm Intermix Media, Inc. ended up settling, by agreeing to pay US$7.5 million and to stop distributing spyware.

The hijacking of Web advertisements has also led to litigation. In June 2002, a number of large Web publishers sued Claria for replacing advertisements, but settled out of court.

Courts have not yet had to decide whether advertisers can be held liable for spyware that displays their ads. In many cases, the companies whose advertisements appear in spyware pop-ups do not directly do business with the spyware firm. Rather, they have contracted with an advertising agency, which in turn contracts with an online subcontractor who gets paid by the number of "impressions" or appearances of the advertisement. Some major firms such as Dell Computer and Mercedes-Benz have sacked advertising agencies that have run their ads in spyware.

Libel Suits by Spyware Developers

Litigation has gone both ways. Since "spyware" has become a common pejorative, some makers have filed libel and defamation actions when their products have been so described. In 2003, Gator (now known as Claria) filed suit against the website PC Pitstop for describing its program as "spyware". PC Pitstop settled, agreeing not to use the word "spyware", but continues to describe harm caused by the Gator/Claria software. As a result, other anti-spyware and anti-virus companies have also used other terms such as "potentially unwanted programs" or greyware to denote these products.

WebcamGate

In the 2010 WebcamGate case, plaintiffs charged two suburban Philadelphia high schools secretly spied on students by surreptitiously and remotely activating webcams embedded in school-issued laptops the students were using at home, and therefore infringed on their privacy rights. The school loaded each student's computer with LANrev's remote activation tracking software. This included the now-discontinued "TheftTrack". While TheftTrack was not enabled by default on the software, the program allowed the school district to elect to activate it, and to choose which of the TheftTrack surveillance options the school wanted to enable.

TheftTrack allowed school district employees to secretly remotely activate the webcam embedded in the student's laptop, above the laptop's screen. That allowed school officials to secretly take photos through the webcam, of whatever was in front of it and in its line of sight, and send the photos to the school's server. The LANrev software disabled the webcams for all other uses (*e.g.*, students were unable to use Photo Booth or video chat), so most students mistakenly believed their webcams did not work at all. In addition to webcam surveillance, TheftTrack allowed school officials to take screenshots, and send them to the school's server. In addition, LANrev allowed school officials to take snapshots of instant messages, web browsing, music playlists, and written compositions. The schools admitted to secretly snapping over 66,000 webshots and screenshots, including webcam shots of students in their bedrooms.

Ghost Push

Ghost Push is a kind of malware which infects the Android OS by automatically gaining root access, downloading malicious software, converting to system app and then losing root access which makes it virtually impossible to remove the infection even by factory reset unless the firmware is reflashed. The malware hogs all system resources making it unresponsive and drains the battery. Advertisements always appears anytime either as a full screen, part of a display, or in status bar. Unnecessary apps are automatically activated and sometimes downloads another malicious software when connected to the internet. It is hard to detect.

History

It was discovered in September 18, 2015 by Cheetah Mobile's CM Security Research Lab.

Further investigation of Ghost Push revealed more recent variants, which, unlike older ones, employ the following routines that make them harder to remove and detect:

- encrypt its APK and shell code,

- run a malicious DEX file without notification,

- add a "guard code" to monitor its own processes,

- rename .APK (Android application package) files used to install the malicious apps,

- and launch the new activity as the payload.

References

- Gralla, Preston (2005). PC Pest Control: Protect Your Computers from Malicious Internet Invaders. Google Books. 1005 Gravenstein Highway North, Sebastopol, CA 95472, US: O'Reilly Media, Inc. p. 237. ISBN 0-596-00926-7. Retrieved 18 January 2014.

- Hantula, Richard (2010). How Do Cell Phones Work?. Google Books. 132 West 31st Street, New York NY 10001, US: Infobase Publishing. p. 27. ISBN 978-1-43812-805-4. Retrieved 18 January 2014.

- "'Ghost Push': An Un-Installable Android Virus Infecting 600,000+ Users Per Day - The world's leading mobile tools provider". cmcm.com. Retrieved 2016-01-09.

- Nicole Perlroth (August 30, 2012). "Software Meant to Fight Crime Is Used to Spy on Dissidents". The New York Times. Retrieved August 31, 2012.

- Cooley, Brian (March 7, 2011). "'Like,' 'tweet' buttons divulge sites you visit: CNET News Video". CNet News. Retrieved March 7, 2011.

- "Gadgets boingboing.net, ''MagicJack's EULA says it will spy on you and force you into arbitration'"". Gadgets. boingboing.net. April 14, 2008. Retrieved September 11, 2010.

- "Initial LANrev System Findings", LMSD Redacted Forensic Analysis, L-3 Services – prepared for Ballard Spahr (LMSD's counsel), May 2010. Retrieved August 15, 2010. Archived June 15, 2010.

- Doug Stanglin (February 18, 2010). "School district accused of spying on kids via laptop webcams". USA Today. Retrieved February 19, 2010.

Mobile Security: An Integrated Study

Mobile security has become an alarming concern in the past few decades. The aspects explained in this text are browser security, wireless security camera and mobile secure gateway. This chapter helps the reader in developing an in depth understanding on the issue of mobile security.

Mobile Security

Mobile security or mobile phone security has become increasingly important in mobile computing. Of particular concern is the security of personal and business information now stored on smartphones.

More and more users and businesses employ smartphones as communication tools, but also as a means of planning and organizing their work and private life. Within companies, these technologies are causing profound changes in the organization of information systems and therefore they have become the source of new risks. Indeed, smartphones collect and compile an increasing amount of sensitive information to which access must be controlled to protect the privacy of the user and the intellectual property of the company.

All smartphones, as computers, are preferred targets of attacks. These attacks exploit weaknesses related to smartphones that can come from means of communication like Short Message Service (SMS, aka text messaging), Multimedia Messaging Service (MMS), Wi-Fi networks, Bluetooth and GSM, the de facto global standard for mobile communications. There are also attacks that exploit software vulnerabilities from both the web browser and operating system. Finally, there are forms of malicious software that rely on the weak knowledge of average users.

Different security counter-measures are being developed and applied to smartphones, from security in different layers of software to the dissemination of information to end users. There are good practices to be observed at all levels, from design to use, through the development of operating systems, software layers, and downloadable apps.

Challenges of Mobile Security

Threats

A smartphone user is exposed to various threats when they use their phone. In just the last two quarters of 2012, the number of unique mobile threats grew by 261%, according to ABI Research. These threats can disrupt the operation of the smartphone, and transmit or modify user data. For these reasons, the applications deployed there must guarantee privacy and integrity of the information they handle. In addition, since some apps could themselves be malware, their functionality and activities should be limited (for example, restricting the apps from accessing location information via GPS, blocking access to the user's address book, preventing the transmission of data on the network, sending SMS messages that are billed to the user, etc.).

There are three prime targets for attackers:

- Data: smartphones are devices for data management, therefore they may contain sensitive data like credit card numbers, authentication information, private information, activity logs (calendar, call logs);

- Identity: smartphones are highly customizable, so the device or its contents are associated with a specific person. For example, every mobile device can transmit information related to the owner of the mobile phone contract, and an attacker may want to steal the identity of the owner of a smartphone to commit other offenses;

- Availability: by attacking a smartphone one can limit access to it and deprive the owner of the service.

The source of these attacks are the same actors found in the non-mobile computing space:

- Professionals, whether commercial or military, who focus on the three targets mentioned above. They steal sensitive data from the general public, as well as undertake industrial espionage. They will also use the identity of those attacked to achieve other attacks;

- Thieves who want to gain income through data or identities they have stolen. The thieves will attack many people to increase their potential income;

- Black hat hackers who specifically attack availability. Their goal is to develop viruses, and cause damage to the device. In some cases, hackers have an interest in stealing data on devices.

- Grey hat hackers who reveal vulnerabilities. Their goal is to expose vulnerabilities of the device. Grey hat hackers do not intend on damaging the device or stealing data.

Consequences

When a smartphone is infected by an attacker, the attacker can attempt several things:

- The attacker can manipulate the smartphone as a zombie machine, that is to say, a machine with which the attacker can communicate and send commands which will be used to send unsolicited messages (spam) via sms or email;

- The attacker can easily force the smartphone to make phone calls. For example, one can use the API (library that contains the basic functions not present in the smartphone) Phone-MakeCall by Microsoft, which collects telephone numbers from any source such as yellow pages, and then call them. But the attacker can also use this method to call paid services, resulting in a charge to the owner of the smartphone. It is also very dangerous because the smartphone could call emergency services and thus disrupt those services;

- A compromised smartphone can record conversations between the user and others and send them to a third party. This can cause user privacy and industrial security problems;

- An attacker can also steal a user's identity, usurp their identity (with a copy of the user's sim card or even the telephone itself), and thus impersonate the owner. This raises security concerns in countries where smartphones can be used to place orders, view bank accounts or are used as an identity card;

- The attacker can reduce the utility of the smartphone, by discharging the battery. For example, they can launch an application that will run continuously on the smartphone processor, requiring a lot of energy and draining the battery. One factor that distinguishes mobile computing from traditional desktop PCs is their limited performance. Frank Stajano and Ross Anderson first described this form of attack, calling it an attack of "battery exhaustion" or "sleep deprivation torture";

- The attacker can prevent the operation and/or starting of the smartphone by making it unusable. This attack can either delete the boot scripts, resulting in a phone without a functioning OS, or modify certain files to make it unusable (e.g. a script that launches at startup that forces the smartphone to restart) or even embed a startup application that would empty the battery;

- The attacker can remove the personal (photos, music, videos, etc.) or professional data (contacts, calendars, notes) of the user.

Attacks Based on Communication

Attack Based on SMS and MMS

Some attacks derive from flaws in the management of SMS and MMS.

Some mobile phone models have problems in managing binary SMS messages. It is possible, by sending an ill-formed block, to cause the phone to restart, leading to denial of service attacks. If a user with a Siemens S55 received a text message containing a Chinese character, it would lead to a denial of service. In another case, while the standard requires that the maximum size of a Nokia Mail address is 32 characters, some Nokia phones did not verify this standard, so if a user enters an email address over 32 characters, that leads to complete dysfunction of the e-mail handler and puts it out of commission. This attack is called "curse of silence". A study on the safety of the SMS infrastructure revealed that SMS messages sent from the Internet can be used to perform a distributed denial of service (DDoS) attack against the mobile telecommunications infrastructure of a big city. The attack exploits the delays in the delivery of messages to overload the network.

Another potential attack could begin with a phone that sends an MMS to other phones, with an attachment. This attachment is infected with a virus. Upon receipt of the MMS, the user can choose to open the attachment. If it is opened, the phone is infected, and the virus sends an MMS with an infected attachment to all the contacts in the address book. There is a real-world example of this attack: the virus Commwarrior uses the address book and sends MMS messages including an infected file to recipients. A user installs the software, as received via MMS message. Then, the virus began to send messages to recipients taken from the address book.

Attacks Based on Communication Networks

Attacks Based on the GSM Networks

The attacker may try to break the encryption of the mobile network. The GSM network encryption algorithms belong to the family of algorithms called A5. Due to the policy of security through obscurity it has not been possible to openly test the robustness of these algorithms. There were originally two variants of the algorithm: A5/1 and A5/2 (stream ciphers), where the former was designed to be relatively strong, and the latter was designed to be weak on purpose to allow easy cryptanalysis and eavesdropping. ETSI forced some countries (typically outside Europe) to use A5/2. Since the encryption algorithm was made public, it was proved it was possible to break the encryption: A5/2 could be broken on the fly, and A5/1 in about 6 hours . In July 2007, the 3GPP approved a change request to prohibit the implementation of A5/2 in any new mobile phones, which means that is has been decommissioned and is no longer implemented in mobile phones. Stronger public algorithms have been added to the GSM standard, the A5/3 and A5/4 (Block ciphers), otherwise known as KASUMI or UEA1 published by the ETSI. If the network does not support A5/1, or any other A5 algorithm implemented by the phone, then the base station can specify A5/0 which is the null-algorithm, whereby the radio traffic is sent unencrypted. Even in case mobile phones are able to use 3G or 4G which have much stronger encryption than 2G GSM, the base station can downgrade the radio communication to 2G GSM and specify A5/0 (no encryption) . This is the basis for eavesdropping attacks on mobile radio networks using a fake base station commonly called an IMSI catcher.

In addition, tracing of mobile terminals is difficult since each time the mobile terminal is accessing or being accessed by the network, a new temporary identity (TMSI) is allocated to the mobile terminal. The TSMI is used as identity of the mobile terminal the next time it accesses the network. The TMSI is sent to the mobile terminal in encrypted messages.

Once the encryption algorithm of GSM is broken, the attacker can intercept all unencrypted communications made by the victim's smartphone.

Attacks Based on Wi-Fi

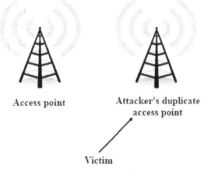

Access Point spoofing

An attacker can try to eavesdrop on Wi-Fi communications to derive information (e.g. username, password). This type of attack is not unique to smartphones, but they are very vulnerable to these attacks because very often the Wi-Fi is the only means of communication they have to access the internet. The security of wireless networks (WLAN) is thus an important subject. Initially wireless networks were secured by WEP keys. The weakness of WEP is a short encryption key which is the same for all connected clients. In addition, several reductions in the search space of the keys have been found by researchers. Now, most wireless networks are protected by the WPA security protocol. WPA is based on the "Temporal Key Integrity Protocol (TKIP)" which was designed to allow migration from WEP to WPA on the equipment already deployed. The major improvements in security are the dynamic encryption keys. For small networks, the WPA is a "pre-shared key" which is based on a shared key. Encryption can be vulnerable if the length of the shared key is short. With limited opportunities for input (i.e. only the numeric keypad) mobile phone users might define short encryption keys that contain only numbers. This increases the likelihood that an attacker succeeds with a brute-force attack. The successor to WPA, called WPA2, is supposed to be safe enough to withstand a brute force attack.

As with GSM, if the attacker succeeds in breaking the identification key, it will be possible to attack not only the phone but also the entire network it is connected to.

Many smartphones for wireless LANs remember they are already connected, and this mechanism prevents the user from having to re-identify with each connection. However, an attacker could create a WIFI access point twin with the same parameters and characteristics as the real network. Using the fact that some smartphones remember the networks, they could confuse the two networks and connect to the network of the attacker who can intercept data if it does not transmit its data in encrypted form.

Lasco is a worm that initially infects a remote device using the SIS file format. SIS file format (Software Installation Script) is a script file that can be executed by the system without user interaction. The smartphone thus believes the file to come from a trusted source and downloads it, infecting the machine.

Principle of Bluetooth-based Attacks

Security issues related to Bluetooth on mobile devices have been studied and have shown numerous problems on different phones. One easy to exploit vulnerability: unregistered services do not require authentication, and vulnerable applications have a virtual serial port used to control the phone. An attacker only needed to connect to the port to take full control of the device. Another example: a phone must be within reach and Bluetooth in discovery mode. The attacker sends a file via Bluetooth. If the recipient accepts, a virus is transmitted. For example: Cabir is a worm that spreads via Bluetooth connection. The worm searches for nearby phones with Bluetooth in discoverable mode and sends itself to the target device. The user must accept the incoming file and install the program. After installing, the worm infects the machine.

Attacks Based on Vulnerabilities in Software Applications

Other attacks are based on flaws in the OS or applications on the phone.

Web Browser

The mobile web browser is an emerging attack vector for mobile devices. Just as common Web browsers, mobile web browsers are extended from pure web navigation with widgets and plug-ins, or are completely native mobile browsers.

Jailbreaking the iPhone with firmware 1.1.1 was based entirely on vulnerabilities on the web browser. As a result, the exploitation of the vulnerability described here underlines the importance of the Web browser as an attack vector for mobile devices. In this case, there was a vulnerability based on a stack-based buffer overflow in a library used by the web browser (Libtiff).

A vulnerability in the web browser for Android was discovered in October 2008. As the iPhone vulnerability above, it was due to an obsolete and vulnerable library. A significant difference with the iPhone vulnerability was Android's sandboxing architecture which limited the effects of this vulnerability to the Web browser process.

Smartphones are also victims of classic piracy related to the web: phishing, malicious websites, etc. The big difference is that smartphones do not yet have strong antivirus software available.

Operating System

Sometimes it is possible to overcome the security safeguards by modifying the operating system itself. As real-world examples, this section covers the manipulation of firmware and malicious signature certificates. These attacks are difficult.

In 2004, vulnerabilities in virtual machines running on certain devices were revealed. It was possible to bypass the bytecode verifier and access the native underlying operating system. The results of this research were not published in detail. The firmware security of Nokia's Symbian Platform Security Architecture (PSA) is based on a central configuration file called SWIPolicy. In 2008 it was possible to manipulate the Nokia firmware before it is installed, and in fact in some downloadable versions of it, this file was human readable, so it was possible to modify and change the image of the firmware. This vulnerability has been solved by an update from Nokia.

In theory smartphones have an advantage over hard drives since the OS files are in ROM, and cannot be changed by malware. However, in some systems it was possible to circumvent this: in the Symbian OS it was possible to overwrite a file with a file of the same name. On the Windows OS, it was possible to change a pointer from a general configuration file to an editable file.

When an application is installed, the signing of this application is verified by a series of certificates. One can create a valid signature without using a valid certificate and add it to the list. In the Symbian OS all certificates are in the directory: c:\resource\swicertstore\dat. With firmware changes explained above it is very easy to insert a seemingly valid but malicious certificate.

Attacks Based on Hardware Vulnerabilities

Electromagnetic Waveforms

In 2015, researchers at the French government agency ANSSI demonstrated the capability to trigger the voice interface of certain smartphones remotely by using "specific electromagnetic waveforms". The exploit took advantage of antenna-properties of headphone wires while plugged into the audio-output jacks of the vulnerable smartphones and effectively spoofed audio input to inject commands via the audio interface.

Juice Jacking

Juice Jacking is a method of physical or a hardware vulnerability specific to mobile platforms. Utilizing the dual purpose of the USB charge port, many devices have been susceptible to having data ex-filtrated from, or malware installed on to a mobile device by utilizing malicious charging kiosks set up in public places, or hidden in normal charge adapters.

Password Cracking

In 2010, researcher from the University of Pennsylvania investigated the possibility of cracking a device's password through a smudge attack (literally imaging the finger smudges on the screen to discern the user's password). The researchers were able to discern the device password up to 68% of the time under certain conditions. Outsiders may perform over-the-shoulder on victims, such as watching specific keystrokes or pattern gestures, to unlock device password or passcode.

Malicious Software (Malware)

As smartphones are a permanent point of access to the internet (mostly on), they can be compromised as easily as computers with malware. A malware is a computer program that aims to harm the system in which it resides. Trojans, worms and viruses are all considered malware. A Trojan is a program that is on the smartphone and allows external users to connect discreetly. A worm is a program that reproduces on multiple computers across a network. A virus is malicious software designed to spread to other computers by inserting itself into legitimate programs and running programs in parallel. However, it must be said that the malware are far less numerous and important to smartphones as they are to computers.

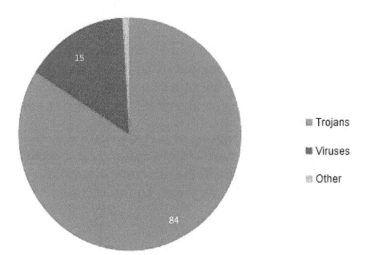

Nonetheless, recent studies show that the evolution of malware in smartphones have rocketed in the last few years posing a threat to analysis and detection.

The Three Phases of Malware Attacks

Typically an attack on a smartphone made by malware takes place in 3 phases: the infection of a host, the accomplishment of its goal, and the spread of the malware to other systems. Malware often use the resources offered by the infected smartphones. It will use the output devices such as

Bluetooth or infrared, but it may also use the address book or email address of the person to infect the user's acquaintances. The malware exploits the trust that is given to data sent by an acquaintance.

Infection

Infection is the means used by the malware to get into the smartphone, it can either use one of the faults previously presented or may use the gullibility of the user. Infections are classified into four classes according to their degree of user interaction:

Explicit Permission

the most benign interaction is to ask the user if it is allowed to infect the machine, clearly indicating its potential malicious behavior. This is typical behavior of a proof of concept malware.

Implied Permission

this infection is based on the fact that the user has a habit of installing software. Most trojans try to seduce the user into installing attractive applications (games, useful applications etc.) that actually contain malware.

Common Interaction

this infection is related to a common behavior, such as opening an MMS or email.

No Interaction

the last class of infection is the most dangerous. Indeed, a worm that could infect a smartphone and could infect other smartphones without any interaction would be catastrophic.

Accomplishment of its Goal

Once the malware has infected a phone it will also seek to accomplish its goal, which is usually one of the following: monetary damage, damage data and/or device, and concealed damage:

Monetary Damages

the attacker can steal user data and either sell them to the same user, or sell to a third party.

Damage

malware can partially damage the device, or delete or modify data on the device.

Concealed Damage

the two aforementioned types of damage are detectable, but the malware can also leave a backdoor for future attacks or even conduct wiretaps.

Spread to Other Systems

Once the malware has infected a smartphone, it always aims to spread one way or another:

- It can spread through proximate devices using Wi-Fi, Bluetooth and infrared;

- It can also spread using remote networks such as telephone calls or SMS or emails.

Examples of Malware

Here are various malware that exist in the world of smartphones with a short description of each.

Viruses and Trojans

- Cabir (also known as Caribe, SybmOS/Cabir, Symbian/Cabir and EPOC.cabir) is the name of a computer worm developed in 2004 that is designed to infect mobile phones running Symbian OS. It is believed to be the first computer worm that can infect mobile phones

- Commwarrior, found March 7, 2005, is the first worm that can infect many machines from MMS. It is sent in the form of an archive file COMMWARRIOR.ZIP that contains a file COMMWARRIOR.SIS. When this file is executed, Commwarrior attempts to connect to nearby devices by Bluetooth or infrared under a random name. It then attempts to send MMS message to the contacts in the smartphone with different header messages for each person, who receive the MMS and often open them without further verification.

- Phage is the first Palm OS virus that was discovered. It transfers to the Palm from a PC via synchronization. It infects all applications that are in the smartphone and it embeds its own code to function without the user and the system detecting it. All that the system will detect is that its usual applications are functioning.

- RedBrowser is a Trojan which is based on java. The Trojan masquerades as a program called "RedBrowser" which allows the user to visit WAP sites without a WAP connection. During application installation, the user sees a request on their phone that the application needs permission to send messages. Therefore, if the user accepts, RedBrowser can send sms to paid call centers. This program uses the smartphone's connection to social networks (Facebook, Twitter, etc.) to get the contact information for the user's acquaintances (provided the required permissions have been given) and will send them messages.

- WinCE.PmCryptic.A is a malicious software on Windows Mobile which aims to earn money for its authors. It uses the infestation of memory cards that are inserted in the smartphone to spread more effectively.

- CardTrap is a virus that is available on different types of smartphone, which aims to deactivate the system and third party applications. It works by replacing the files used to start the smartphone and applications to prevent them from executing. There are different variants of this virus such as Cardtrap.A for SymbOS devices. It also infects the memory card with malware capable of infecting Windows.

- Ghost Push is a malicious software on Android OS which automatically root the android device and installs malicious applications directly to system partition then unroots the device to prevent users from removing the threat by master reset (The threat can be removed only by reflashing). It cripples the system resources, executes quickly, and harder to detect.

Ransomware

Mobile ransomware is a type of malware that locks users out of their mobile devices in a pay-to-unlock-your-device ploy, it has grown by leaps and bounds as a threat category since 2014. Specific to mobile computing platforms, users are often less security-conscious, particularly as it pertains to scrutinizing applications and web links trusting the native protection capability of the mobile device operating system. Mobile ransomware poses a significant threat to businesses reliant on instant access and availability of their proprietary information and contacts. The likelihood of a traveling businessman paying a ransom to unlock their device is significantly higher since they are at a disadvantage given inconveniences such as timeliness and less likely direct access to IT staff.

Spyware

- Flexispy is an application that can be considered as a trojan, based on Symbian. The program sends all information received and sent from the smartphone to a Flexispy server. It was originally created to protect children and spy on adulterous spouses.

Number of Malware

Below is a diagram which loads the different behaviors of smartphone malware in terms of their effects on smartphones:

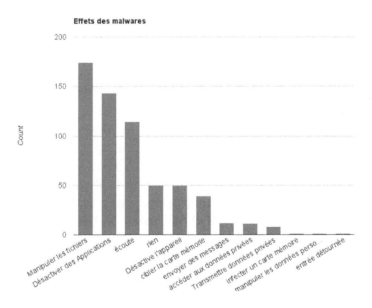

Effects of Malware

We can see from the graph that at least 50 malwares exhibit no negative behavior, except their ability to spread.

Portability of Malware Across Platforms

There is a multitude of malware. This is partly due to the variety of operating systems on smartphones. However attackers can also choose to make their malware target multiple platforms, and malware can be found which attacks an OS but is able to spread to different systems.

To begin with, malware can use runtime environments like Java virtual machine or the .NET Framework. They can also use other libraries present in many operating systems. Other malware carry several executable files in order to run in multiple environments and they utilize these during the propagation process. In practice, this type of malware requires a connection between the two operating systems to use as an attack vector. Memory cards can be used for this purpose, or synchronization software can be used to propagate the virus.

Countermeasures

The security mechanisms in place to counter the threats described above are presented in this section. They are divided into different categories, as all do not act at the same level, and they range from the management of security by the operating system to the behavioral education of the user. The threats prevented by the various measures are not the same depending on the case. Considering the two cases mentioned above, in the first case one would protect the system from corruption by an application, and in the second case the installation of a suspicious software would be prevented.

Security in Operating Systems

The first layer of security within a smartphone is at the level of the operating system (OS). Beyond the usual roles of an operating system (e.g. resource management, scheduling processes) on a smartphone, it must also establish the protocols for introducing external applications and data without introducing risk.

A central idea found in the mobile operating systems is the idea of a sandbox. Since smartphones are currently being designed to accommodate many applications, they must put in place mechanisms to ensure these facilities are safe for themselves, for other applications and data on the system, and the user. If a malicious program manages to reach a device, it is necessary that the vulnerable area presented by the system be as small as possible. Sandboxing extends this idea to compartmentalize different processes, preventing them from interacting and damaging each other. Based on the history of operating systems, sandboxing has different implementations. For example, where iOS will focus on limiting access to its public API for applications from the App Store by default, Managed Open In allows you to restrict which apps can access which types of data. Android bases its sandboxing on its legacy of Linux and TrustedBSD.

The following points highlight mechanisms implemented in operating systems, especially Android.

Rootkit Detectors

The intrusion of a rootkit in the system is a great danger in the same way as on a computer. It is important to prevent such intrusions, and to be able to detect them as often as possible. Indeed, there is concern that with this type of malicious program, the result could be a partial or complete bypass of the device security, and the acquisition of administrator rights by the attacker. If this happens, then nothing prevents the attacker from studying or disabling the safety features that were circumvented, deploying the applications they want, or disseminating a method of intrusion by a rootkit to a wider audience. We can cite, as a defense mechanism, the Chain of trust in iOS. This mechanism relies on the signature of the different applications required to start the operating system, and a certificate signed by Apple. In the event that the signature checks are inconclusive, the device detects this and stops the boot-up. If the Operating System is compromised due to Jailbreaking, root kit detection may not work if it is disabled by the Jailbreak method or software is loaded after Jailbreak disables Rootkit Detection.

Process Isolation

Android uses mechanisms of user process isolation inherited from Linux. Each application has a user associated with it, and a tuple (UID, GID). This approach serves as a sandbox: while applications can be malicious, they can not get out of the sandbox reserved for them by their identifiers, and thus cannot interfere with the proper functioning of the system. For example, since it is impossible for a process to end the process of another user, an application can thus not stop the execution of another.

File Permissions

From the legacy of Linux, there are also filesystem permissions mechanisms. They help with sandboxing: a process can not edit any files it wants. It is therefore not possible to freely corrupt files necessary for the operation of another application or system. Furthermore, in Android there is the method of locking memory permissions. It is not possible to change the permissions of files installed on the SD card from the phone, and consequently it is impossible to install applications.

Memory Protection

In the same way as on a computer, memory protection prevents privilege escalation. Indeed, if a process managed to reach the area allocated to other processes, it could write in the memory of a process with rights superior to their own, with root in the worst case, and perform actions which are beyond its permissions on the system. It would suffice to insert function calls are authorized by the privileges of the malicious application.

Development Through Runtime Environments

Software is often developed in high-level languages, which can control what is being done by a running program. For example, Java Virtual Machines continuously monitor the actions of the execution threads they manage, monitor and assign resources, and prevent malicious actions. Buffer overflows can be prevented by these controls.

Security Software

Above the operating system security, there is a layer of security software. This layer is composed of individual components to strengthen various vulnerabilities: prevent malware, intrusions, the identification of a user as a human, and user authentication. It contains software components that have learned from their experience with computer security; however, on smartphones, this soft-ware must deal with greater constraints.

Antivirus and Firewall

An antivirus software can be deployed on a device to verify that it is not infected by a known threat, usually by signature detection software that detects malicious executable files. A firewall, meanwhile, can watch over the existing traffic on the network and ensure that a malicious application does not seek to communicate through it. It may equally verify that an installed application does not seek to establish suspicious communication, which may prevent an intrusion attempt.

Visual Notifications

In order to make the user aware of any abnormal actions, such as a call they did not initiate, one can link some functions to a visual notification that is impossible to circumvent. For example, when a call is triggered, the called number should always be displayed. Thus, if a call is triggered by a malicious application, the user can see, and take appropriate action.

Turing Test

In the same vein as above, it is important to confirm certain actions by a user decision. The Turing test is used to distinguish between a human and a virtual user, and it often comes as a captcha.

Biometric Identification

Another method to use is biometrics. Biometrics is a technique of identifying a person by means of their morphology(by recognition of the eye or face, for example) or their behavior (their signature or way of writing for example). One advantage of using biometric security is that users can avoid having to remember a password or other secret combination to authenticate and prevent malicious users from accessing their device. In a system with strong biometric security, only the primary user can access the smartphone.

Resource Monitoring in the Smartphone

When an application passes the various security barriers, it can take the actions for which it was designed. When such actions are triggered, the activity of a malicious application can be sometimes detected if one monitors the various resources used on the phone. Depending on the goals of the malware, the consequences of infection are not always the same; all malicious applications are not intended to harm the devices on which they are deployed. The following sections describe different ways to detect suspicious activity.

Battery

Some malware is aimed at exhausting the energy resources of the phone. Monitoring the energy consumption of the phone can be a way to detect certain malware applications.

Memory Usage

Memory usage is inherent in any application. However, if one finds that a substantial proportion of memory is used by an application, it may be flagged as suspicious.

Network Traffic

On a smartphone, many applications are bound to connect via the network, as part of their normal operation. However, an application using a lot of bandwidth can be strongly suspected of attempting to communicate a lot of information, and disseminate data to many other devices. This observation only allows a suspicion, because some legitimate applications can be very resource-intensive in terms of network communications, the best example being streaming video.

Services

One can monitor the activity of various services of a smartphone. During certain moments, some services should not be active, and if one is detected, the application should be suspected. For example, the sending of an SMS when the user is filming video: this communication does not make sense and is suspicious; malware may attempt to send SMS while its activity is masked.

The various points mentioned above are only indications and do not provide certainty about the legitimacy of the activity of an application. However, these criteria can help target suspicious applications, especially if several criteria are combined.

Network Surveillance

Network traffic exchanged by phones can be monitored. One can place safeguards in network routing points in order to detect abnormal behavior. As the mobile's use of network protocols is much more constrained than that of a computer, expected network data streams can be predicted (e.g. the protocol for sending an SMS), which permits detection of anomalies in mobile networks.

Spam Filters

As is the case with email exchanges, we can detect a spam campaign through means of mobile communications (SMS, MMS). It is therefore possible to detect and minimize this kind of attempt by filters deployed on network infrastructure that is relaying these messages.

Encryption of Stored or Transmitted Information

Because it is always possible that data exchanged can be intercepted, communications, or even information storage, can rely on encryption to prevent a malicious entity from using any data obtained during communications. However, this poses the problem of key exchange for encryption algorithms, which requires a secure channel.

Telecom Network Monitoring

The networks for SMS and MMS exhibit predictable behavior, and there is not as much liberty compared with what one can do with protocols such as TCP or UDP. This implies that one cannot predict the use made of the common protocols of the web; one might generate very little traffic by consulting simple pages, rarely, or generate heavy traffic by using video streaming. On the other hand, messages exchanged via mobile phone have a framework and a specific model, and the user does not, in a normal case, have the freedom to intervene in the details of these communications. Therefore, if an abnormality is found in the flux of network data in the mobile networks, the potential threat can be quickly detected.

Manufacturer Surveillance

In the production and distribution chain for mobile devices, it is the responsibility of manufacturers to ensure that devices are delivered in a basic configuration without vulnerabilities. Most users are not experts and many of them are not aware of the existence of security vulnerabilities, so the device configuration as provided by manufacturers will be retained by many users. Below are listed several points which manufacturers should consider.

Remove Debug Mode

Phones are sometimes set in a debug mode during manufacturing, but this mode must be disabled before the phone is sold. This mode allows access to different features, not intended for routine use by a user. Due to the speed of development and production, distractions occur and some devices are sold in debug mode. This kind of deployment exposes mobile devices to exploits that utilize this oversight.

Default Settings

When a smartphone is sold, its default settings must be correct, and not leave security gaps. The default configuration is not always changed, so a good initial setup is essential for users. There are, for example, default configurations that are vulnerable to denial of service attacks.

Security Audit of Apps

Along with smart phones, appstores have emerged. A user finds themselves facing a huge range of applications. This is especially true for providers who manage appstores because they are tasked with examining the apps provided, from different points of view (e.g. security, content). The security audit should be particularly cautious, because if a fault is not detected, the application can spread very quickly within a few days, and infect a significant number of devices.

Detect Suspicious Applications Demanding Rights

When installing applications, it is good to warn the user against sets of permissions that, grouped together, seem potentially dangerous, or at least suspicious. Frameworks like such as Kirin, on Android, attempt to detect and prohibit certain sets of permissions.

Revocation Procedures

Along with appstores appeared a new feature for mobile apps: remote revocation. First developed by Android, this procedure can remotely and globally uninstall an application, on any device that has it. This means the spread of a malicious application that managed to evade security checks can be immediately stopped when the threat is discovered.

Avoid Heavily Customized Systems

Manufacturers are tempted to overlay custom layers on existing operating systems, with the dual purpose of offering customized options and disabling or charging for certain features. This has the dual effect of risking the introduction of new bugs in the system, coupled with an incentive for users to modify the systems to circumvent the manufacturer's restrictions. These systems are rarely as stable and reliable as the original, and may suffer from phishing attempts or other exploits.

Improve Software Patch Processes

New versions of various software components of a smartphone, including operating systems, are regularly published. They correct many flaws over time. Nevertheless, manufacturers often do not deploy these updates to their devices in a timely fashion, and sometimes not at all. Thus, vulnerabilities persist when they could be corrected, and if they are not, since they are known, they are easily exploitable.

User Awareness

Much malicious behavior is allowed by the carelessness of the user. From simply not leaving the device without a password, to precise control of permissions granted to applications added to the smartphone, the user has a large responsibility in the cycle of security: to not be the vector of intrusion. This precaution is especially important if the user is an employee of a company that stores business data on the device. Detailed below are some precautions that a user can take to manage security on a smartphone.

A recent survey by internet security experts BullGuard showed a lack of insight into the rising number of malicious threats affecting mobile phones, with 53% of users claiming that they are unaware of security software for Smartphones. A further 21% argued that such protection was unnecessary, and 42% admitted it hadn't crossed their mind ("Using APA," 2011). These statistics show consumers are not concerned about security risks because they believe it is not a serious problem. The key here is to always remember smartphones are effectively handheld computers and are just as vulnerable.

Being Skeptical

A user should not believe everything that may be presented, as some information may be phishing or attempting to distribute a malicious application. It is therefore advisable to check the reputation of the application that they want to buy before actually installing it.

Permissions Given to Applications

The mass distribution of applications is accompanied by the establishment of different permissions mechanisms for each operating system. It is necessary to clarify these permissions mechanisms to users, as they differ from one system to another, and are not always easy to understand. In addition, it is rarely possible to modify a set of permissions requested by an application if the number of permissions is too great. But this last point is a source of risk because a user can grant rights to an application, far beyond the rights it needs. For example, a note taking application does not require access to the geolocation service. The user must ensure the privileges required by an application during installation and should not accept the installation if requested rights are inconsistent.

Be Careful

Protection of a user's phone through simple gestures and precautions, such as locking the smartphone when it is not in use, not leaving their device unattended, not trusting applications, not storing sensitive data, or encrypting sensitive data that cannot be separated from the device.

Ensure Data

Smartphones have a significant memory and can carry several gigabytes of data. The user must be careful about what data it carries and whether they should be protected. While it is usually not dramatic if a song is copied, a file containing bank information or business data can be more risky. The user must have the prudence to avoid the transmission of sensitive data on a smartphone, which can be easily stolen. Furthermore, when a user gets rid of a device, they must be sure to remove all personal data first.

These precautions are measures that leave no easy solution to the intrusion of people or malicious applications in a smartphone. If users are careful, many attacks can be defeated, especially phishing and applications seeking only to obtain rights on a device.

Centralized Storage of Text Messages

One form of mobile protection allows companies to control the delivery and storage of text messages, by hosting the messages on a company server, rather than on the sender or receiver's phone. When certain conditions are met, such as an expiration date, the messages are deleted.

Limitations of Certain Security Measures

The security mechanisms mentioned in this article are to a large extent inherited from knowledge and experience with computer security. The elements composing the two device types are similar, and there are common measures that can be used, such as antivirus and firewall. However, the implementation of these solutions is not necessarily possible or at least highly constrained within a mobile device. The reason for this difference is the technical resources offered by computers and mobile devices: even though the computing power of smartphones is becoming faster, they have other limitations than their computing power.

- Single-task system: Some operating systems, including some still commonly used, are single-tasking. Only the foreground task is executed. It is difficult to introduce applications such as antivirus and firewall on such systems, because they could not perform their monitoring while the user is operating the device, when there would be most need of such monitoring.

- Energy autonomy: A critical one for the use of a smartphone is energy autonomy. It is important that the security mechanisms not consume battery resources, without which the autonomy of devices will be affected dramatically, undermining the effective use of the smartphone.

- Network Directly related to battery life, network utilization should not be too high. It is indeed one of the most expensive resources, from the point of view of energy consumption. Nonetheless, some calculations may need to be relocated to remote servers in order to preserve the battery. This balance can make implementation of certain intensive computation mechanisms a delicate proposition.

Furthermore, it should be noted that it is common to find that updates exist, or can be developed or deployed, but this is not always done. One can, for example, find a user who does not know that there is a newer version of the operating system compatible with the smartphone, or a user may discover known vulnerabilities that are not corrected until the end of a long development cycle, which allows time to exploit the loopholes.

Next Generation of Mobile Security

There is expected to be four mobile environments that will make up the security framework:

Rich Operating System

In this category will fall traditional Mobile OS like Android, iOS, Symbian OS or Windows Phone. They will provide the traditional functionaity and security of an OS to the applications.

Secure Operating System (Secure OS)

A secure kernel which will run in parallel with a fully featured Rich OS, on the same processor core. It will include drivers for the Rich OS ("normal world") to communicate with the secure kernel ("secure world"). The trusted infrastructure could include interfaces like the display or keypad to regions of PCI-E address space and memories.

Trusted Execution Environment (TEE)

Made up of hardware and software. It helps in the control of access rights and houses sensitive applications, which need to be isolated from the Rich OS. It effectively acts as a firewall between the "normal world" and "secure world".

Secure Element (SE)

The SE consists of tamper resistant hardware and associated software. It can provide high levels of security and work in tandem with the TEE. The SE will be mandatory for hosting proximity payment applications or official electronic signatures.

Browser Security

Browser security is the application of Internet security to web browsers in order to protect networked data and computer systems from breaches of privacy or malware. Security exploits of browsers often use JavaScript - sometimes with cross-site scripting (XSS) - sometimes with a secondary payload using Adobe Flash. Security exploits can also take advantage of vulnerabilities (security holes) that are commonly exploited in all browsers (including Mozilla Firefox, Google Chrome, Opera, Microsoft Internet Explorer, and Safari).

Security

Web browsers can be breached in one or more of the following ways:

- Operating system is breached and malware is reading/modifying the browser memory space in privilege mode

- Operating system has a malware running as a background process, which is reading/modifying the browser memory space in privileged mode

- Main browser executable can be hacked

- Browser components may be hacked

- Browser plugins can be hacked

- Browser network communications could be intercepted outside the machine

The browser may not be aware of any of the breaches above and may show user a safe connection is made.

Whenever a browser communicates with a website, the website, as part of that communication, collects some information about the browser (in order to process the formatting of the page to be delivered, if nothing else). If malicious code has been inserted into the website's content, or in a worst-case scenario, if that website that has been specifically designed to host malicious code, then vulnerabilities specific to a particular browser can allow this malicious code to run processes within the browser application in unintended ways (and remember, one of the bits of information that a website collects from a browser communication is the browser's identity- allowing specific vulnerabilities to be exploited). Once an attacker is able to run processes on the visitor's machine, then exploiting known security vulnerabilities can allow the attacker to gain privileged access (if the browser isn't already running with privileged access) to the "infected" system in order to perform an even greater variety of malicious processes and activities, on the machine or even the victim's whole network.

Breaches of web browser security are usually for the purpose of bypassing protections to display pop-up advertising collecting personally identifiable information (PII) for either Internet marketing or identity theft, website tracking or web analytics about a user against their will using tools such as web bugs, Clickjacking, Likejacking (where Facebook's like button is targeted), HTTP cookies, zombie cookies or Flash cookies (Local Shared Objects or LSOs); installing adware, viruses, spyware such as Trojan horses (to gain access to users' personal computers via cracking) or other malware including online banking theft using man-in-the-browser attacks.

Vulnerabilities in the web browser software itself can be minimized by keeping browser software updated, but will not be sufficient if the underlying operating system is compromised, for example, by a rootkit. Some subcomponents of browsers such as scripting, add-ons, and cookies are particularly vulnerable ("the confused deputy problem") and also need to be addressed.

Following the principle of defence in depth, a fully patched and correctly configured browser may not be sufficient to ensure that browser-related security issues cannot occur. For example, a rootkit can capture keystrokes while someone logs into a banking website, or carry out a man-in-the-middle attack by modifying network traffic to and from a web browser. DNS hijacking or DNS spoofing may be used to return false positives for mistyped website names, or to subvert search results for popular search engines. Malware such as RSPlug simply modifies a system's configuration to point at rogue DNS servers.

Browsers can use more secure methods of network communication to help prevent some of these attacks:

- DNS: DNSSec and DNSCrypt, for example with non-default DNS servers such as Google Public DNS or OpenDNS.

- HTTP: HTTP Secure and SPDY with digitally signed public key certificates or Extended Validation Certificates.

Perimeter defenses, typically through firewalls and the use of filtering proxy servers that block malicious websites and perform antivirus scans of any file downloads, are commonly implemented as a best practice in large organizations to block malicious network traffic before it reaches a browser.

The topic of browser security has grown to the point of spawning the creation of entire organizations, such as The Browser Exploitation Framework Project, creating platforms to collect tools to breach browser security, ostensibly in order to test browsers and network systems for vulnerabilities.

Plugins and Extensions

Although not part of the browser per se, browser plugins and extensions extend the attack surface, exposing vulnerabilities in Adobe Flash Player, Adobe (Acrobat) Reader, Java plugin, and ActiveX that are commonly exploited. Malware may also be implemented as a browser extension, such as a browser helper object in the case of Internet Explorer. Browsers like Google Chrome and Mozilla Firefox can block—or warn users of—insecure plugins.

Conversely, extensions may be used to harden the security configuration. US-CERT recommends to block Flash using NoScript. Charlie Miller recommended "not to install Flash" at the computer security conference CanSecWest. Several other security experts also recommend to either not install Adobe Flash Player or to block it.

Password Security Model

The contents of a web page is arbitrary, but controlled by the entity owning the domain named displayed in the address bar. If HTTPS is used, then encryption is used to secure against attackers with access to the network from changing the page contents. For normal password usage on the WWW, when the user is confronted by a dialog asking for their password, they are supposed to look at the address bar to determine whether the domain name in the address bar is the correct place to send the password. For example, for Google's single sign-on system (used on e.g. youtube. com), the user should always check that the address bar says "https://accounts.google.com" before inputting their password.

The browser guarantees that the address bar is correct. This guarantee is a reason that browsers will generally display a warning when entering fullscreen mode, on top of where the address bar would normally be, so that a fullscreen website cannot make a fake browser user interface with a fake address bar.

Privacy

Flash

An August 2009 study by the Social Science Research Network found that 50% of websites using Flash were also employing flash cookies, yet privacy policies rarely disclosed them, and user controls for privacy preferences were lacking. Most browsers' cache and history delete functions do not affect Flash Player's writing Local Shared Objects to its own cache, and the user community is much less aware of the existence and function of Flash cookies than HTTP cookies. Thus, users having deleted HTTP cookies and purged browser history files and caches may believe that they have purged all tracking data from their computers when in fact Flash browsing history remains. As well as manual removal, the BetterPrivacy addon for Firefox can remove Flash cookies. Adblock Plus can be used to filter out specific threats and Flashblock can be used to give an option before allowing content on otherwise trusted sites.

Hardware Browser

A hardware-based solution which runs a non-writable, read-only file system and web browser based on the LiveCD approach. Brian Krebs recommends the use of a LiveCD to be protected from organized cybercrime. The first such hardware browser was the ZeusGard Secure Hardware Browser which was released in late 2013. Each time the bootable media is started the browser starts in a known clean and secure operating environment. Data is never stored on the device and the media cannot be overwritten, so it's clean each time it boots.

Browser Hardening

Browsing the Internet as a least-privilege user account (i.e. without administrator privileges) limits the ability of a security exploit in a web browser from compromising the whole operating system.

Internet Explorer 4 and later allows the blacklisting and whitelisting of ActiveX controls, add-ons and browser extensions in various ways.

Internet Explorer 7 added "protected mode", a technology that hardens the browser through the application of a security sandboxing feature of Windows Vista called Mandatory Integrity Control. Google Chrome provides a sandbox to limit web page access to the operating system.

Suspected malware sites reported to Google, and confirmed by Google, are flagged as hosting malware in certain browsers.

There are third-party extensions and plugins available to harden even the latest browsers, and some for older browsers and operating systems. Whitelist-based software such as NoScript can block JavaScript and Adobe Flash which is used for most attacks on privacy, allowing users to choose only sites they know are safe - AdBlock Plus also uses whitelist ad filtering rules subscriptions, though both the software itself and the filtering list maintainers have come under controversy for by-default allowing some sites to pass the pre-set filters.

Wireless Security Camera

Wireless security cameras are closed-circuit television (CCTV) cameras that transmit a video and audio signal to a wireless receiver through a radio band. Many wireless security cameras require at least one cable or wire for power; "wireless" refers to the transmission of video/audio. However, some wireless security cameras are battery-powered, making the cameras truly wireless from top to bottom.

Digital wireless camera

Wireless cameras are proving very popular among modern security consumers due to their low installation costs (there is no need to run expensive video extension cables) and flexible mounting options; wireless cameras can be mounted/installed in locations previously unavailable to standard wired cameras. In addition to the ease of use and convenience of access, wireless security camera allows users to leverage broadband wireless internet to provide seamless video streaming over-internet.

Types

Analog Wireless

Analog wireless is the transmission of audio and video signals using radio frequencies. Typically, analog wireless has a transmission range of around 300 feet (91 meters) in open space; walls, doors, and furniture will reduce this range.

Analog wireless is found in three frequencies: 900 MHz, 2.4 GHz, and 5.8 GHz. Currently, the majority of wireless security cameras operate on the 2.4 GHz frequency. Most household routers, cordless phones, video game controllers, and microwaves operate on the 2.4 GHz frequency and may cause interference with your wireless security camera. 900 MHz is known as Wi-Fi Friendly because it will not interfere with the Internet signal of your wireless network.

Advantages include:

- Cost effective: the cost of individual cameras is low.

- Multiple receivers per camera: the signal from one camera can be picked up by any receiver; you can have multiple receivers in various locations to create your wireless surveillance network

Disadvantages:

- Susceptible to interference from other household devices, such as microwaves, cordless phones, video game controllers, and routers.

- No signal strength indicator: there is no visual alert (like the bars on a cellular phone) indicating the strength of your signal.

- Susceptible to interception: because analog wireless uses a consistent frequency, it is possible for the signals to be picked up by other receivers.

- One-way communication only: it is not possible for the receiver to send signals back to the camera.

Digital Wireless Cameras

Digital wireless is the transmission of audio and video analog signals encoded as digital packets over high-bandwidth radio frequencies.

Advantages include:

- Wide transmission range—usually close to 450 feet (open space, clear line of sight between camera and receiver)

- High quality video and audio

- Two-way communication between the camera and the receiver

- Digital signal means you can transmit commands and functions, such as turning lights on and off

- You can connect multiple receivers to one recording device, such as security DVR

Uses and Applications

Wireless security cameras on a lamp post deployed by NYPD

Home Security Systems

Wireless security cameras are becoming more and more popular in the consumer market, being a cost-effective way to have a comprehensive surveillance system installed in a home or business for an often less expensive price. Wireless cameras are also ideal for people renting homes or apartments. Since there is no need to run video extension cables through walls or ceilings (from the camera to the receiver or recording device) one does not need approval of a landlord to install a wireless security camera system. Additionally, the lack of wiring allows for less "clutter," avoiding damage to the look of a building.

A wireless security camera is also a great option for seasonal monitoring and surveillance. For example, one can observe a pool or patio

Barn Cameras

Wireless cameras are also very useful for monitoring outbuildings as wireless signals can be sent from one building to another where it is not possible to run wires due to roads or other obstructions. One common use of these is for watching animals in a barn from a house located on the same property. One such example of this can be seen in this story of one of the first BarnCam in the New York Times.

Law Enforcement

Wireless security cameras are also used by law enforcement agencies to deter crimes. The cameras can be installed in many remote locations and the video data is transmitted through government-only wireless network. An example of this application is the deployment of hundreds of wireless security cameras by New York City Police Department on lamp posts at many streets throughout the city.

Wireless Range

Wireless security cameras function best when there is a clear line of sight between the camera(s) and the receiver. Outdoors, and with clear line of sight, digital wireless cameras typically have a range between 250 and 450 feet. Indoors, the range can be limited to 100 to 150 feet. Cubical walls, drywall, glass, and windows generally do not degrade wireless signal strength. Brick, concrete floors, and walls degrade signal strength. Trees that are in the line of sight of the wireless camera and receiver may impact signal strength.

The signal range also depends on whether there are competing signals using the same frequency as the camera. For example, signals from cordless phones or routers may affect signal strength. When this happens, the camera image may freeze, or appear "choppy". Typical solution involves locking the channel that wireless router operates on.

Mobile Secure Gateway

Mobile secure gateway (MSG) is an industry term for the software or hardware appliance that provides secure communication between a mobile application and respective backend resources typically within a corporate network. It addresses challenges in the field of mobile security.

MSG is typically composed of two components - Client library and Gateway. The client is a library that is linked with the mobile application. It establishes secure connectivity to Gateway using cryptographic protocol typically SSL/TLS. This represents a secured channel used for communication between the mobile application and hosts. Gateway separates internal IT infrastructure from the Internet, allowing only an authorized client requests to reach a specific set of hosts inside restricted network.

Client Library

The Client library is linked with the corresponding mobile application, and that provides secure access via the Gateway to the set of Hosts. The Client library exposes public API to the mobile application, mimicking platform default HTTP client library. The application uses this API to communicate with the desired hosts in a secure way.

Gateway

Gateway is a server or daemon typically installed onto physical or virtual appliance placed into DMZ. Gateway public interface is exposed to the Internet (or other untrusted network) and accepts TCP/IP connections from mobile applications. It operates on IPv4 and/or IPv6 networks. Incoming client connections typically use SSL/TLS to provide security for the network communication and a mutual trust of communicating peers. Communication protocol is typically based on SPDY or HTTP.

Host

Gateway forwards requests from connected apps to a collection of configured hosts. These are typically HTTP or HTTPS servers or services within an internal network. The response from a host is sent back to the respective mobile app.

References

- Dunham, Ken; Abu Nimeh, Saeed; Becher, Michael (2008). Mobile Malware Attack and Defense. Syngress Media. ISBN 978-1-59749-298-0.

- Dixon, Bryan; Mishra, Shivakant (June–July 2010). On and Rootkit and Malware Detection in Smartphones (PDF). 2010 International Conference on Dependable Systems and Networks Workshops (DSN-W). ISBN 978-1-4244-7728-9.

- Roth, Volker; Polak, Wolfgang; Rieffel, Eleanor (2008). Simple and Effective Defense Against Evil Twin Access Points. ACM SIGCOMM HotNets. doi:10.1145/1352533.1352569. ISBN 978-1-59593-814-5.

- Thirumathyam, Rubathas; Derawi, Mohammad O. (2010). Biometric Template Data Protection in Mobile Device Using Environment XML-database. 2010 2nd International Workshop on Security and Communication Networks (IWSCN). ISBN 978-1-4244-6938-3.

- Smith, Dave. "The Yontoo Trojan: New Mac OS X Malware Infects Google Chrome, Firefox And Safari Browsers Via Adware". IBT Media Inc. Retrieved 21 March 2013.

- "How to Create a Rule That Will Block or Log Browser Helper Objects in Symantec Endpoint Protection". Symantec.com. Retrieved 12 April 2012.

- Guo, Chuanxiong; Wang, Helen; Zhu, Wenwu (November 2004). Smart-Phone Attacks and Defenses (PDF). ACM SIGCOMM HotNets. Association for Computing Machinery, Inc. Retrieved March 31, 2012.

- Albanesius, Chloe (19 August 2011). "German Agencies Banned From Using Facebook, 'Like' Button". PC Magazine. Retrieved 24 August 2011.

- "Local Shared Objects -- "Flash Cookies"". Electronic Privacy Information Center. 2005-07-21. Archived from the original on 16 April 2010. Retrieved 2010-03-08.

- Töyssy, Sampo; Helenius, Marko (2006). "About malicious software in smartphones". Journal in Computer Virology. Springer Paris. 2 (2): 109–119. doi:10.1007/s11416-006-0022-0. Retrieved 2010-11-30.

Permissions

Index

CPSIA information can be obtained
at www.ICGtesting.com
Printed in the USA
BVHW02*0437020218
506942BV00003B/139/P